EAI/Springer Innovations in Communication and Computing

Series Editor

Imrich Chlamtac, European Alliance for Innovation, Ghent, Belgium

Editor's Note

The impact of information technologies is creating a new world yet not fully understood. The extent and speed of economic, life style and social changes already perceived in everyday life is hard to estimate without understanding the technological driving forces behind it. This series presents contributed volumes featuring the latest research and development in the various information engineering technologies that play a key role in this process.

The range of topics, focusing primarily on communications and computing engineering include, but are not limited to, wireless networks; mobile communication; design and learning; gaming; interaction; e-health and pervasive healthcare; energy management; smart grids; internet of things; cognitive radio networks; computation; cloud computing; ubiquitous connectivity, and in mode general smart living, smart cities, Internet of Things and more. The series publishes a combination of expanded papers selected from hosted and sponsored European Alliance for Innovation (EAI) conferences that present cutting edge, global research as well as provide new perspectives on traditional related engineering fields. This content, complemented with open calls for contribution of book titles and individual chapters, together maintain Springer's and EAI's high standards of academic excellence. The audience for the books consists of researchers, industry professionals, advanced level students as well as practitioners in related fields of activity include information and communication specialists, security experts, economists, urban planners, doctors, and in general representatives in all those walks of life affected ad contributing to the information revolution.

Indexing: This series is indexed in Scopus, Ei Compendex, and zbMATH.

About EAI

EAI is a grassroots member organization initiated through cooperation between businesses, public, private and government organizations to address the global challenges of Europe's future competitiveness and link the European Research community with its counterparts around the globe. EAI reaches out to hundreds of thousands of individual subscribers on all continents and collaborates with an institutional member base including Fortune 500 companies, government organizations, and educational institutions, provide a free research and innovation platform.

Through its open free membership model EAI promotes a new research and innovation culture based on collaboration, connectivity and recognition of excellence by community.

More information about this series at http://www.springer.com/series/15427

M. Kathiresh • G. R. Kanagachidambaresan
Sheldon S. Williamson
Editors

E-Mobility

A New Era in Automotive Technology

Springer

EAI
RESEARCH MEETS INNOVATION

Editors
M. Kathiresh
Department of Electrical and Electronics Engineering
PSG College of Technology
Coimbatore, India

G. R. Kanagachidambaresan
Department of Computer Science and Engineering
Vel Tech Rangarajan Dr Sagunthala R&D Institute of Science and Technology
Chennai, India

Sheldon S. Williamson
Electric Energy Storage Systems for Transportation Electrification
Ontario Tech University
Oshawa, ON, Canada

ISSN 2522-8595 ISSN 2522-8609 (electronic)
EAI/Springer Innovations in Communication and Computing
ISBN 978-3-030-85423-2 ISBN 978-3-030-85424-9 (eBook)
https://doi.org/10.1007/978-3-030-85424-9

This Springer imprint is published by the registered company Springer Nature Switzerland AG
The registered company address is: Gewerbestrasse 11, 6330 Cham, Switzerland

To students, scholars and family members

Preface

One of the major reasons for Global Warming is the emission of hazardous gases like carbon dioxide from Internal Combustion Engines used in Automobiles. It is anticipated that the world's human population will reach 10 billion in the year 2050. With this trend, the number of vehicles on road is also expected to be around 2.3 billion by 2050. It is also found that 36% of the total harmful gases emitted into the environment are due to the use of gasoline engines in automobiles. In spite of many guidelines and laws proposed by the government to regulate the emission of these pollutants, the increased number of vehicles with internal combustion engines continues to be a significant factor for environmental pollution. Researchers focus on green technologies that can prevent or reduce these adverse effects happening to the environment. The solution to this problem is to make use of Low- or Zero-Emission Vehicles like Electric Vehicles (EV) and Hybrid Electric Vehicles (HEV) that use electricity as the energy to drive the wheels of the automobiles. These categories of automobiles cause less environmental pollution, and their dependency on fossil fuels is also less compared to conventional vehicles.

Electric vehicles make use of electric motors instead of internal combustion engines, in order to drive the wheels of the vehicles. EV uses a large pack of batteries that store electrical energy to power an electric motor. These batteries must be plugged in to a charging station to get recharged. Hybrid Electric Vehicles are driven by the combination of internal combustion engines and electric motors. HEV combine the benefits of high fuel economy and low tailpipe emissions with the power and range of conventional vehicles. Thus, it is very imperative to learn the technology behind these kinds of alternative fuel vehicles so as to cope up with the upcoming trend of vehicles.

This book starts with an introduction to EV and HEV and their impact on the environment. The book elucidates the various electrical machines used to drive the powertrain of EV and HEV along with the converters, inverters, and control mechanisms. The book elucidates the power flow mechanism in EV and HEV followed by a detailed study on energy storage devices, battery management system, and also regenerative braking system which plays a vital role in the energy management of EV. The role of artificial intelligence in the energy management of EV and HEV is also illustrated in this book. The book further discusses Vehicle to Grid technology and issues pertaining to the process of smart charging. Various innovative applications involving the Internet of Things are also presented in this book. This book is intended for academicians, researchers, and industrialists.

Coimbatore, India — M. Kathiresh
Chennai, India — G. R. Kanagachidambaresan
Oshawa, ON, Canada — Sheldon S. Williamson

Acknowledgements

We would like to thank Dr. Imrich Chlamtac, the Series Editor, and Ms. Eliska, the Managing Editor at European Alliance for Innovation, for giving us this opportunity to edit a book in the series Innovations in Communication and Computing, Springer. We would also like to thank Ms. Mary James for providing this opportunity to edit a book in this series. We express our sincere thanks to the authors for their contribution and whole hearted cooperation in the making of this book. We also thank the reviewers for their constructive criticism and comments which enriched this work. We are grateful to PSG College of Technology, Coimbatore, Vel Tech Rangarajan Dr. Sagunthala R&D Institute of Science and Technology, Avadi, and Ontario Tech University Canada for the constant support and encouragement. Finally, we would like to acknowledge with gratitude the support and love of our family members and colleagues, without whom this book would not have been possible.

Contents

About the Editors

M. Kathiresh is a Faculty in the Department of Electrical and Electronics Engineering, PSG College of Technology, Coimbatore, India. He has completed his undergraduate degree in the stream of Electronics and Communication Engineering, postgraduate degree in Embedded and Real-Time Systems and Doctorate in the area of Automotive Embedded Systems at Anna University, Chennai. He has done several innovative projects which are now published in reputed journals, international conferences, and books, and some are patented. He has also carried out many consultancy projects for various leading industries, to help the industries to adapt Industry 4.0. He is currently working in a research project funded by the Department of Science and Technology, India. His research areas include Automotive Embedded Systems, Internet of Things, and Real-Time Embedded Systems. He is a life member of the Institution of Engineers (India) and Indian Society of Technical Education. He is a recipient of the Young Achiever Award from IEI in the year 2020.

G. R. Kanagachidambaresan is an Associate Professor in the Department of Computer Science and Engineering in Vel Tech Rangarajan Dr. Sagunthala R&D Institute of Science and technology. He completed his BE in Electrical and Electronics Engineering and ME in Pervasive Computing Technologies in 2010 and 2012. He completed his Ph.D. in 2017 from PSG College of Technology, Anna University, on Wireless Body Area

Networks domain. His research interest includes Internet of things, wireless sensor network, real-time embedded systems and single-board computers, and Industry 4.0. He is experienced in developing tailor-made IoT solutions for industries and other client applications. He has edited several books published by Springer; he is an associate editor of *Wireless Networks*, a journal from Springer. He is also an editor in chief of the Next-Generation Computing and Communication Engineering series from Wiley. He has published more than 20+ SCI journals, edited 8+ books, published 10+ patents, developed 10+ software, and copyrighted the same. He has more than 8 years of teaching experience. Presently he is working on ISRO, DBT, and NSTL-funded projects worth 1,20,000 USD. He has also made consultancy for companies like AMCO WABCO on Industry 4.0. He is presently managing director for Eazythings Technology Private Limited, Chennai.

Sheldon S. Williamson is Professor and Canada Research Chair in Electric Energy Storage Systems for Transportation Electrification at Ontario Tech University, Canada. His research interests include novel research to shift Canada's transportation industry to smart electric energy storage systems and fast-charging infrastructures. He has authored and co-authored several books. He is a member of the IEEE and the Association of Professional Engineers Ontario. He holds a PhD Degree in Electrical Engineering from the Illinois Institute of Technology, USA.

Chapter 1
Introduction to Electric Vehicles and Hybrid Electric Vehicles

K. Latha Maheswari, S. Kavitha, and M. Kathiresh

1 Introduction to EV and HEV

Air pollution, global warming, and the rapid decline of the world's petroleum resources are important challenges faced by the automobile industry which led to the shift of their focus on the electric vehicle. In recent decades, research and development in the automobile had highlighted the future for the development of high-efficiency, secure, and sustainable electric transportation. It is anticipated that electric vehicles (EVs), hybrid electric vehicles (HEVs), and fuel cell vehicles will replace traditional vehicles in the future. An electric vehicle (EV) is operated by an electric motor rather than an internal combustion engine that burns a mixture of fuel and gases to generate power. Electric Vehicles (EVs) provide a smooth and swift acceleration without producing any atmospheric pollutants. The electric vehicle (EV), which is fueled by renewable energy sources and is enabled by high-efficiency electric motors and controllers, provides an efficient, reliable, and environmentally sustainable urban transportation system. Indeed, electrification of the transportation fleet is a done deal for modern countries, and it is viewed as a significant feature of the intelligent transportation systems paradigm.

Exploring different facets of the presence of EVs will therefore be a useful reference for transportation infrastructure planners, policymakers, and decision-makers to become familiar with the characteristics of this emerging technology

K. L. Maheswari (✉) · M. Kathiresh
Department of Electrical and Electronics Engineering, PSG College of Technology, Coimbatore, India
e-mail: klm.eee@psgtech.ac.in; mkh.eee@psgtech.ac.in

S. Kavitha
Department of Electronics and Instrumentation Engineering, Dr. Mahalingam College of Engineering and Technology, Pollachi, India

M. Kathiresh et al. (eds.), *E-Mobility*, EAI/Springer Innovations in Communication and Computing, https://doi.org/10.1007/978-3-030-85424-9_1

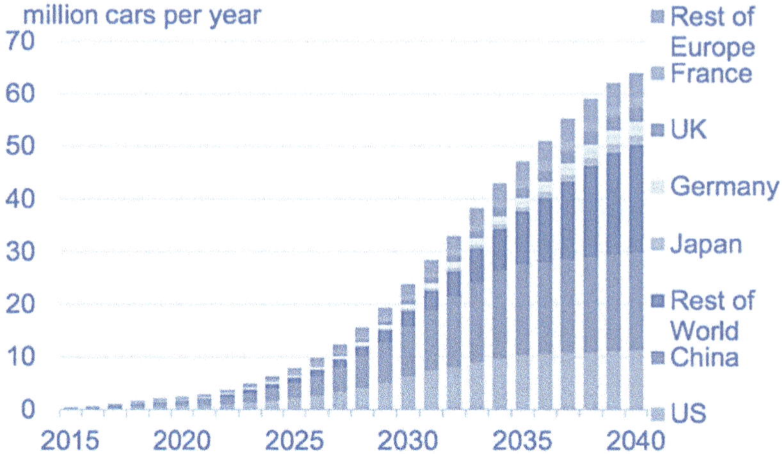

Fig. 1.1 Annual global electric vehicle sales by market [1]

and guide it in the right direction. The subject of electric and hybrid vehicles is becoming particularly significant as a result of demand from governments, environmental campaigners, and related industries. Figure 1.1 gives the statistics about the annual global electric vehicle sale by the market for the year 2015–2040. China continues to lead the way in electric vehicle adoption, followed by Europe and the United States, while India's worldwide stock of electric vehicles was just five million in 2018 [2]. Light Electric Vehicles [LEV] are one of the fastest expanding categories of the EV industry. The majority of LEVs are electric two-wheelers, followed by electric three-wheelers. Electric two-wheelers and three-wheelers provide an opportunity to raise the percentage of cleaner and efficient vehicles in Asian nations, such as India and China, where air pollution is a persistent problem that is predicted to intensify with rapid urbanization [3, 4]. As a result, policymakers are advocating the Clean Mobility Initiative to reduce vehicular emissions, reduce carbon emissions, and improve energy security by reducing reliance on petroleum imports [5]. According to the Electric Power Research Institute (EPRI), by 2020, 2030, and 2050, EV demand is projected to reach 35 percent, 51 percent, and 62 percent, respectively [6]. As the number of electric vehicles on the road increases, the grid's load during peak hours will need to be supplemented by solar energy to provide uninterrupted service. This trend indicates that proper training should be provided to understand the technological aspects of these vehicles. Electric and Hybrid Vehicle teaching materials with cutting-edge technologies are in even greater demand in today's world.

2 Adverse Effect of Atmospheric Pollutants

The case studies demonstrate the adverse effects of hazardous air pollutants, with combustion engine vehicles (CEV) accounting for the substantial proportion of air pollution. Climate change and air pollution are strongly intertwined [7]. The amount of incoming sunlight is affected by air pollutants, which results in increasing the temperature of the Earth. Food safety issues, iceberg melting, animal extinction, and plant damage are all significant consequences of climate change and the impact of global planetary warming on diverse ecosystems [3, 8]. To achieve the 1.5 °C objective established in the Paris Agreement, G-20 nations will need to advance their 2030 emission targets along with significantly enhanced mitigation, adaptation, and funding during the next decade [9]. Particulate matter (PM) is generated in the atmosphere mostly as a consequence of chemical reactions between pollutants [3]. The ultrafine Particulate Matter (PM2.5) generated by gasoline-powered cars is dangerous. Their concentrations in the air have grown in past decades, causing about 1,640,000 fatalities in India each year [10].

China, United States, India, Russia, and Japan are the five nations with poor air quality in the environmental health category, according to the Environmental Pollution Index (EPI) 2020 [7]. Figure 1.2 illustrates the annual CO_2 emission for different areas across the world for the past two decades. The government of India has Bharat Stage (BS) VI fuel standards to mitigate the environmental degradation, growing pollution levels, and health hazards caused by vehicular pollution. With the implementation of BS VI, India will be on par with the United States and other developed European countries in the automobile sector. To minimize air pollution, the governments have planned to shift from regular fuel-based vehicles to electric vehicles, along with enacting a number of strict regulations.

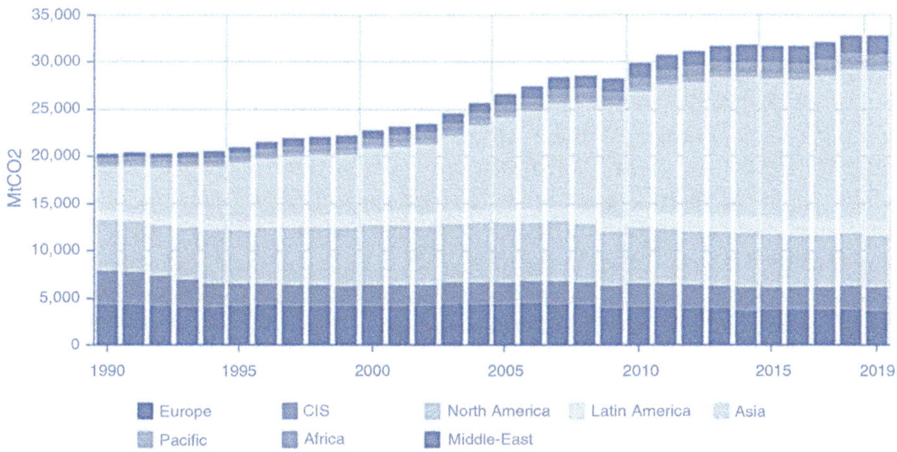

Fig. 1.2 Annual CO_2 emissions by region [9]

3 Need for Electric Vehicles

Crude oil price fluctuations and increasing import bills, along with significant expenditures on oil refineries and related distribution infrastructure, make the conventional vehicle uneconomical for developing countries. Several studies show that the adoption of electric vehicles in fuel-importing, service-dominated countries has a beneficial influence on GDP [8]. Electric vehicle (EV) technology development has progressed to satisfy the world's increasingly vital need for environmental preservation and energy conservation [11]. Electric vehicles (EVs) can help to alleviate the current predicament by lowering local pollution concentrations in cities. They can lead to significant savings while reducing pollution and have various benefits which are non-existent in the traditional internal combustion engine vehicles [12]. EVs are extremely resilient and have a huge amount of rapid torque produced by electric engines, allowing them to accelerate and decelerate smoothly and quickly. With the launch of electric vehicles, the global automobile market is undergoing a significant shift toward clean mobility options. As a result of stringent restrictions linked to global warming, fuel economy, and conventional energy limits, electric vehicle manufacturers and governments have been paying attention to electric, hybrid, and fuel-cell hybrid electric vehicle (FCEV) technologies.

4 History of Electric Vehicles

Although the rapid evolution of electric vehicles (EV) has been accelerated in the last decade, its early invention has a fascinating history. The very first electric motor for cars was designed by a Hungarian, Anyos Jedlik, in 1828. Following that, two inventors named Robert Anderson and Thomas Davenport from Scottish and America developed an electric car in the 1830s. In 1865, Gaston Plante, a French Physicist, substituted the non-rechargeable EV batteries with rechargeable batteries. EVs became viable for car users when Camille Faure from France reformatted and improved lead acid batteries formerly designed by Plante. This steered the way for electric cars in Europe. William Morrison of Des Moines, Iowa, built the first effective electric car in the United States in 1891. In 1899, a new land speed record of 100 km/h (62 mph) was set by the electric car named "La Jamais Contente" by Belgian race car driver Camille Jenatzy. The first real-world electric car for use was designed by Thomas Parker in the year 1895. EVs found their peak sales in 1912. The cost of basic cars was on average about $3000 (about $84,000 today). About 3200 commercial EVs were produced and used in Europe by 1914 [5]. Figure 1.3 shows the first inventor of the electric motor car and his prototype.

The bloom for electric cars and steam-powered vehicles was emerging in the nineteenth century, but due to falling limitations such as poor speed, battery limits, and high cost, the expansion of EVs gradually began to decline. The first hybrid car was invented by Ferdinand Porsche, a German automotive engineer, in 1900. The

Fig. 1.3 Image of first inventor of electric motor for cars and the prototype [12]

lead-acid batteries in EVs were replaced by Nickel-alkaline batteries by Thomas Edition in 1907. The commercial development of EVs took place after 2002. The credit of enjoying EVs and PHEV in this contemporary era is due to the work of EV. Lohner-Porsche Mixte Hybrid produced the first hybrid electric car in 1899. Surprisingly, the first automobile driven on the moon was also an electric car, the Lunar Roving Vehicle, which was launched on July 31, 1971, as part of the Apollo 15 mission. Between 1996 and 1999, famous and well-known GM mass-produced their electric car named "EV1," which had a range of 80 miles and could accelerate to 50 miles per hour in 7 s. The introduction of the Toyota Prius using nickel metal hydride battery (World's first mass-produced HEV) in 1997 and its launch in 2000 worldwide made car users to look back at electric cars. Also, the increased gasoline prices and growing concern about carbon pollution have helped make the Prius the best-selling hybrid worldwide during the past decade. Before that, Honda released the Insight hybrid in the United States in 1999. The third major event that increased EV anxiety was Tesla Motors' luxury electric sports car that covered 200 miles on a single charge.

In 1994, a joint venture between India and California led to the establishment of REVA Electric Car Company in Bangalore, which sold about 4000 vehicles around the world till 2011 and it is now established in more than 26 countries. The U.S. market paved the way for the Chevy Volt and Nissan Leaf companies in 2010. GEM vehicles sold 45,000 EVs in December 2010, and over 1000 Think City vehicles were sold in the United States and Europe. The Tesla Roadster (Sporty two-seater) is the first electric car suited for the Highways of America and saw sales of about 2100 vehicles in 31 countries. The Tesla S model was a luxurious four-seater produced for the whole family. Six excellent stations for charging about half the battery capacity in 30 min are provided by Tesla in North America in main traffic zones. In 2009, Mitsubishi launched the i-Mi EV for users in Japan and Europe at the end of 2010. The first modern electric car was launched on the Japanese and American market at the end of 2010 with zero tailpipe emission by the Nissan Leaf. In 2011, the Nissan Leaf was launched in the markets of France, Ireland, the Netherlands, Norway, Portugal, Spain, Switzerland, and Great Britain. At the end of

2012, more than 30,000 models were sold worldwide. It became the best-selling electric car in the world at that time. By the end of 2011 and at the beginning of 2012, Renault started launching its series of electric vehicles, Kangoo Z. E., Fluence Z. E., Twizy Z. E., and Zoe Z. E. models, respectively [13].

Rather than four-wheelers, Charger E-Bicycles were the first in the United States that designed the Pedelec model in 1997. In China, there were six million e-bike users, and Jiangsu Yadea was given credit for this. European countries like Norway, the Netherlands, and France have taken great interest in using PHEVs. Tesla introduced the Model 3 E-car in 2016 and continued to stand top in sales in 2017. In September 2018, the Norwegian market share of all-electric vehicles was 45.3%, with plug-in hybrids accounting for 14.9%. Tesla produced one million E-cars in March 2020, and in August 2020, its sales reached 645,000 units. China consistently stood first in utilizing E-Bikes and, on average, they used 22.2 million e-motorbikes around the twentieth century and the count reached 120 million by 2010. Geoby is the world's leading manufacturer of e-bikes [14].

5 Classification of EV

Electric Vehicles in general are classified based on their charging method into any one of the following types [13]:

- Battery electric vehicles (pure battery/electric)—Use rechargeable electric battery source only for its entire operation.
- Hybrid electric vehicles (Plug-in hybrid electric)—Use both battery and gasoline engines.
- Fueled electric vehicles—Fuel cell (made of hydrogen/methanol and oxygen) or metal air battery replaces the rechargeable electric battery.
- Supply-line electric vehicles—Heavy-duty vehicles such as buses and trams, as well as trolley buses, that charge their batteries via main power lines.
- EVs with flywheels and supercapacitors—Normal BEV along with additional charge generation components by regenerative action.
- Solar-powered vehicles—Use solar panels as back up to recharge main batteries.
- Vehicles using linear motors—Instead of creating rotary motion by torque, an electric motor with its stator and rotor "unrolled" produces linear motion due to force along its length to run the vehicle.

Electric vehicles are categorized as follows based on their range, speed, and acceleration performance:

- Neighborhood electric vehicles (NEVs)—small vehicles with a short range (less than 25 km).
- City electric vehicles are compact and have a limited range (less than 50 km).
- E-REV (extended range EVs)—Also known as full-performance battery-electric cars, used as traditional passenger automobiles, ranging from 100 to 600 km.

6 Battery Electric Vehicles (BEV)

BEVs are propelled by one or more electric motors. The typical configuration of a battery electric vehicle is shown in Fig. 1.4. The batteries can be recharged using grid electricity, house outlets, and non-grid sources such as solar panels, or on-board recuperative energy devices [11, 14]. In comparison to hybrid vehicles, which we shall examine next, the configuration of an electric car is more versatile [15, 16]. The following are the reasons for this flexibility:

- In an electric vehicle, energy is transferred mostly through flexible electrical lines rather than bolted flanges or hard shafts. As a result, distributed subsystems in the EV are a reality.
- EVs support a variety of propulsion systems, including independent four-wheel drive and in-wheel drive.

To manage the torque of the motor, the clutch assembly and transmission system are frequently replaced with an electric motor drive system. An accelerator pedal, an electric motor drive/controller, batteries, and traction electric motors are the essential components of an electric vehicle's drive system.

6.1 Major Components of EV

It is very essential to identify and analyze various components used in EVs. Some manufacturers use slightly different nomenclature, but the following are the key components [17]:

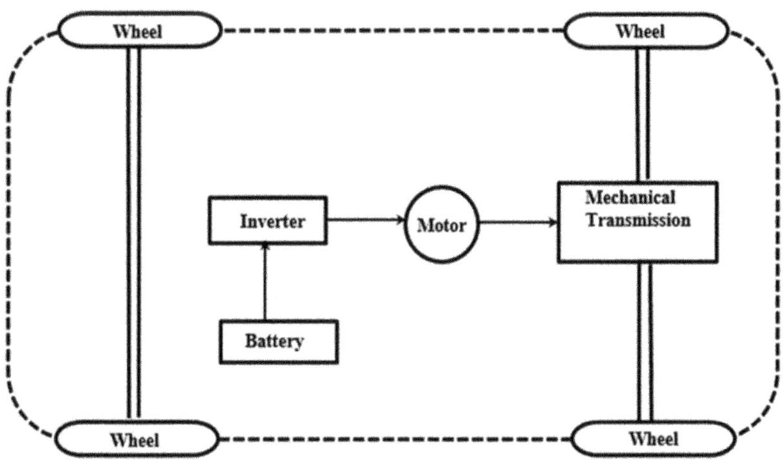

Fig. 1.4 Typical configuration of battery-operated electric vehicles

6.1.1 Energy Storage Devices

Despite the creation and design of several types of batteries for BEVs, they are still unable to meet performance standards. Furthermore, trade-offs between energy density, power density, cycle life, and cost limit each battery type for specific BEV applications [18]. As a result, there is no obvious winner in terms of the best battery technology for all BEVs. The Waste Batteries and Accumulators Regulations 2009 apply to all battery suppliers. This is a legal requirement, which means that producers must return batteries to customers to be reused, repurposed, or properly disposed of [15]. The various energy storage devices used in EV and HEV are listed below

- Lead-acid batteries (Pb–Pb02): Due to advantages such as low cost, fast recharge capability, high specific power, robustness against severe temperature variation, and availability in a variety of sizes and designs, the lead-acid battery is still an attractive candidate. A nominal 12 V lead-acid battery is made up of six cells that are connected in series. Due to the weight of their lead collectors, they have low specific energy, low energy density, limited cycle life, high cutoff voltage, and lack of long-term storage [17, 18].
- Alkaline batteries (Ni–Cad): The essential components of a Ni–Cad cell for the vehicle are: positive plate—Nickel Hydrate (NiOOH), negative plate—cadmium (Cd), and electrolyte—potassium hydroxide (KOH) and water (H_2O). This battery has a long cycle life, as well as rapid recharge capability, a wide operating temperature range, low self-discharge rate, great long-term storage, and a variety of sizes and designs, despite its high initial cost and poor specific energy. Despite the battery's recyclability, if it is not properly disposed of, cadmium in the battery can harm the environment.
- Nickel metal hydride battery (NiMH): The high specific energy and energy density of this battery, which are double that of lead-acid batteries, are its key advantages. Furthermore, due to their recyclability, these batteries have a fast recharge capability, a long cycle life, a wide operating temperature range, and are environmentally friendly. This technology's biggest disadvantage is its high starting cost.
- Nickel zinc (Ni–Zn) battery: Due to its high specific energy and power densities, cheap cost materials, deep cycle capability, and reasonably wide operating temperature range, this type of battery is appropriate for BEV. Its low cycle life, on the other hand, limits its applicability in BEV applications.
- Lithium-ion (Li-ion) battery: This provides a high energy density, great performance at high temperatures, high specific power, high specific energy, and a long life cycle. Li-ion batteries have two drawbacks: high prices and rapid self-discharge rates. Lithium-ion batteries are non-hazardous and contain useful materials that may be recycled, so they can be discovered [15, 16].
- Sodium–nickel chloride (Na–$NiCl_2$): The use of sodium (Na) for the negative electrodes in rechargeable batteries has seen substantial advancement. Because of its high potential of 2.71 V, low weight, nontoxic nature, relative abundance and

quick availability, and inexpensive cost, sodium is appealing. Sodium must be utilized in liquid form to build effective batteries. Sodium has a melting point of 98 °C (208 °F).

- Sodium–sulfur (Na–S): A liquid sodium cathode, an alumina electrode, and a metal in contact with the anode (a sulfur electrode) surround the entire assembly in a Na–S battery. The main issue with this system is that it requires a running temperature of 300–350 °C. This battery's cells are relatively tiny, requiring just roughly 15 g of sodium per cell. Each cell has a capacity of roughly 10 Ah.
- Fuel cells: In a fuel cell, the energy of oxidation of common fuels, which is generally represented as heat, can be directly turned into electricity. An anode, a cathode, and an electrolyte are required to separate these processes in a fuel cell. The electrolyte is mixed with the fuel immediately. It has been discovered that combining hydrogen with oxygen produces the most efficient design. Fuel cells are extremely dependable and quiet in operation, but they are rather costly to build.
- Super-capacitors: Super- or ultra-capacitors have extremely high capacitance yet with a small size. They can be used in traditional vehicles to eliminate the requirement for huge alternators. They can be utilized to recover braking energy that has been dissipated as heat and to reduce losses in electric power steering. A hybrid bus system stores 1600 kJ of electrical energy in 30 ultra-capacitors (20 farads at 400 V). The capacitor bank weighs 950 kg.
- Flywheels: Recovering the energy lost when a vehicle breaks is a highly effective approach to increase fuel economy and cut pollutants. The flywheel is connected to the tram by an infinitely variable cone and ball gearbox. The tram is decelerated by accelerating the flywheel using the gearbox, switching the vehicle's kinetic energy to the flywheel's kinetic energy—a type of regenerative braking that works well. Using this stored energy instead of engine energy to reaccelerate, the vehicle decreases engine fuel usage and CO_2 emissions. Flywheel systems are a fun alternative to batteries and super-capacitors. They are less complex, more compact, and less weight when compared directly.

6.1.2 Electric Traction Motors

Electric Motors are mainly used for powering the mechanical movement of wheels. To deliver the requisite traction force for vehicle motion, electric motors are used to transform electrical energy from energy sources into mechanical energy [19]. A BEV's electric motors must handle a variety of driving tasks, including frequent starting and stopping, rapid acceleration and deceleration, low-torque high-speed cruising, high-torque low-speed hill climbing, and propelling the vehicle from a standstill. An electric motor's type, size, weight, and performance in a BEV are determined by the vehicle's overall powertrain specifications. Single or multiple motor configurations, fixed or variable transmission, and whether the motor is geared or gearless are all included in these specifications. However, the following are the essential needs and parameters for the right selection of electric motors for a BEV:

- To produce sufficient maximum torque, often four or five times the rated torque, for temporary acceleration and hill-climbing;
- To provide high efficiency throughout a wide speed and torque range to reduce overall vehicle weight and increase driving range;
- Provide great controllability, steady-state precision, and dynamic performance;
- Provide appropriate robustness against high temperatures, severe weather, and frequent vibration;
- Provide excellent regenerative braking efficiency.

The electric motor generates constant torque at low speeds, while crossing the base speed, the motor generates constant power, and the torque begins to decrease with speed. An electric motor with a higher speed ratio can provide more maximum torque, which results in faster beginning acceleration and better gradeability. There are several ways to categorize the traction electric motors used in BEVs:

- Series wound DC motors, Shunt wound DC motors, and Independently excited DC motors are the most popular DC motors used in BEVs.
- Induction motors (IM), Synchronous motors (PM brushless motor), and Switched reluctance motors are the AC motor choices.

6.1.3 Power Electronic Modules

The power control unit or a motor control unit regulates the power electronics devices (inverter) in EV [20, 21]. It responds to the driver's signals (brake, acceleration, etc.) by switching the power electronics accordingly. The control instructs the motor to either propel the vehicle or to act as a generator, charging the battery [21]. It may also be in charge of air conditioning and PAS. The power electronics control module is made up of various subsystems like Rectifiers, Filters, regulators, and Inverters/Converters [20]. The rectifier transforms alternating current (AC) into direct current (DC), which is sent into the high voltage battery when the car is charged. The DC–DC converter is in charge of converting a high-voltage (for example, 400 V) to a low voltage network (12 V) to fuel lights, horns, radios, power windows, and so on. The inverter regulates the speed and torque of the electric machine by converting direct current from the battery to alternating three-phase current. When the vehicle is in energy recovery mode (braking), the inverter converts three-phase AC to DC in the opposite direction. The power flows between power sources, loads, and power buses are controlled by this subsystem. Power electronic modules ensure effective power management between various energy sources and storage elements, which plays an important role. The key characteristics of converters for electric vehicles are their small size, lightweight, dependability, and high efficiency.

6.1.4 Battery Management Controller

It monitors and manages the battery and determines the state of charge of the cells. It keeps the cells cool and protects them from overcharging and deep discharge. When the battery system is idle or in a severe scenario such as a fire, electronic switches disconnect the battery system [22].

6.1.5 Displays /Indicators (HMI Touch Panels)

A variety of strategies like displays alarms and indicators are utilized to keep the driver informed. A touch screen interface is now the most frequent method of delivering information and allowing the driver to regulate parameters [22].

6.1.6 Power Bus

The power bus is a DC connection between sources and loads. Two direct current buses are included in such vehicles: one is labeled as the high-voltage power bus and the other is a low-voltage power bus. The high-voltage bus gives power to propulsion motors and other high-power loads, while the low-voltage bus gives electricity to low-power accessory loads including lighting, microcontrollers, and small motors. To transport energy back and forth, the two DC power buses are connected to one or more DC–DC converters. It is worth noting that the "high"- and "low"-voltage buses do not have defined voltages; instead, they are determined by the vehicle configuration. A third power bus is occasionally included to provide power for external plug-in appliances or tools. The utilization of numerous power buses at different voltages not only answers power requirements but also resolves safety concerns [23].

6.1.7 Suspension Systems of EVS and HEVS

When developing a suspension system for EVs and HEVs, there are a few things to keep in mind. Heavy-duty construction is a major parameter of concern. The weight of EVs and HEVs with storage systems can be up to 25% more than that of conventional automobiles. This added weight has a substantial impact on the vehicle's ride and handling characteristics. Apart from employing proper tires with optimal cornering coefficients, the suspension system should also be able to accommodate the above-said weight issue. The load-carrying capability of the suspension mechanism should be improved. As a result, the suspension arms and linkages will be stronger, with stiffer bushings. Reducing the suspension system's weight can help reduce the vehicle's unsprung weight and overall weight. When manufacturing suspension arms and links, modern lightweight materials such as aluminum alloys are advantageous. The Toyota Prius' suspension system, for example, is constructed of aluminum alloys. This prominent trend should continue in the future.

6.1.8 Regenerative Suspension System

Regenerative suspension systems have the capability to harvest the energy generated by suspension vibration while also managing it. The key component of this system is a regenerating shock absorber, which recovers vibrational energy from the vehicle and turns it into useful electricity. Extending the harvesting capacity of regenerative suspension systems is the subject of ongoing study.

6.1.9 Steering System of EVS and HEVS

Electric power-assisted steering is a sort of rack and pinion steering system found in most modern EVs and HEVs. Conventional automobiles, as previously stated, have limited amounts of onboard electric power and a low level of electric voltage. As a result, electric power-assisted steering systems are limited to smaller traditional automobiles. Electric and hybrid-electric vehicles' energy storage devices, on the other hand, can easily offer higher power and voltages, which are suited for electric actuators in larger and heavier cars. As a result, electric power-assisted steering systems can be used in EVs and HEVs of any size or type. More advanced steering methods, such as the steer-by-wire system, are used in modern electric and hybrid electric concept automobiles. This technology is used in the steering system of the Audi A2 [24].

6.1.10 EVS and HEVS' Braking Systems

EVs and HEVs, like conventional vehicles, are equipped with sophisticated braking systems such as ABS, EBD, TCS, and ECS. Additionally, EV and HEV braking systems provide the following special features:

- Electronically controlled electric vacuum pump: EVs and HEVs include this module to provide vacuum energy to the brake booster. To keep the vacuum pump from constantly cycling, you will need to use a vacuum reservoir.
- Electro-hydraulic power boosters are an alternative to electric vacuum pumps in terms of design. The hydraulic pressure amplifies the braking force supplied by the driver's hydraulic pump, which is powered by electricity. The hydraulic pressure for the electro-hydraulic boosters comes from the steering system hydraulic pump.
- Regenerative braking system: EVs currently have a battery pack and an electric motor/generator, allowing them to use regenerative braking. Because regenerative braking systems cannot generate enough braking torque on their own,

especially during emergency braking, all modern EVs and HEVs use mechanical friction braking systems in addition to regenerative braking.

- Brake-by-wire (BBW) system: As previously stated, BBW systems are not yet found in mass-produced automobiles. This is because this technology is insufficiently trustworthy for use in important safety systems such as braking. Brake-by-wire technology, like steer-by-wire, can be energy-efficient if a higher level of electric power is available. Due to their on-board high voltage circuitry, EVs and HEVs offer a good testing platform for this new technology.

6.2 Universal Operating Modes of EV

The standard Electric vehicle is envisioned to operate in any one of the following modes:

- Mode1—Start-up: Under typical circumstances, the motor will start the engine at 1000 rpm right away. The engine will be cranked if the state of charge (SOC) of the high-voltage battery module is too low or if the temperature is too low or when the motor system fails.
- Mode 2—Acceleration: An inverter converts current from the battery module to AC and supplies it to the motor during acceleration. The output of the motor is used to complement the engine's output, resulting in increased power. The amount of time available for acceleration is maximized. The battery module's current is also converted to 12 V DC for use in the vehicle's electrical system. This minimizes the load created by a standard alternator, allowing for faster acceleration. When the battery's charge level is lowered to the bare minimum, no assistance is supplied.
- Mode 3—Cruising: The engine drives the motor, which now works as a generator when the car is cruising. The output current is then used to charge the battery before being used to power the vehicle's electrical system.
- Mode 4—Deceleration: The engine is operated by the wheels during deceleration (during fuel cut), allowing regeneration to occur. The high-voltage battery is charged using the generated power. A higher quantity of regeneration will be allowed during braking (brake switch on). More charge is transferred to the battery module in this mode. An "ABS-busy" signal is issued to the motor control module if the ABS system is managing wheel locking. To avoid interfering with the ABS system, this will immediately turn off the generator.
- Mode 5—Idle: The energy flow while idling is identical to that of cruising. If the battery module's state of charge is very low, the motor control system will send a signal to the engine control module (ECM) to increase the idle speed to around 1100 rpm [17].

6.3 Basic Configuration of Battery Electric Vehicles (BEV)

There are three key subsystems in the EV: electric propulsion, energy source, and auxiliary device. The energy source (Battery), the energy management unit, and the energy refilling unit make up the energy source subsystem. The electronic controller, the power converter, the electric motor (EM), the mechanical transmission, and the driving wheels make up the electric propulsion subsystem. The power steering unit and temperature control unit and auxiliary power supply make up the auxiliary subsystem [15]. The following configurations are used by electric powertrain system for propulsion of Electric Vehicles

- Out-wheel motor rear-wheel drive: An electric motor, fixed gearing, and a differential make up this setup. Modern EVs, unlike traditional ICE vehicles, can easily obtain the appropriate torque-speed characteristics for vehicle motion by combining an electric motor with fixed gearing. The electric motor, in general, has continuous power over a wider speed range. Due to the absence of the complex transmission system and its clutch or torque converter, this arrangement has a simpler structure, lighter weight, and smaller dimensions than converted electric powertrains. This powertrain design is depicted in Fig. 1.5b.
- Out-wheel motor front-wheel drive: An electric motor, fixed gearing, and a differential are all integrated into a single assembly on the front axle in this setup. This powertrain design is depicted in Fig. 1.5c.
- Dual out-wheel motors front-wheel drive: The mechanical differential separates the action of the left and right wheels in the configurations stated previously. The differential has been removed in this configuration, and the wheels are instead driven by separate electric motors and fixed gearings. Because electric motors can precisely control individual wheel torques, this design improves vehicle stability. The installation of an electric motor, a power converter, and fixed gearing, on the other hand, raises the system's initial cost. This powertrain setup is depicted in Fig. 1.5d.
- In-wheel motor drive: Mechanical components of an electric powertrain system can be reduced or even eliminated using in-wheel motor technology. Electric motors are housed inside a pair or all of the wheels in this design, with or without fixed gearing. By decreasing the motor speed while using fixed gearing and a high-speed inner-rotor electric motor, the desired wheel speed can be attained. The wheel torque is controlled directly by the electric motor in this setup (and hence, the vehicle speed). Figure 1.5e depicts an in-wheel motor all-wheel drive system. Despite its benefits, an in-wheel motor will almost certainly result in a heavier combined tire-wheel-suspension system. In terms of vehicle dynamics and stability control, this attribute is undesirable. Furthermore, the electric motors must be resistant to lateral and longitudinal stress as well as water infiltration [25].

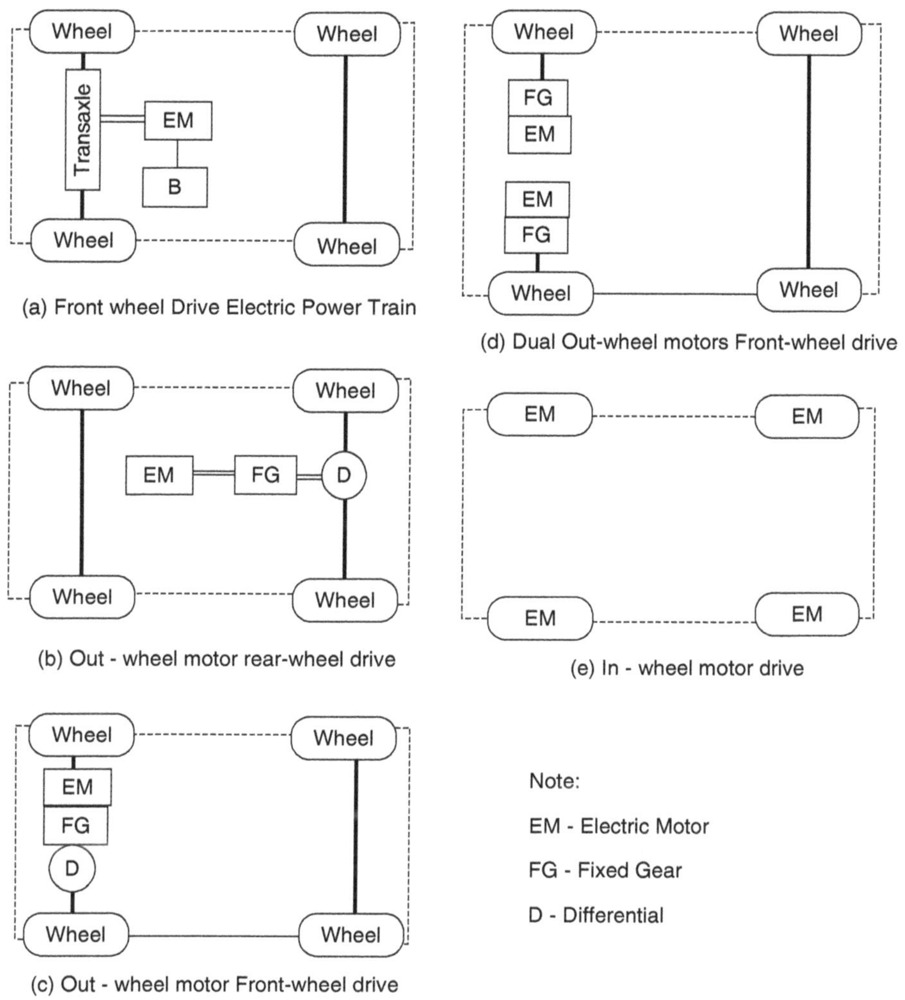

Fig. 1.5 Common EV configurations

6.4 Working of EV Subsystems

The working of EV along with regenerative power flow is shown in Fig. 1.6. The electronic controller gives suitable control signals to switch on or off the power converter, which regulates the power flow between the electric motor and the energy source, based on the control inputs from the brake and accelerator pedals. While the car is being driven, the battery charge can be extended further by using the regenerative braking system. The otherwise wasted kinetic energy from braking is stored in a storage battery and used to power the motor when needed later. The EV's regenerative braking causes the reverse power flow, and this regenerative energy can

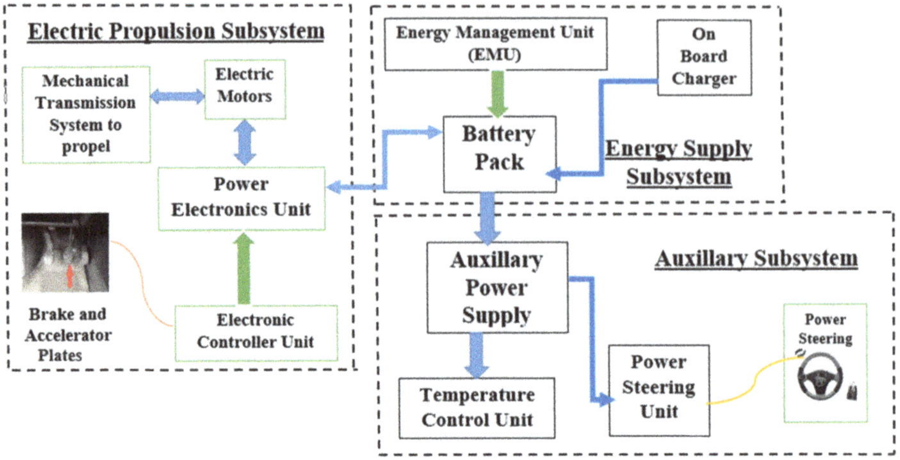

Fig. 1.6 Working of EVs along with regenerative power flow
[Note: Blue Color denotes Electric supply lines (Double side Blue arrow indicates Regenerative Action); Green color represents Control signals; Orange color signifies motions from and to the vehicle parts]

be retained if the energy source is receptive. To govern regenerative braking and energy recovery, the energy management unit works with the electronic controller. It also communicates with the energy-refueling unit to govern refueling and monitor the energy source's usage. The auxiliary power supply provides all EV auxiliaries with the necessary power at various voltage levels, particularly the temperature control and power steering units. The controller will generally govern the power supplied to the motor in forward and reverse directions and hence the vehicle speed [15].

6.5 Advantages of Battery Electric Vehicle (BEV)

Electric vehicles provide a number of advantages over conventional ICE vehicles, in addition to environmental benefits [16]

1. No tailpipe emissions: BEV keeps the environment free from toxic pollutants, unlike conventional gasoline engines, as they use alternative energy sources. Although BEVs have the potential to produce zero greenhouse gases, the kind of energy generation influences the well-to-wheel emission. Even if the electricity used to charge the batteries originates from a CO_2-emitting source like a coal-fired power plant, the amount of CO_2 emitted by BEV is around half to one-third less than IC vehicles.
2. No gas or oil changes: The electric vehicle battery needs to be recharged and hence the oil replacement process is not required here since the engine system is

redesigned with electronic systems. Ability to conveniently charge at home at Low cost—EVs can be conveniently charged at Home overnight or at charge stations at the cheapest rate, whereas conventional IC engine vehicles need demanding gasoline refilling at high cots.
3. Fast and smooth Acceleration—Since the internal systems are remodeled in EVs, the acceleration system works very smoothly when compared to IC engines.
4. Higher "tank-to-wheels" efficiency: Compared to traditional gasoline vehicles, they can convert roughly 59–62% of the electrical energy given by the grid to power the wheels, whereas a normal gasoline car can only convert around 17–21% of the energy stored in gasoline to practical power.
5. Cost Effective: Electric vehicles are also less expensive to maintain because they have fewer mechanical and emission control components. A muffler, catalytic converter, tailpipe, and gas tank, for example, are all missing from BEVs.

6.6 Disadvantages of Battery Electric Vehicle (BEV)

Despite these advantages, there are a few drawbacks to consider with BEVs that are listed subsequently,

1. EV initial production cost is comparatively more expensive than that of IC engine vehicles.
2. The low energy and power density of batteries in comparison to liquid fuels is now a major challenge for BEVs.
3. Another issue is the time it takes to charge the device. Although recharging a battery pack is less expensive than refueling a tank, the time it takes to recharge (about 4–8 h at home and a minimum of 30 min at a charging station) might be long.
4. Furthermore, due to the restricted capacity of EV battery packs, the current driving range achievable on a single charge is less than that of a conventional car running on a full tank of gasoline.
5. Battery packs are still relatively expensive and heavy, despite increases in battery storage capacity and lifespan.

7 Hybrid Electric Vehicle (HEV)

A hybrid electric vehicle (HEV) employs one or more electric motors along with an internal combustion engine (ICE) to propel the vehicle. It has two or more power sources, offering a wide variety of possibilities. The most prevalent hybrid vehicles combine an internal combustion engine (IC engine) with a battery, along with an electric motor and generator [26, 27]. The electric motor in a hybrid vehicle is powered by batteries that are constantly replenished by a generator that is driven by the ICE. The main advantage of an electric drive used in HEV is that it produces

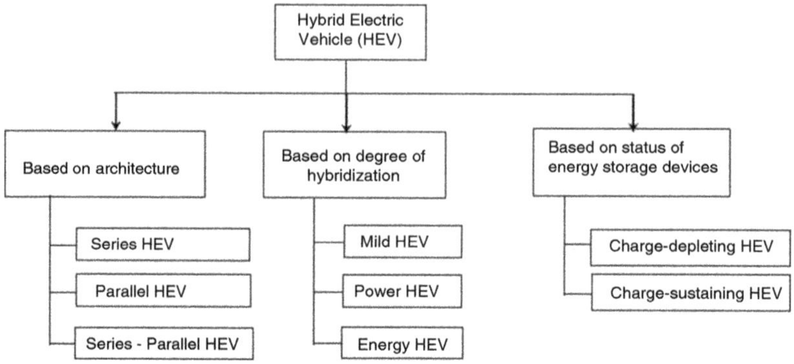

Fig. 1.7 Classification of hybrid electric vehicle (HEV)

great torque at low speeds, making it an ideal counterpart to an ICE that produces torque at higher speeds. The benefit of combining a motor and an engine makes HEV to run at its optimum speed for reducing emissions and consumption while still delivering necessary torque (with appropriate electronic control) [28, 29]. Figure 1.7 shows the various classification of Hybrid Electric Vehicle (HEV) based on architecture, degree of hybridization, and status of energy storage devices. The most popular method of categorizing hybrid cars is to look at the path of energy flow from the battery to the wheels through the power transmission lines. The power transmission line can be either a mechanical path that includes an IC engine and mechanical transmission or an electrical path that includes an energy storage system, converter, propulsion motor, and mechanical transmission or both. The vehicle's powertrain is engineered to satisfy both the vehicle's base load and peak load requirement during acceleration and starting. Architecture-based hybrids, such as series, parallel, and series–parallel hybrids, are formed by the arrangement of hybrid powertrain components in the power transmission line [30].

Hybrid vehicles can also be categorized into mild, power, and energy hybrids depending on their degree of hybridization. It is important to determine a variable termed the "degree of hybridization"in hybrid systems as shown in Eq. (1.1).

$$\text{Degree of Hybridization (DOH)} = \frac{\text{Electric motor power}}{\text{Electric motor power} + \text{IC engine power}} \tag{1.1}$$

The degree to which the engine is downsized while the electrical and energy storage components are upsized determines the evolution from mild to energy hybrids. Depending on whether the energy storage device needs to be charged from an external source or is self-sustaining with its onboard electricity generation capabilities, hybrid cars are categorized as charge-depleting or charge-sustaining hybrids. Plug-in hybrid electric vehicle is an example for charge-depleting type of HEV as it uses rechargeable batteries that can be charged at a charging station [31].

7.1 Series Hybrid Electric Vehicle

The mechanical output from the ICE is turned into electrical energy using a generator in the series hybrid powertrain, as shown in Fig. 1.8. The electrical energy from the generator is either utilized to charge the battery or transferred from the battery to the electric traction motor that drives the wheels. An AC–DC converter for battery charging and a DC–AC inverter for traction motor propulsion are both required in terms of power electronic converters used in series hybrid vehicles [32].

The operating modes of a series hybrid electric powertrain are classified as

- Engine-alone traction mode: The battery remains idle, and the combustion engine only delivers the total power for vehicle motion through the generator and power converter in this mode, and the electric motor serves as an electric transmission between the engine and the driving wheels.
- Electric-alone traction mode: The engine is turned off in this mode, and the electric source provides the entire amount of power essential for vehicle mobility.
- Hybrid mode: The engine–generator and batteries provide traction power to the wheels simultaneously in this mode.
- Engine traction and battery charging mode: The engine–generator generates traction power while also charging the batteries in this mode of operation.
- Regeneration mode: During braking, deceleration, and downhill driving, the regenerative braking system charges the batteries. The traction motor acts as a generator in this mode, and the engine generator is switched off.
- Battery-charging mode: The traction motor is turned off in this mode, and the engine generator is used to charge the batteries.
- Hybrid battery-charging mode: Both the engine generator and the traction motor work as generators, simultaneously charging the batteries.

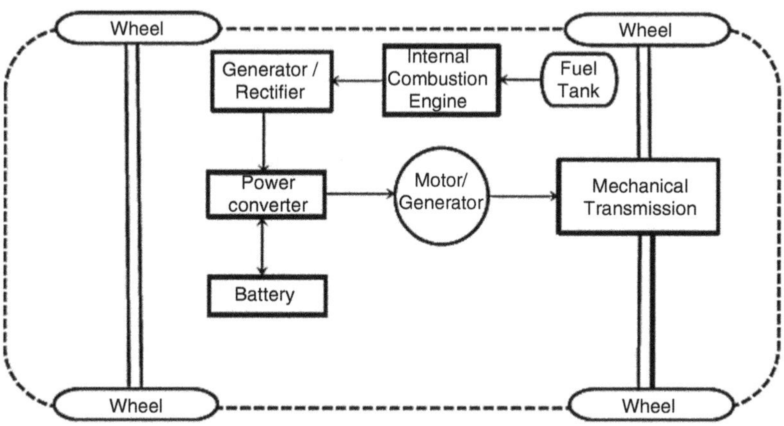

Fig. 1.8 Architecture of Series Hybrid Electric Vehicle (HEV)

7.1.1 Advantages

The advantages of a series hybrid electric vehicles design are listed below

- There is flexibility in the placement of engine generator set
- The drivetrain is much simpler and easy to design
- The engine is mechanically isolated from the wheels of the vehicle, allowing it to run at its most economical speed and torque, resulting in less idling time and lower emissions.

7.1.2 Disadvantages

The disadvantages in a series hybrid electric vehicles design are listed below

- It requires a longer chain of energy transmission and is less efficient than parallel hybrid electric vehicles
- Although these vehicles operate at a fraction of their maximum power, the motor used in them should be designed for maximum power accounting for climbing steep slopes
- All three drivetrain components must be sized for maximum power for long-distance, sustained, and high-speed driving; otherwise, the batteries will quickly deplete, forcing the ICE to supply power through the generator.

7.2 Parallel Hybrid Electric Vehicle

A parallel hybrid electric vehicle can provide propulsion power to the wheels through several energy conversion devices. The mechanical and electrical powers are mechanically coupled in this design, allowing them to be used simultaneously or independently. Mechanical devices such as torque-couplers and speed couplers are typically used in parallel hybrid powertrains to integrate mechanical and electrical power outputs. A gearbox unit or a pulley/chain assembly is used in a mechanical torque-coupler to combine the torques of the combustion engine and the electric motor [25, 33]. During regenerative braking or when the ICE output power is more than the needed power at the wheels, the electric motor can act as an electric generator to charge the battery. Figure 1.9 shows the architecture of one type of parallel hybrid vehicle, which has an IC engine for driving the front axle and an electric motor for driving the rear axle of the vehicle. In a parallel hybrid system, the ICE may be turned off and just the electric motor is powered by the battery pack. The operating modes of a parallel hybrid electric powertrain are classified as

- Engine-alone traction mode: In this mode, the electric motor is turned off and the combustion engine produces all of the power needed for driving. It is used when the engine is operating at or around its optimum operating conditions.

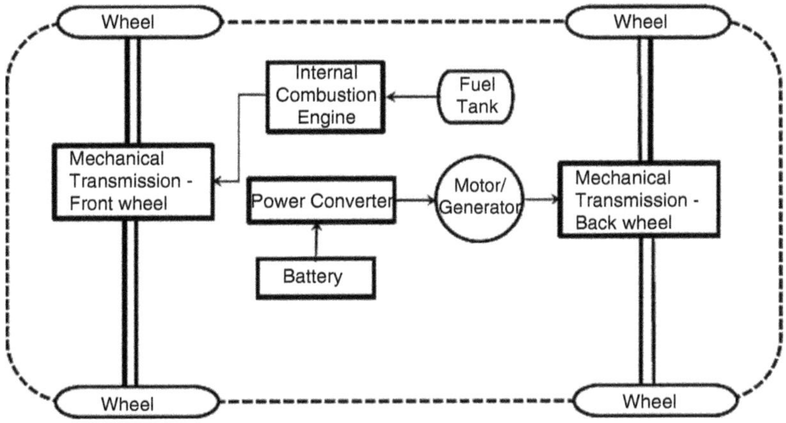

Fig. 1.9 Architecture of parallel hybrid electric vehicle (HEV)

- Electric-alone traction mode: The combustion engine is turned off in this mode, and the electric source provides the entire amount of power required for vehicular mobility. It is used for low-speed operation and starting of the vehicle.
- Hybrid mode: This mode allows the wheels to receive power from both sources and is used when additional power is required, such as while accelerating or driving at high speeds.
- Engine traction and battery charging mode: This mode happens when the ICE generates more power than the vehicle requires for motion. The surplus power is used to recharge the battery by switching the electric motor to generator mode.
- Regeneration mode: The kinetic energy lost by the vehicle when braking or moving downhill is recovered in this mode by the powertrain's regenerative system.

7.2.1 Advantages

The advantages of a parallel hybrid architecture are

- It has increased efficiency and performance when compared to series hybrid electric vehicle
- It is economic and has lower cost as compared to series hybrid electric vehicles and can be used for long-distance trips
- Due to the omission of a generator and the use of a smaller traction motor, the vehicle is more compact

7.2.2 Disadvantages

The disadvantages of a parallel hybrid architecture are

- As power flow must be controlled and blended from two parallel sources, the control complexity increases dramatically

- Because of the mechanical connection between the engines and the driven wheels, the operating points of the engines cannot be fixed in a specific speed range

7.3 Series-Parallel Hybrid Electric Vehicle

Power-split hybrid or series-parallel hybrid powertrains integrate the strongest features of both series and parallel hybrids to provide a more efficient hybrid powertrain design [33]. Figure 1.10 illustrates the configuration of the series-parallel architecture-based hybrid electric vehicle with planetary gear power split.

In contrast to a series hybrid powertrain, the series-parallel connects the engine to the end drives through a mechanical link, allowing the engine to directly drive the wheels [22, 29]. This design distributes the power from the engine into two paths: one uses a mechanical gear system to transmit the power to the wheels, while the other uses a generator and an electric motor to transmit the power to the wheels. This implies that the electric motor and the combustion engine can maximize efficiency by operating independently or in unison. The major devices employed to combine the powers are a planetary-geared, electronically operated continuous variable transmission (E-CVT) and a magnetic continuous variable transmission (M-CVT) [34]. In hybrid powertrains, E-CVTs are being used as a power split device, integrating engine power with that of the electric motor/generator through a planetary gear set [35]. Magnetic materials without mechanical contact are used in a magnetic continuous variable transmission (M-CVT) to provide torque transfer and adjustable gear ratios.

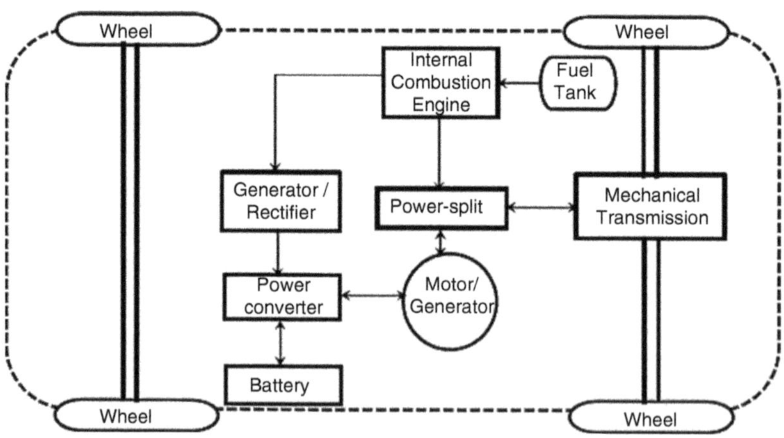

Fig. 1.10 Architecture of series-parallel hybrid electric vehicle (HEV)

7.3.1 Advantages

The advantages of a Series-Parallel hybrid architecture are

- It is capable of producing continuous high output power when compared to either a series or a parallel HEV.
- IC engine used in Series-Parallel HEV is smaller and more efficient compared to conventional ICE
- It uses an electric motor at low traction requirements and improves efficiency

7.3.2 Disadvantages

The disadvantages of a Series-Parallel hybrid architecture are

- A mechanical coupling and a planetary gear unit present in series-parallel HEV result in significant power loss
- The power management system in series-parallel HEV is complex and complicated.

7.4 Plug-in Hybrid Electric Vehicle (PHEV)

PHEVs are similar to charge-sustaining hybrids, with the exception that they include a larger energy storage system and a power electronic interface for grid connection [36].

By driving the automobile with grid power, plug-in hybrid electric vehicles (PHEVs) have the potential to reduce transportation fuel consumption. PHEVs can be driven for the first few miles using only the electric energy stored in the onboard battery, and then the range may be extended by using the internal combustion engine. A PHEV's drivetrain can be designed in a parallel, series, or power-split configuration, with the ability to charge the batteries both on-board and via a power outlet. Figure 1.11 shows the architecture of the Series Plug-in Hybrid Electric Vehicle (PHEV). The plug-in hybrid electric vehicle is rated as "PHEVX" based on zero-emission distance (X) traveled by the vehicle in electricity-only mode. Table 1.1 shows the comparison of characteristics and features between various categories of electric vehicles.

7.5 Various Charging Units and Methods

On pure EVs and PHEVs, this device is usually positioned near where the external power supply is connected. It converts and regulates the mains voltage to a level

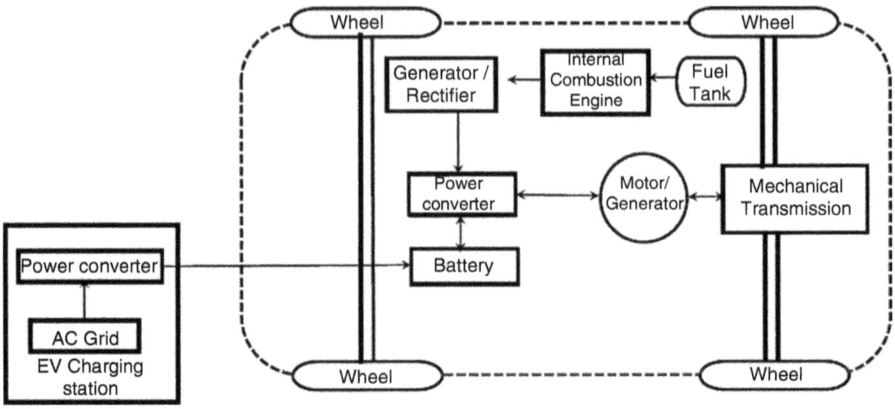

Fig. 1.11 Architecture of series plug-in hybrid electric vehicle (PHEV)

suitable for charging the battery cells. Although most of the EVs may be charged at home, national infrastructure is still being developed. The International Electro-Technical Commission (IEC) publishes international standards for the Standardization of charging cables, plugs, and techniques.

7.5.1 AC Charging

The most common charging method is alternating current. With relatively cheap investments, it is conceivable in the private sector as well as at charging stations in the semi-public and public sectors. As a result, this charging method has a bright future. It is the most popular and adaptable charging method. Charging can be done at home or CEE plugs in charging modes 1 and 2. A car can be charged at a charging station in charging mode 3.

7.5.2 DC Charging

DC charging stations are much more expensive than AC charging stations. In practice, automobiles with a DC charging connection also include a normal charging connection, allowing the vehicle to be charged at home. There are two means of DC charging:

- DC low charging: up to 38 kW with type 2 plugs
- DC high charging: up to 170 kW

Table 1.1 Comparison of characteristics and features of various electrified vehicles [29]

S. No.	Type of EV	Propulsion	Energy storage system	Energy source	Advantages	Disadvantages	Major Issues
1.	Battery electric vehicle (BEV)	Electric motor drive	Battery Super-capacitor	Grid electricity On-board electricity charging	Zero local emission High energy efficiency Independent of fossil fuels Commercially available	Relatively short range High initial cost	Battery size and management Charging facilities Cost Battery lifetime
2.	Fuel cell electric vehicle (FCEV)	Electric motor drive	Battery Super-capacitor Hydrogen tank	Hydrogen On-board electricity Charging	Zero/low local emission High energy efficiency Independent of fossil fuels (if not using fossil fuel to Produce hydrogen)	Relatively short range High initial cost Under development	Fuel-cell cost Fuel-cell life cycle and reliability Hydrogen distribution and Infrastructure Cost
3.	Hybrid electric vehicle (HEV)	Internal combustion engine (primarily) Electric motor drive (secondary)	Battery Super-capacitor Fuel tank	Gasoline/diesel On-board electricity Charging	Low local emissions High fuel economy Long driving range Commercially available	Dependence on fossil fuels Higher cost than ICEVs	Battery sizing and management Control, optimization, and management of multiple energy sources
4.	Plug-in hybrid electric vehicle (PHEV)	Electric motor drive (primarily) Internal combustion engine (secondary)	Battery Super-capacitor Fuel tank	Grid electricity Gasoline/diesel On-board electricity Charging	Minimum local emissions Higher fuel economy than ICEVs Longer driving range than BEVs Commercially available	Slightly dependence on fossil fuels Higher cost than ICEVs	Battery sizing and management Control, optimization, and management of multiple energy sources

7.5.3 Inductive Charging

Inductive loops are used to charge without using any connections. For both the charging station and the car, the technical complexity and expenses are significant. This system is not yet marketable or suitable for large-scale production.

7.5.4 Battery Replacement

At the charge station, the vehicle's rechargeable battery is swapped with a fully charged battery. You can continue driving after a few minutes in this instance. Charge stations, on the other hand, would have to keep a variety of battery types on hand, which would be as challenging in practice.

7.5.5 Vehicle-to-Grid Technology

V2G (vehicle-to-grid) is a system that employs bidirectional power from the automobile to the grid as well as the standard grid-to-car charging procedure. The car battery can be used as a power backup for the home or company if this method is deployed. Mostly charged from renewable sources such as PV panels or wind and hence, this production method is not only environmentally supportive, but it is also an excellent way to stabilize demand variations in the grid. This concept is vastly intensifying at the time of writing.

7.5.6 TESLA'S Power Wall

While the Tesla Power Wall is not an EV, it is a spin-off that, when combined with home solar charging, could have a substantial impact on EV use [14].

7.5.7 Wireless Power Transfer (WPT)

This is a technique for extending the range of an EV while reducing its weight and expense. WPT is a cutting-edge method for wirelessly charging electric vehicle batteries. WPT does not necessitate the use of charging poles or accompanying infrastructure. Electric vehicles just park over an induction pad and the charging process begins immediately. Energy is delivered from a road-side primary coil system of limited length to the secondary coil of a non/slow-moving vehicle. Energy is transferred to a secondary coil of a high-speed vehicle via a specialized driving lane equipped with a high-power primary coil system with dynamic wireless charging.

There are three types of classifications:

- Stationary WPT: Charged when the vehicle is parked and no one is inside.
- Quasi-dynamic WPT: Charged when the car is stopped and the driver is inside.
- Dynamic WPT: Charged when the vehicle is in motion.

Three WPT power classes (SAE J2954) are also available:

- High Duty: 200–250 kW
- Light Duty Home: 3.6 kW
- Light Duty Fast Charge: 19.2 kW

7.5.8 Solar Charging

Many experiments are being carried out to generate electricity for electric vehicles using solar panels, and the results are being evaluated and studied. Figure 1.12 shows the electric vehicle that is charged at the charging station using the renewable energy source.

8 Future of Electric Vehicles

The current situation of oil gas price increases and pollution hazards seriously evidence us that only EVs or hybrid vehicles can save us from the two big issues. Clinically proven results show that toxic pollutants in air badly affects the various system of the human body and produce serious side effects and threats for human life [2, 3]. Also, increased oil prices mean that humans cannot afford much for their daily petrol allowances that too in this pandemic economy. So, no way, EVs are the right

Fig. 1.12 Charging stations of Tesla [14]

option for the people to adhere with. U.S. Ex-president Obama in 2012 rightly took the initiative to bring together all automakers to start designing and see that 2022 see major usage of Plug-win Hybrid vehicles.

References

1. EIA. (2011). *Annual energy outlook 2011 with projections to 2035*, Outlook.
2. Kettles, D., et al. (2016). Electric vehicle sales and future projections. *Applied Energy*.
3. Manisalidis, I., Stavropoulou, E., Stavropoulos, A., & Bezirtzoglou, E. (2020). Environmental and health impacts of air pollution: A review. *Frontiers in Public Health*. https://doi.org/10.3389/fpubh.2020.00014
4. McMillan, A. (March 11, 2018). *Global warming 101—Definition, facts, causes and effects of global warming* | NRDC.org.
5. Handy, G. (2015). *History of electric cars*. Edison Tech Cent..
6. Article by McCluskey, R. K. (1980). "*Electric vehicle—A petroleum independence alternative*". https://www.energy.gov/
7. Yale. (2020). *Environmental performance index 2020*. Yale University.
8. Juyal, S., et al. (2018). *ZERO emission vehicles (ZEVs): Towards a policy framework*. NITI Aayog.
9. Climate Transparency. (2019). *The G20 transition towards a net-zero emissions economy*, Climate Transparency Report. https://germanwatch.org/en/climate-transparency-report
10. Akhbarizadeh, R., Dobaradaran, S., Amouei Torkmahalleh, M., Saeedi, R., Aibaghi, R., & Faraji Ghasemi, F. (2021). Suspended fine particulate matter (PM2.5), microplastics (MPs), and polycyclic aromatic hydrocarbons (PAHs) in air: Their possible relationships and health implications. *Environmental Research*. https://doi.org/10.1016/j.envres.2020.110339
11. Gogoi, B. J. (2017). Need for electric vehicle: An Indian scenario & factors influencing electric vehicle purchase. *International Journal of Economic Research, 14*, 189–202.
12. Wolf, S., & Korzynietz, R. (2019). Innovation needs for the integration of electric vehicles into the energy system. *World Electric Vehicle Journal, 10*, 76. https://doi.org/10.3390/wevj10040076
13. Carley, D.(2014). The beginners guide to electric vehicles (EV). *Emotive The Electric Vehicle Experience*. https://pluginbc.ca/wp/wp-content/uploads/2014/07/EV-Beginners-Guide_Final_Sept2_2014.pdf
14. Tesla. (2020). About Tesla, *Tesla.com*.
15. Larminie, J., & Lowry, J. (2012). *Electric vehicle technology explained* (2nd ed.). Wiley.
16. Besselink, I. J. M., Van Oorschot, P. F., Meinders, E., & Nijmeijer, H. (2010). *Design of an efficient, low weight battery electric vehicle based on a VW Lupo 3L*.
17. Todd, J., Chen, J., & Clogston, F. (2013). *Creating the clean energy economy: Analysis of electric vehicle industry*. International Economic Development Council.
18. Khaligh, A., & Li, Z. (2010). Battery, ultracapacitor, fuel cell, and hybrid energy storage systems for electric, hybrid electric, fuel cell, and plug-in hybrid electric vehicles: State of the art. *IEEE Transactions on Vehicular Technology, 59*, 2806–2814. https://doi.org/10.1109/TVT.2010.2047877
19. Zeraoulia, M., Benbouzid, M. E. H., & Diallo, D. (2006). Electric motor drive selection issues for HEV propulsion systems: A comparative study. *IEEE Transactions on Vehicular Technology*. https://doi.org/10.1109/TVT.2006.878719
20. Emadi, A., Williamson, S. S., & Khaligh, A. (2006). Power electronics intensive solutions for advanced electric, hybrid electric, and fuel cell vehicular power systems. *IEEE Transactions on Power Electronics*. https://doi.org/10.1109/TPEL.2006.872378

21. Bellur, D. M., & Kazimierczuk, M. K. (2007). DC-DC converters for electric vehicle applications. *IEEE*. https://doi.org/10.1109/EEIC.2007.4562633
22. Ehsani, M., Gao, Y., & Emadi, A. (2009). Modern electric, hybrid electric, and fuel cell vehicles: Fundamentals, theory, and design. In *Fundamentals, theory, and design*, 2nd ed. CRC Press.
23. Denton, T. (2016). *Electric and hybrid vehicles*. CRC Press.
24. Christlein, J., & Maar, E. (2001). *The Audi space frame on the A2: A trendsetting all-aluminium car body concept in a compact class car*. SAE Transactions.
25. Amir Khajepour, A., Saber Fallah, M., Goodarzi, A. (2014). Electric and hybrid vehicles: technologies, modeling, and control: a mechatronic approach, Wiley; 1st edition – 11 April 2014[ISBN: 978-1-118-34151-3].
26. Husain, I. (2010). *Electric and hybrid vehicles: Design fundamentals* (2nd ed.). CRC Press.
27. Mi, C., Masrur, M. A., & Gao, D. W. (2011). *Hybrid electric vehicles: Principles and applications with practical perspectives*. Wiley.
28. Erjavec, J., & Arias, J. (2006). *Hybrid, electric and fuel-cell vehicles*. Cengage Learning.
29. Chan, C. C., Bouscayrol, A., & Chen, K. (2010). Electric, hybrid, and fuel-cell vehicles: Architectures and modeling. *IEEE Transactions on Vehicular Technology, 59*(2), 589–598. https://doi.org/10.1109/TVT.2009.2033605
30. Adlard, E. R. (2019). Mehrdad Ehsani, Yimin Gao, Stefano Longo and Kambiz M. Ebrahimi: Modern electric hybrid electric and fuel cell vehicles (3rd Edn). *Chromatographia*. https://doi.org/10.1007/s10337-018-3600-7
31. Björnsson, L. H., & Karlsson, S. (2017). Electrification of the two-car household: PHEV or BEV? *Transportation Research Part C: Emerging Technologies*. https://doi.org/10.1016/j.trc.2017.09.021
32. Joshi, A. S., Shah, N. P., & Mi, C. (2009). Modeling and simulation of a dual clutch hybrid vehicle powertrain. *IEEE*. https://doi.org/10.1109/VPPC.2009.5289682
33. Grammatico, S., Balluchi, A., & Cosoli, E. (2010). *A series-parallel hybrid electric powertrain for industrial vehicles*. doi: https://doi.org/10.1109/VPPC.2010.5729045.
34. Jian, L., & Chau, K. T. (2009). A novel electronic-continuously variable transmission propulsion system using coaxial magnetic gearing for hybrid electric vehicles. *Journal of Asian Electric Vehicles*. https://doi.org/10.4130/jaev.7.1291
35. Miller, J. M. (2006). Hybrid electric vehicle propulsion system architectures of the e-CVT type. *IEEE Transactions on Power Electronics*. https://doi.org/10.1109/tpel.2006.872372
36. Singh, H., Ambikapathy, A., Logavani, K., Arun Prasad, G., & Thangavel, S. (2021). Plug-in hybrid electric vehicles (PHEVs). *Green Energy and Technology*.

Chapter 2
Environmental Impact of Electric Vehicles

P. Sweety Jose, P. Subha Hency Jose, G. Jims John Wessley, and P. Rajalakshmy

1 Introduction

One of the acute issues in the future market uptake of EVs is the improvements in battery technology and other EV components. There is a rapid increase in the technologies, the views of professionals in battery manufacturers, car manufacturers, and research institutes play an important role in the elaborate prediction of the upcoming growth of these technologies.

The market expansion for electric vehicles is quickly increasing since 2008. The reason is that the technology is becoming more extensively used by car manufacturers and decrease in cost. Another reason for its reputation is due to the growing care for the environmental impact that gas emissions lead to Global Warming. Global warming is a very important concern for the entire world. Few scientists have reported that the main reason for global warming is deforestation and ocean tides. But the report of the U.S. Environmental Protection Agency released in January 2017 has proved this argument as incorrect. Another view is that global warming is a natural planetary process. The Environmental Protection Act (2017) agrees that global warming is a process of maintaining an acceptable temperature on the earth. In the nineteenth century, vehicle emission has resulted in 1.5°Fahrenheit increase in temperature [1]. The air pollution contributors in India are shown in Fig. 2.1 which shows that transport contributes to about 14% in air pollution.

Factors that lead to the rise in emission are as stated subsequently:

P. S. Jose (✉)
Department of EEE, PSG College of Technology, Coimbatore, India
e-mail: psj.eee@psgtech.ac.in

P. S. H. Jose · G. J. J. Wessley · P. Rajalakshmy
Karunya University, Coimbatore, India
e-mail: hency20002000@karunya.edu; jims_john@karunya.edu; rajalakshmy@karunya.edu

M. Kathiresh et al. (eds.), *E-Mobility*, EAI/Springer Innovations in Communication and Computing, https://doi.org/10.1007/978-3-030-85424-9_2

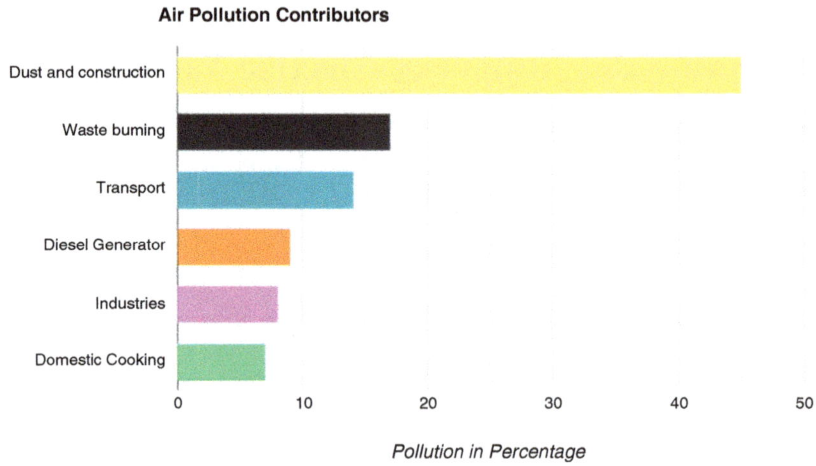

Fig. 2.1 Air pollution Contributors in India
Source: https://en.wikipedia.org/wiki/Air_pollution_in_India

- Increase in population
- Increase in production
- Increase in consumption of energy
- Increase in transportation.

From 1990 to 2012, in just 22 years, global greenhouse gas emissions increased by 41% [2]. In the forth coming years, if it continues, it will lead to increase in temperature and will affect the global climate. Although the United States accounts for 4.5% of the world's population, it consumes 19.2% of the world's energy and is the second-largest energy consumer in the world after China [3]. Although smaller countries take efforts to decrease their emissions and to produce energy through renewable technologies, the impact will be good only if the high-energy consumers are reducing their emissions [4]. The U.S. Department of Transportation is responsible for 28% of the country's greenhouse gas emissions, which means that the more electric vehicles are used, the greater the possibility of reducing emissions [5]. Meet higher requirements than ever before, which will increase your overall impact. In 2020, the number of electric cars on the road increased from 740,000 in 2015 to several million [6]. The most relevant scientific papers from 2013 to the present have been used to analyze the environmental impact of all-electric vehicles. Since the debut of the Tesla Roadster, interest in the electric vehicle industry has skyrocketed. Therefore, the corresponding article was not published until a few years later. Many articles used have received many references in the practice and discussion of electric vehicles. The main databases for finding articles are Science Direct, Environmental Science and Technology, and American Chemical Society Publications, and the U.S. Department of Energy.

2 Significant Features of Electric Vehicles

The purchase of electric vehicles (EV) is increasing over the last decade. The prevalent aims for this increase are as follows:

- Electric cars are more efficient than gasoline cars
- Can reduce your dependence on fossil fuels.
- Compared to most vehicles, they require less maintenance.

One of the most maintainable forms of transportation is Electric Vehicles as they depend only on electric power. This electric power generation can be dependent on sustainable, renewable resources. The four important factors to be pondered for the evaluation of the impact of electric cars on the environment are as follows:

- tailpipe release
- well-to-wheel release
- the energy to charge the battery
- vehicle's effectiveness.

The most important benefit of electric vehicles is their impact on improving urban air quality. While driving, electric cars do not emit carbon dioxide and hence it reduces pollution This reduces air pollution significantly. In other words, electric vehicles are making roads greener and making our communities healthier places for walking and cycling. An electric car on the road can save 1.5 million grams of carbon dioxide on average.

In addition to air pollution, electric vehicles reduce noise pollution particularly in towns where speeds are commonly low. Electric vehicles are extreme noiseless than conventional vehicles and create a more peaceful environment. A comparison between them is shown in Fig. 2.2.

To compare the features of electric cars with that of other cars in terms of environmental effects, the Luxembourg Institute of Science and Technology has developed a wonderful methodology. The following were the objectives of the methodology:

- Performance of electro-mobility by cutting per-km emissions from its fossil-fuel-based counterpart.
- Analysis of the conditions in which electric vehicles may "counter perform":

What happens when the battery size changes? or the background electricity mix? the battery lifetime? what about winter conditions affecting battery performances?

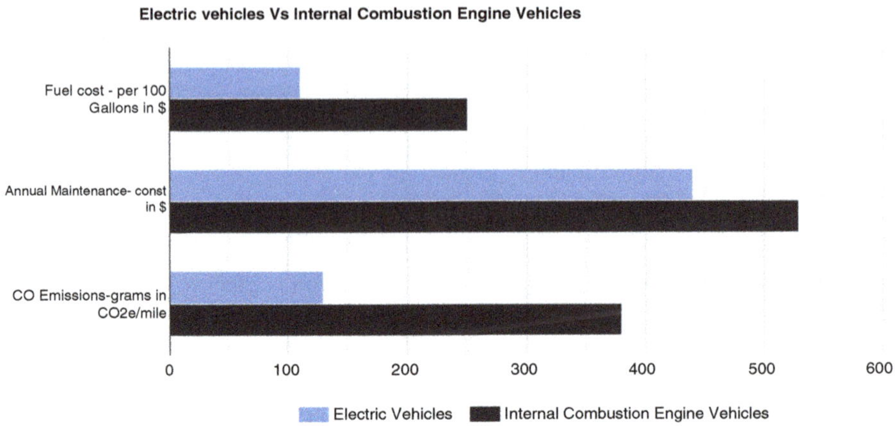

Fig. 2.2 Comparison between Electric Vehicles and Internal Combustion Vehicles
Sources: Applied, Energy, Vol.209, 2018, pages 108–119, CO_2 emissions data from the union of concerned scientists, 2018

3 Power Source

The least-known parts of electric vehicles are the energy used to charge these vehicles and the emissions associated with the production of that energy. In 2016, the U.S. Department of Energy announced that 64% of the energy produced in the United States comes from gaseous gasoline and coal [7]. Flammable gas creates generally 50% of the measure of outflows that coal does; however, on the off chance that 66% of the nation's energy is coming from these non-inexhaustible sources, the measure of discharges is still high. The development in the environmentally friendly power field is very cheerful considering the measure of energy delivered in 2012 was just 13.2% [3]. That expanded to 34% in 2016 with the utilization of atomic, hydro, wind, biomass, and sun-powered energy [8]. The nation has enormous potential for considerably more development in light of the all the way open spaces in the west for sun-oriented boards and wind turbine ranches.

Indeed, even with the energy coming from this 66% non-environmentally friendly power grid, electric vehicles actually created not exactly a large portion of the discharges of gas-fueled vehicles. Electric vehicles yearly produce 4587 pounds of CO_2 while gas and hybrids produce 23,885 pounds of CO_2 [8]. This measure only applies to emissions from vehicle manufacturing, power plants, ignition systems, early fuel production, and grid losses that are not taken into account. With each one of those parts considered, the assessed U.S. normal discharges are 202 g CO_2e/km, and the gauge for a gas vehicle is about 300 g CO_2e/km [4]. The more energy provided by the environmental protection system, the more environmentally friendly the all-electric vehicles will be, and their "zero-emission" examples are closer to reality. The 14 most populated cities on Earth are in India which are listed below in Fig. 2.3.

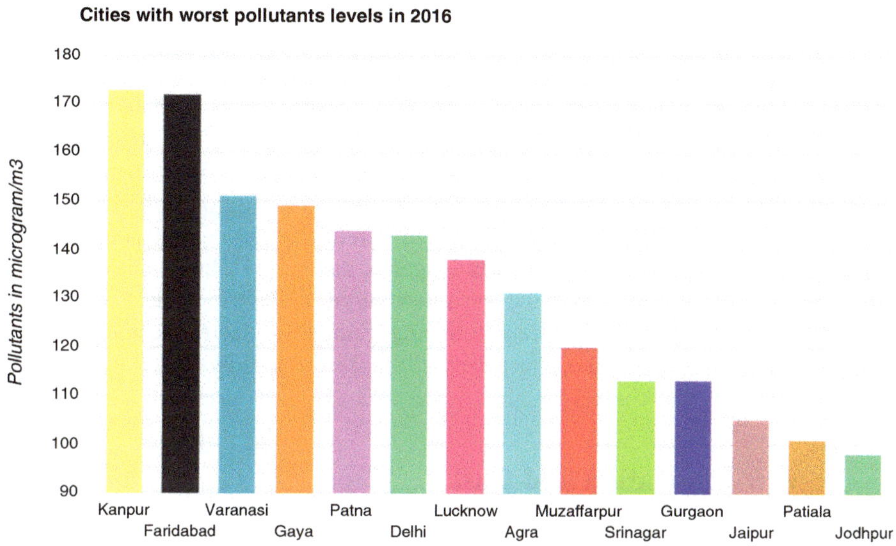

Fig. 2.3 Cities with worst pollutants levels in 2016 in India
Source: WHO report, 2018: Baoding (China) ranks 15th in the list

4 Batteries

Lithium-ion batteries completely save energy for electric vehicles. Lithium-ion batteries can also allow a vehicle to travel up to 335 miles on a single charge; the actual mileage depends on the model [9]. Since lithium-ion batteries were first provided to consumers in 1992, lithium-ion batteries have made significant progress. Their current weight performance is twice that, and the cost per kilowatt is only $300 [6]. When Tesla delivered the Roadster in 2008, the cost of lithium-ion batteries was as high as $900. Although it is still expensive, this reduction helps reduce the cost of all-electric vehicles. These batteries are the preferred solution because they "have high power-to-weight ratio, high energy efficiency, excellent high-temperature performance, and low self-dilution" [10]. These components make lithium batteries the most promising type of onboard energy storage and actuators for all-electric vehicles.

The chemical design of various lithium-ion batteries is currently being tested and developed. Nissan Leaf and BMW i3 use lithium nickel manganese cobalt batteries with an energy output of 140–180 Wh/kg [6]. The challenge is to find a synthetic compound that can extend the energy limit but remains stable at high temperatures and continuous load cycles. The new Tesla Model S uses lithium aluminum, nickel, and cobalt batteries, "helping to achieve an energy thickness of more than 240 Wh/kg" [6]. Experts began to test new materials for lithium batteries to increase the thickness of their energy used in electric vehicles; vehicles can choose to travel an extra mile at a one-time fee, which is cheaper for car owners who commute 20–80 miles a day.

The applications for lithium-particle batteries stretch out a long way past electric vehicles. This market became 73% somewhere in the range of 2010 and 2014 [11]. This could restrict the development of the electric vehicle industry due to the generally low flexibility of excellent lithium for these batteries. In December 2015, the cost of 99% lithium brought into China multiplied to 13,000 dollars for every ton; with practically 50% of an electric vehicle's creation costs being made up by the expense of their batteries, this represents a gigantic danger [11]. In the long run, costs are expected to fall, which may trigger a new surge in electric vehicle production similar to the year 2008. The more noteworthy number of electric vehicles out and about, the more prominent the effect they will make on outflows creation.

5 Battery Innovation and Cost

The scope of FEVs and the All-Electric Range (AER) of PHEVs and EREVs keep on being a significant determinant of expenses as it drives the size of the battery, and the expense of energy stockpiling keeps on being moderately high. Battery cost execution is the single most prominent test to the commercialization of all sorts of EV models.

Battery makers demonstrate that every battery age is probably going to be underway for 4–5 years in any event to recover capital speculations and R&D costs, so 2011/2012 presentation of the original of car lithium-ion batteries infers that the second-generation batteries could be popularized in 2016/17 and third-generation batteries in the mid-2020 time span.

In the view of a review of battery innovation improvements, the accompanying advancements comparative with a 2010 battery are foreseen. Upgrades of 20% to 25% in specific energy with a comparative decrease in cost by 2016 are principally because of improved battery design and packaging. Upgrades of 70% to 75% in specific energy and half decrease in cost per kWh by 2020 to 2022 with the presentation of cutting-edge materials for the anodes and cathodes, for example, silicon anodes. Potential for a significant increase of specific energy and 70% cost decrease per kWh by 2030 with the presentation of lithium-sulfur batteries.

In light of accessible examination and current battery information, apparently, current battery life ought to surpass 7 years and might associate with 10 years for "normal" use. There is still a lot of vulnerability with respect to battery schedule life at more serious surrounding temperatures while more moderate temperatures may permit certifiable battery life to associate with 10 years by and large. It is envisioned to proceed with progress to 2020 by which time, a normal life may be from 13 years to long-term range.

It is commonly seen that dissimilar to cadmium and toxic batteries, current known definitions of the Li-Ion battery materials do not present critical natural worries past fire well-being and landfill use. It is accepted that there are no significant worries that would recognize reusing Li-Ion batteries comparative with current lead corrosive and nickel-metal hydride batteries. Battery reusing financial aspects appears to be troublesome and hard to Electric vehicle emissions: tailpipe and well-to-wheel.

An electric vehicle does not emit any exhaust gas (also known as direct exhaust gas) when it starts. Because of this factor alone, electric vehicles are more environmentally friendly than conventional gas vehicles available today.

For the environmental protection of electric vehicles, the introduction of "from well to bicycle" must also be taken into consideration. This is a general term that includes ozone-depleting substances and toxins in the air that are irradiated to generate and distribute energy for control. Energy production creates another type of system-related emissions. The sustainability of an electric car driving demonstration is just the beginning, but if the main goal of buying an electric car is to reduce ozone-depleting substances and pollutants, making sure to use zero-emission energy as much as possible.

When considering admirably-to-wheel discharges, all-electric vehicles produce a normal of around 4450 pounds of CO_2 comparable every year. Therefore, traditional gasoline vehicles more than double every year. Your electric car is responsible for the flow of fuel from the oil well to the wheels usually depending on your geographic area and the fuel source most commonly used for power generation.

Flammable gas gives most of the power in the United States, followed intently by coal. Generally considered to be the "cleanest" petroleum product, it emits 50–60% less carbon dioxide than coal. In the United States, coal accounts for approximately 65% of carbon dioxide vapor. Regardless of whether your energy is mainly used in coal-fired power plants, when driving an electric car, there may still be less or less exhaust gas from the pit to the wheel. The combination of multiple power generation methods means that driving an electric car produces less well-to-wheel emissions than a traditional car.

6 Environmental Effects of Electric Car Production

There is a lot of energy consumption involved in the production of electric vehicles. The emission of toxic pollutants during the manufacturing of electric vehicles is much higher than that of traditional fuel-powered vehicles. This is because the manufacturing process involves the production of lithium-ion batteries that form an integral part of electric vehicles. Statistics reveal that the emission generated during the manufacturing of the electric vehicle amounts to more than one-third of the CO_2 emissions generated during the entire life cycle of the vehicle [12]. However, recent advancements in technology and the highly efficient manufacturing techniques in place have resulted in a drastic decrease in the amount of emissions generated during the production of batteries.

7 Growing Demand for Reusable and Recyclable Batteries

Extensive research is being carried out to explore various methods to reuse batteries for electric storage. The day is not far when every home will have batteries being used to store the energy required for their consumption. Such initiative will go a long way in reducing the ill effects of vehicle emissions on the environment. Despite the environmental issues involved in battery manufacturing, electric vehicles still remain a greener option due to the drastic reduction in emissions created during the entire lifecycle of the electric vehicle.

8 Need for Electricity

There are many questions that arise among the public regarding the strategies adopted to produce green electricity to energize an electric vehicle. Research findings by the European Energy Agency state that despite limitations arising due to the emissions during electricity generation, the CO_2 emission produced by electric cars is approximately 17–30% less than that emitted from a conventional fuel-operated vehicle. It is also found that there is a remarkable decrease in the emissions due to electricity generation when low carbon electricity is used. The electric tariffs are normally generated from 100% renewable sources to enable the electric car drivers aware of the charging methods so that they wisely operate the vehicle to minimize the environmental impact while driving the vehicle.

Nowadays, a hybrid combination of an electrically operated engine and the conventional fuel-powered engines is quite popular since they are found to reduce environmental pollution to a certain extent. The positive impacts of the hybrid combination depend on the proportion of distance covered in electrical miles and the method used to charge the vehicle. Therefore, it is important that the drivers of electric vehicles be aware of the method used to generate the electric power that energizes the vehicle. The discussions mentioned above clearly bring out the fact that electrical vehicles have a major role to play not only in reducing the emissions due to transport but also in making the world pollution-free and the air, clean enough to breathe.

9 The Efficiency of Electric Cars Defined

The primary reason for the sustainable nature of electric cars is the resource of power used to energize the vehicle. However, another factor that makes it a preferred choice for car buyers is its high efficiency. Efficiency is generally the ratio of output produced for a given input. In the case of electric vehicles, it is observed that the technology used can convert 59–62% of the electric input to usable power to energize the vehicle, whereas its counterpart traditional fuel-operated vehicles can convert only 17–21% of the energy obtained by the combustion of gasoline to usable power for the vehicle. The performance index that helps car buyers to evaluate the operational efficiency of any vehicle, namely, traditional, electric, or hybrid, is the MPGe (Miles Per Gallon equivalent). It is the distance a car can travel while consuming the same energy as a gallon of gasoline, measured in miles. The average MPGe of conventional fuel vehicles is about 24.7 MPGe. Although this is much better than its earlier performance, it is found to be inferior to the MPGes of the electrical vehicles in the market, which is around 100 MPGe, more than four times the efficiency of the traditional vehicles.

10 Positive Impact of Electric Cars on Environment

It is a known fact that electric vehicles generate lesser emissions than traditional fuel-powered vehicles. However, the benefits expected by the user determine how the vehicle is operated. If it is expected to attain zero emissions and ensure green energy, then, it should be kept in mind that not all sources of electric power are similar. Instead, it is a wise option to energize the car using renewable sources of energy like solar, wind, and so on. When purchasing an electric car, the users prefer to integrate the vehicle charging with the rooftop solar panel used for domestic purposes. On average, residential buildings with 5 kW installation spend about US$10,465 for the solar panel. The amount retrieval period may vary depending on the location. But it is found that most of the houses that have installed solar panels break even on the expenses after 7 years. It is promising to note that the solar savings may even increase with a faster retrieval time when combined with an electric vehicle application. This is because the installation of a solar panel will lead the user to do away with the expenses for gasoline and energize the vehicle with the electricity from the solar panels at no cost. It may also be required to expand the size of the solar panel to meet the need for increased capacity to charge the electric vehicle. The additional panels required to power the electric vehicle depend on the efficiency of the vehicle, the frequency of usage of the vehicle, and the solar potential of the area of interest. In case it is not possible to generate the required electrical energy from solar resources at one's property, it is possible to subscribe to a community solar charging system that is based on a sharing approach. This is rapidly gaining momentum across the country, and most of the utility organizations choose to purchase electricity from such renewable energy resources.

11 Direct Environmental Impacts of Electric Vehicles

It is interesting to note that electric vehicles produce almost zero tailpipe emissions. They utilize the energy stored in their batteries to energize their wheels to move. Experiments have proved that this conversion is extremely efficient in that only a very small amount of heat is lost during the conversion process. It is important to consider the environmental impacts resulting from the mining of battery material and its processing. These emissions are primarily due to the mining of coal and the extraction and refining of raw materials for the battery manufacturing process. However, these impacts are considerably lesser than those resulting from the usage of a gasoline engine.

12 Indirect Environmental Impacts of Electric Vehicles

With all the advantages of electric vehicles in place, there are a few factors of concern too. The worst impact of electric vehicles is revealed when we look up in the supply chain. It has been found that the concentration of particulate matter increases to a large extent. This is the result of coal power generation. Studies show that the grid average has been varying recently, with a shift toward renewable and natural gas. However, this shift is not going to happen overnight. On considering the effect of Electric Vehicles on climate change, it has been found that the electric vehicles powered by the current grid average reduces climate change-related impacts, but increase particulate pollution that results in a net higher impact on the environment than the existing conventional techniques. Hybrid Vehicles reduce the impact of both techniques considerably, and the buyers need not worry about range or recharging. Surprisingly, the existing technology is found to operate much cleaner than the electric vehicles powered by the grid average. Hybrid Technology is available ready to use from manufacturers like Toyota.

13 Economic Policies in World Nations

As the world's second-largest energy consumer, the United States has formulated a legal policy as a measure to reduce the country's vehicle emissions. In 1975, the American industries developed guidelines for average commercial fuel consumption. Over the years, the bill has been further revised, and the U.S. Environmental Protection Agency estimates that by 2050, 500 million tons of carbon dioxide emissions will be reduced each year. Ultimately, these initiatives will help the environment, but solutions must be found immediately. Due to the rapid increase in emissions, the 2005 Energy Policy Act was amended to immediately provide tax relief and loans to companies engaged in important research and production of

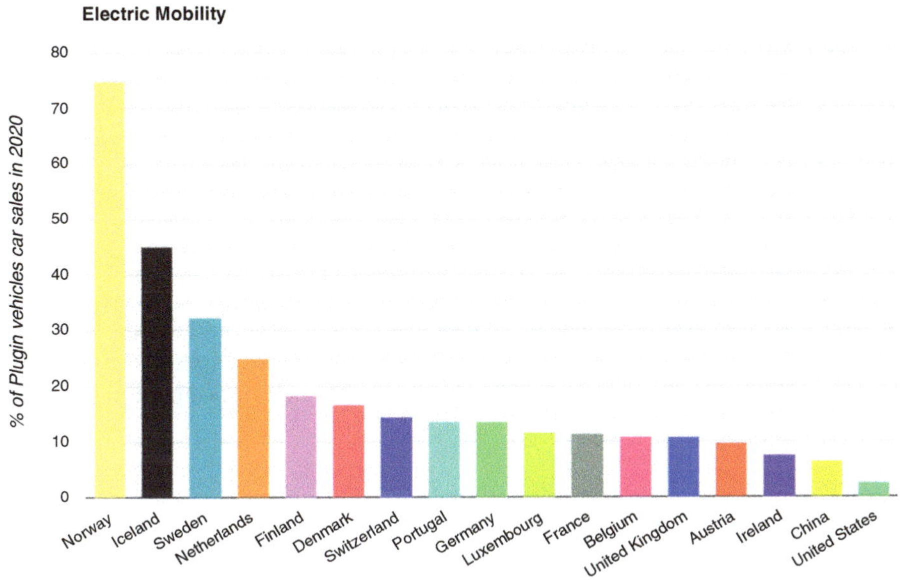

Fig. 2.4 Countries with the highest share of plug-in-vehicles
Sources: ACEA, CAAM, EV-Volumes

renewable energy such as solar and wind energy [13]. Each state should have its own guidelines for emission standards and renewable energy research funding.

The economic downfall in 2008 adversely affected the renewable energy domain due to the huge unemployment rate and state of the economy that dominated the media. Now, the economy has recovered, and more people have come forward to fight for a change in environmental policies. Such efforts will certainly go a long way to enhance the research activities in the renewable energy domain, thereby making the grid and electric vehicle usage have a huge impact on the reduction of greenhouse gases.

In the case of Electric Mobility, United Kingdom races ahead. Countries with the highest share of plug-in vehicles in the new passenger car sales in 2020 are shown in Fig. 2.4.

14 Summary

It is clear that the usage of electric vehicles will certainly make a remarkable impact in reducing greenhouse gas emissions. Although two-thirds of the energy production in the United States is from renewable energy sources, electric vehicles are found to produce less than half the share of CO_2 emissions compared to the traditional gasoline-powered vehicles annually. The objective of the study was to compile the pieces of information to create a complete picture of the potential significance of this technology. Electrical vehicles have gradually begun to make a difference now. But their impact could be increased further by considerable efforts taken by the governments to amend legal policies, encourage research activities in this direction and manufacture more electric vehicles. The major limitation of this study was the lack of sufficient research carried out on cradle to grave emissions involved in the production of lithium-ion batteries. This field being a less trodden one and since each battery has its own chemical property, it is difficult to estimate the emission during the production process of the batteries. This is important because the batteries are the most expensive part of the car. Moreover, it is also the most energy-consuming part of the production. Considerable research needs to be done in this domain, namely, electric vehicles, renewable energy, and lithium-ion batteries. Nations all over the world are on the right track to kickstart a significant improvement in this area, thus causing reduction in greenhouse gases making the planet earth a better place to live in.

References

1. Sheppard, K. (2014, April 22). *Congress hasn't passed a major environmental law in 1,894 days*. Retrieved April 18, 2017, from http://www.huffingtonpost.com/2014/04/22/congress-environmental-laws_n_5193815.html
2. Samimi, A., & Zarinabadi, S. (2012). Reduction of greenhouse gases emission and effect on the environment. *Journal of American Science*, 1011–1015.
3. Aslani, A., & Wong, K. V. (March 2014). Analysis of renewable energy development to power generation in the United States. *Renewable Energy Journal, 63*, 15–161. https://doi.org/10.1016/j.renene.2013.08.047
4. Wilson, L. (2013). *The 'electric cars aren't green' myth debunked*. Retrieved February 10, 2017, from http://shrinkthatfootprint.com/electric-cars-green
5. Jenn, A., Azevedo, I. M., & Michalek, J. J. (2016). Alternative fuel vehicle adoption increases Fleet gasoline consumption and greenhouse gas emissions under United States corporate average fuel economy policy and greenhouse gas emissions standards. *Environmental Science & Technology, 50*(5), 2165–2174. https://doi.org/10.1021/acs.est.5b02842
6. Patel, P. (2015, July 13). *Improving the lithium-ion battery*. Retrieved March 25, 2017, from http://pubs.acs.org/doi/full/10.1021/acscentsci.5b00223
7. Schmid, A. (May 2017). An analysis of the environmental impact of electric vehicles. *Missouri S&T's Peer to Peer, 1*(2).
8. Car emissions and global warming. *Reports and multimedia*. Retrieved July18, 2014, from http://www.ucsusa.org/clean-vehicles/car-emissions-and-global-warming#.WPbPS4jytPY
9. Lambert, F. (2017, April 13). *Tesla Model S 100D officially takes top spot for longest range EV with EPA 335-mile rating*. Retrieved April 14, 2017, from https://electrek.co/2017/04/13/tesla-model-s-100d-longest-range-ev-epa/
10. *Emissions from hybrid and plug-in electric vehicles*. Retrieved February 10, 2017, from http://www.afdc.energy.gov/vehicles/electric_emissions.php
11. Narins, T. P. (2017). The battery business: Lithium availability and the growth of the global electric car industry. *The Extractive Industries and Society*. https://doi.org/10.1016/j.exis.2017.01.013
12. Frequently asked questions about climate change. (2017, January 19). Retrieved March 25, 2017, from https://www.epa.gov/climatechange/frequently-asked-questions-about-climate-change
13. Pg Abas, A. E., Yong, J., Mahlia, T. M. I., Hannan, M. A. Techno-economic analysis and environmental impact of electric vehicle. https://doi.org/10.1109/ACCESS.2019.2929530

Chapter 3
Analysis of the Different Types of Electric Motors Used in Electric Vehicles

K. C. Ramya, J. Geetha Ramani, A. Sridevi, Rajakumar S. Rai, and D. Ruth Anita Shirley

1 Introduction

The principle supply of electricity in an electric vehicle is the electric powered strength that proves to be an impetus part of the development of electric vehicles in the automobile industry. This electrical strength is focused on the electric-powered vehicle wherein it is converted into mechanical energy in the form of rotations. This strength is carried forward to the wheels of the vehicle using an optimal transmitting device, enabling movement of the vehicle. The electric motor serves as the backbone of the electric vehicle, providing the power to drive. Depending on the type of electric vehicle and its functionality, an electric motor is chosen. Electric vehicles might ignite DC/AC [1] motors based on the configuration or dependent on the utilization of the electric vehicle. An electric motor is a mandatory part of any electric vehicle. This has paved the way to creating a large market for electric vehicle parts [2], primarily the motors [3], which contribute toward improving the performance and affordability of the vehicle. In 2013, 44.6 million electric motors were required for building electric vehicles while in 2023, about 129 million electric vehicles are estimated to be used and 158.8 million

K. C. Ramya · D. R. A. Shirley (✉)
Department of Electrical and Electronics Engineering, Sri Krishna College of Engineering and Technology, Coimbatore, India

J. G. Ramani
Department of Electronics and Communication Engineering, SNS College of Technology, Coimbatore, India

A. Sridevi
Department of Electrical Sciences, Adithya Institute of Technology, Coimbatore, India

R. S. Rai
Department of Mechanical Engineering, Karunya Institute of Technology and Sciences, Coimbatore, India

M. Kathiresh et al. (eds.), *E-Mobility*, EAI/Springer Innovations in Communication and Computing, https://doi.org/10.1007/978-3-030-85424-9_3

are expected to be used in 2023 [4]. The motors that are generally used today in air and through water are mostly brushless since the exit of DC brushed motors. The importance of brushless traction motors is indicated by the sinusoidal output with permanent magnet AC [5] and the trapezoidal output in the brushless DC motor. However, both motors show outstanding performance with the inclusion of additive measures like regenerative and reverse braking. These dynamics have however been shaken due to the in-wheel motors that are used in today's large vehicles (electric vehicles) [6] to make movement simple, comfortable, and adaptable. In today's scenario, the average money used on electric motors in electric vehicles deal with the use of smaller motors in smaller vehicles such as the mobile household robots built in Japan, power chairs for the disabled, and mobility scooters that are commonly used in the United States and Europe, two-wheelers in China, sea scooters, motorized lifters and stair walkers, walkies (popularly used in Japan as golf caddies), golf-cars, go-karts, all-terrain vehicles (ATVs), and tiny quad bikes [7]. These small vehicles require about 60% traction motors that are available in the market. However, with the increase in larger electric vehicles, the market for electric motors [8] has also increased substantially. In the near future, these electric vehicles powered by electric motors will dominate the streets, replacing the age-old fuel-based vehicles [9].

For example, as the military decides to buy battlefield hybrids instead of the stereotype electric tools, the value for military electric vehicles begins to increase many-fold [10]. Another example is executed in China where hybrid versions of buses are bought, in line with the national transportation system resulting in a higher market value. Similarly, the electric vehicle market is bound to grow leaps and bounds in the years ahead, leading to a demand for traction motors [11]. The electric motors that are used in electric vehicles are reaching higher values of torque and power. A low-cost electric bike will be utilizing a 0.25 kW motor, cars will require a power of 70 kW per motor while a large bus or forklift will need a motor with 250–350 kW and an AUV will require 400 kW [11]. By and large, the torque for the motor will be about 6000 Nm though only about 0.4 Nm of torque will be required by most vehicles used by the disabled. Based on the analysis of asynchronous motors, it has been observed that about 52% of buses use this motor apart from about 65% of military vehicles. One of the biggest vehicle builders [12], Toyota, uses asynchronous motors for many of its buses and forklifts.

An electric vehicle motor works based on a physical process introduced during the 1900s [13]. It is made up of a stator (fixed part of the motor) and rotor (rotating part of the motor) which operates in a magnetic field created, using current. There are two terms that are used in an interchangeable fashion: "Motor" and "Engine." It is crucial to know the difference between the two as used in the automotive industry. An engine refers to a machine that converts thermal energy into mechanical energy while a motor refers to the machine that converts energy into mechanical energy [14]. Since thermal energy is involved in an engine, it refers to the process of combustion. This means that an engine is also one type of motor although not all motors are engines. In motors, electricity is used to create mechanical energy [15] that drives the electric vehicle. Similarly, electric motors also find application in many devices. When it uses Direct Current (DC) [16], it can perform certain basic operations. In this case, the motor is directly connected to the source of energy and subsequently, the speed of rotation will depend on the intensity of the current. Though these motors enable easy rotation, they cannot produce the required amount of power, are unreliable, and relatively large to be used in an electric vehicle [17]. However, they find application in devices such as windows, windshield wipers, and other similar simpler mechanisms used in the car.

2 Electric Motor in Electric Vehicle

2.1 Progress of the Electric Motors

In [18], Thanh Anh Huynh analyzed some traction motors to determine the thermal execution and electromagnetic field generation. The motor examination was performed using two driving cycles: one for freeway driving and the other for urban driving. Assessment of electromagnetic execution took place using the flux weakening capacity and the estimation of torque output for accelerating vehicle speed. Similarly, in [19], the authors have discussed the different machine techniques that are used such as DC, switched reluctance, permanent magnet, and induction machine techniques for high-speed traction implementation. Based on the HST zone, correlation varies with respect to power density, fault torque, proficiency, quality, cost, and adaptability to non-critical failure capacity. Dr. Sab Safi in [20] has determined that critical research is carried out in HEVs and EVs due to the need for ecological vehicles to effectively utilize power. In his research work, Dr. Sab Safi has contemplated the idea of using induction machines instead of PMSMs. He also found that some portions of the IM are related to SRM and PM synchronous machines. Authors in [21] observed the capacity of the motor to save energy by decreasing the power loss using wheel torque designation. Loss of power in slope steer move as well as when driving straight occurs for different motors along with power loss for different wheel torque are investigated using identical methods. In [22] Juan de Santiago and other researchers have examined the drivelines of the various electric vehicles (EVs). In his work, the author has discussed the different EVs and the choice of the motor based on the type of framework used. The authors of [23] show the use of electric motors and their applications in hybrid electric vehicles by means of the outcome. In 2009, Gianmario Pellegrino [24] introduced three electric traction motors that had high proficiency and power with respect to inverter estimate and drift measurements. Based on the vehicle, internal permanent magnet (IPM), surface mounted permanent magnet, and inductor motors are analyzed.

2.2 Working of an Electric Motor

For easy understanding of the electric motor consider a 4-pole 3-phase AC induction motor [25]. The motor is connected to a battery which supplies electrical energy to the stator. A typical magnet is concocted with the help of a stator using the coils arranged in the core's opposite sides. Hence, when the battery supplies electrical energy the coils in the motor create a magnetic field that is capable of tugging the conducting rod at the rotor outside. This enables the rotor to spin, creating the necessary mechanical energy that turns the gears resulting in a tire rotation. In general, a non-electric vehicle will contain both an alternator and an engine. It works in a cycle wherein the wheels and gears are powered by the engine which in turn is powered by the battery [26]. The alternator present in non-electric vehicles is powered by the rotation of the wheels and will contribute to recharge the battery. This is also the reason why a small electric vehicle like a car that remains idle for a prolonged period of time needs to be jumped, i.e., the battery needs to be recharged for the car to start properly. However, there is no alternator in an electric car. Then how does the battery recharge? Though an electric vehicle does not have an alternator, the electric motor will act as both the alternator and a motor. An alternating current characterizes electricity using a variation of current and voltage with respect to time. The alternating nature of a typical AC signal results in stepped-up or stepped-down voltage at different time intervals. This makes the use of an AC motor unique in an electric vehicle. As mentioned earlier the motor is started using the battery which passes this energy to the gear, enabling rotation of the tires. This process is initiated when the driver steps on the accelerator, initiating the rotor to move with respect to the rotating magnetic field. However, when the foot is removed from the accelerator, the rotating magnetic field stops and the rotor begins to decrease its speed of rotation, contrary to stopping along with the magnetic field [27]. Thus the battery is recharged by the motion of the rotor to spin quicker than that of the magnetic field, making it behave as an alternator.

2.3 Alternating Current or Direct Current

The basic difference between the two currents is a known fact: while one is consistent (DC), the other is intermittent (AC) [28]. However, when choosing an electric motor, there are more factors that need to be taken into consideration.

- Direct current: The direct current is a unidirectional and constant flow of electricity when the polarity of the voltage is maintained with respect to time. For example, the positive and negative poles are marked in a battery. A unidirectional current is produced using the constant potential difference. Apart from batteries, solar and fuel cells also produce direct current.
- Alternating current: An alternating current characterizes electricity based on the variation of current and voltage following a sinusoidal pattern. Because of the

shape of the waveform, both current as well as voltage alternates between negative and positive polarity, with respect to time. This shape of the waveform also indicates the generation of electricity.

The best choice among these two forms of current for transferring useable energy over a large distance is AC electricity.

2.4 Hierarchy of Electric Motors

The hierarchy of electric motors is represented in Fig. 3.1. wherein the EVs are generally classified according to the presence and absence of commutators.

2.5 Electric Vehicle Industries and the Environment

As the performance of the electric motors improves, the performance of electric vehicles also elevates substantially so much so that they outperform their counterparts. Though the use of electric cars is not as prominent as expected, the exploration of various companies like Toyota and Tesla has instilled hope for the future of transportation with electric motors at the core. Companies like Alphabet, US Cobalt, Tesla Motors Inc., Albermarle, and Ford motors are some of the renowned companies that have revolutionized electric vehicles all around the world. Electric engines

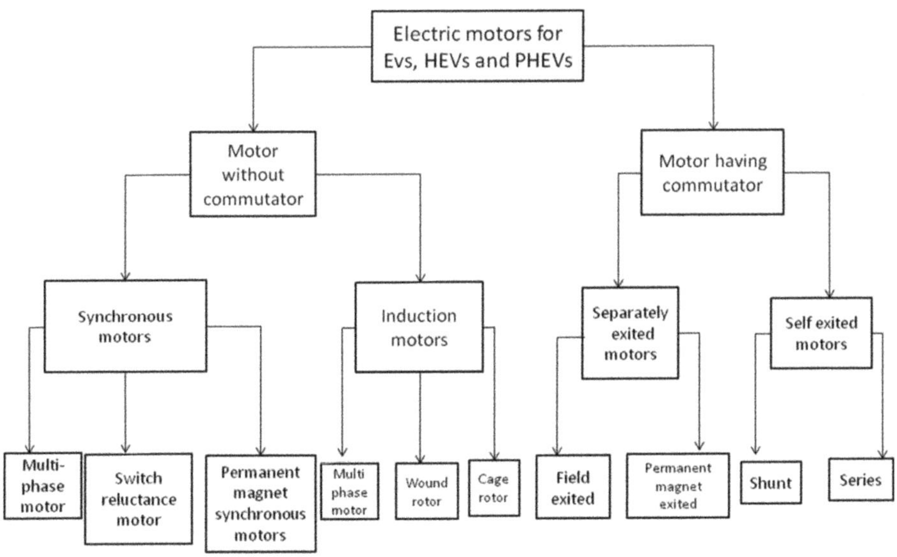

Fig. 3.1 Hierarchy of electric motors

have a direct as well as an indirect impact on the environment. The invention of electric cars has liberated buses from the use of gasoline and fuel-based cars resulting in a pollution-free atmosphere across the highways and cities. It resolves the strain of toxins in the air due to harmful emissions [29]. However, this in turn has created an additional problem of electricity production demand. Since electric engines do not emit much noise when compared with other gas or fuel-powered engines, it results in a significant reduction in noise pollution. Similarly, maintenance, as well as lubricants expense for a typical motor, is not essential for an electric motor, saving oils and chemicals used at the auto-shops.

3 Motors in Electric Vehicles

In general, most of the electric cars that are in use today operate using DC motors with a power of 4 kW or lesser. For Electric Vehicles that use advanced power, the induction motor serves as an apt choice of AC motor. Normally a vector drive is used to provide acceleration and torque management. A Brushless DC (BLDC) motor is commonly used in low power applications. Almost all EVs use batteries as the primary energy depository system. In recent years, a vast amount of research is being carried out in using in-battery improvements (such as lithium particles) for electric vehicle. The motors used, show diverse characteristics which make it crucial to assess the different motors based on certain guidelines to help determine the apt motor that can be used for the EV [30]. Electric vehicles use electric motors that possess important characteristics such as great control, low upkeep cost, high vitality, and easy plan. In general, the motors that are used by the EV manufacturers are Permanent Magnet Brushless Motors, Switched reluctance motors, Synchronous motors, Induction Motors, and DC motors. Table 3.1 gives an overview of the motors used in different electric vehicles built over the years.

Table 3.1 Motors used in electric vehicles

Product Name	Manufacturer	Types	Year
Toyota Prius	Toyota	Permanent magnet motor	1997
Tesla Model S	Tesla	Induction motor	2012
Fiat 500e	Fiat	Permanent magnet motor	2014
Volkswagen Golf Electric	Volkswagen	Permanent magnet motor	2014
Nissan Leaf	Nissan	Permanent magnet synchronous motor	2017
Mitsubishi i-MiEV ES	Mitsubishi	Permanent magnet synchronous motor	2017
Focus Electric	Ford	Permanent magnet motor	2018
Chevrolet Bolt EV	Chevrolet	Permanent magnet motor	2018
Audi E-Tron	Audi	Permanent magnet synchronous motor	2019
Kia Niro	Kia	AC synchronous permanent magnet	2020
Nissan Leaf	Nissan	Permanent magnet motor	2021

3.1 Direct Current Motors

The DC motors or Brushed DC motors were widely used during the 1990s. Brushed DC motors have the ability to have a high torque value at low velocities, resulting in appropriate output. But, this brushed DC motor has poor power density making it unsuitable to be used in electric vehicles. On the other hand, the use of a brushless DC motor will prove to be less maintenance and better efficiency. A Direct Current (DC) drive is used conspicuously in electric vehicles because of its impeccable torque speed and control over speed. There are two sections aligned using excursion motor namely: motors without communication and switching motors. A typical switching motor is a DC motor that uses both shunt and series excitation. They are also great contenders for applications that require low power. In a DC motor, the commutator acts as a strong inverter and is hence recommended to be used in gadgets, as they are economic and simple to build. However, the major drawback when using this motor is the high maintenance caused by both the commutators and brushes (hence also knows as Brushed DC Motors). The DC motors are still used in the Indian Railways.

3.2 Induction Motors

In general electric vehicles make use of three-phase induction motors because of their great speed control, high proficiency, and absence of a commutator. The three-phase AC supply is related to the winding of the stator which results in the development of a rotating magnetic field. The lack of a commutator has resulted in the maintenance-free tasks as well as high reliability and hence this motor is ideal for electric vehicles. Moreover, to uplift the dynamic execution of the electric drive framework, vector control is utilized in this motor. This inclusion provides a variety of speed range when using this motor. Overall, the three-phase AC induction motor (IM) is low accelerated and leads to better productivity and quality.

3.3 Permanent Magnet Synchronous (PMS) Motors

In general, the rotor moves at a synchronous speed in a synchronous motor. Though the stator uses a three-phase AC supply the rotor is powered by a DC supply. Brushless AC motors are commonly called permanent magnet synchronous (PMS) motors. In terms of vitality productivity, the PMS motor is found to be more effective and is comparable to that of the induction motor. Many renowned electric vehicle makers like Toyota, Honda, and Nissan have chosen PMS motors for their efficiency and effectiveness. Powerful dissemination of warmth, higher proficiency, and higher power thickness are some of the promising aspects of this motor. Due to the immense energy thickness of the magnets, agitation is confined to a smaller space resulting in higher profits for the machine. Since excitation current is not necessary it is possible to improve the speed proficiently. The losses involved while using the PMS motors are iron losses which can be easily overcome using an air conditioning framework. Thus PMS motors are found to be more efficient than induction motors in terms of productivity and control thickness. The biggest drawback is the cost of NdFeB in the magnets used in the motor. Another aspect of weakness is the requirement for additional current segment used for field debilitating, which results in lower productivity at large speeds and losses at the stator.

3.4 Permanent Magnet Brushless DC and AC Motors

Another type of motor that can be used in electric vehicles is the permanent magnet brushless DC (PM-BLDC) motor. These motors are used for many purposes by transforming the rotor and stator of the permanent magnet DC motor. Though this motor's setup is similar to that of a permanent magnet synchronous motor, the BLDC motor is powered by an AC supply with a waveform that is rectangular instead of the typical sinusoidal shape. The major advantage of the PM-BLDC motor is its ability to provide higher torque in line with voltage and current level. Because of these characteristics, the PM BLDC motor proves to be a compatible electric vehicle motor.

3.5 Switched Reluctance Motor (SRM)

The SRMs make use of the switching of rotor position to align the different phase windings in a sequence. It is possible to develop a wide speed using this methodology. Torque is induced in this manner based on the movement of the rotor toward the slightest reluctance. SRM is used in electric vehicles due to its ability to adapt in times of failure and large beginning torque.

3.6 Comparison of All the Motors Used in EVs and HEVs

The motors that are used in the electric motors are compared in Table 3.2 based on their power, efficiency, and other characteristics to determine the best electric motor based on the vehicle.

Table 3.2 A comparison of the motors using EVs and HEVs

	Induction motor	Switched reluctance motors	Permanent magnet motors	Brushless DC motors
Type	AC	AC	DC	AC
Family	Induction slip ring squirrel cage	Synchronous unexcited	Separately excited	Synchronous excited PM
Power to stator	AC	Pulsed DC	PM	Pulsed DC
Power to rotor	Induced	Induced	DC	PM
Weight	Medium	Medium	Medium	Low
Overall	Medium	Medium	Medium	High
Commutation method	External electronic	External electronic	Mechanical commutation	Internal electronic
Controller cost	High	High	Medium	Very high
Speed range	Controllable	Controllable	Limited by brushes, easy control	Excellent
Starting torque	High	Up to 200% of the rated torque	>200% of the rated torque	>175% of the rated torque
Speed control method	Frequency dependent	Frequency dependent	PWM	Frequency dependent
Maintenance requirement	Low	Low	Brushes wear	Low
Efficiency	High	Less than PMDC	High	High
Application	ICVs, EVs, and HEVs	ICVs	ICVs, EVs, and HEVs	ICVs, EVs, and HEVs
Efficiency with motor and power	85	86	91	79
Efficiency with power electronic devices only	94	91	94	98.5
Efficiency with motors only	91	95	97.5	81
Pros	High efficiency	Low inertia that can be modified according to the application	High starting torque	Long-life, tremendous power, fast responses, outstanding speed, and torque
Cons	Expensive controller	Requires power sensing, has ripples in torque, and is not very powerful	Limited rotation speed, bulky, requires maintenance, susceptible to damage if dropped	High cost, limited economy to small motor size
Examples	Chevrolet/ Siverado (USA)	Holden/ ECOmmodore	Honda/insight (Japan)	Peugeot Citroen/ Berlingo (France)

4 Parameters Considered in Selecting the Electric Motors

This chapter aims to establish a correlation between the various electric motors that are available in the market and the various factors considered by the electric vehicle manufacturers to find the apt electric motor. The following are specific parameters that have been observed while choosing the electric motor.

4.1 Power Density

The motor apex power is used to calculate the power density of an electric motor in relation to capacity-to-weight proportion. The apex power yield (kW) divided by mass (kg) gives the power density of any motor. Hence power density holds the measuring unit kW/kg. As observed in Fig. 3.2, the power density appears stronger in PM motors due to the newness of the permanent magnets' high power density. It has been observed by Thomas Finken et al. [31] that the highest power thickness is possible in a PMS machines, enabling them to establish the essential field within the confined motor cell. Putting this advantage to better use, the PM brushless motor provides the highest power density while closely followed by SRM and induction motors. On the other hand, the DC motors are said to exhibit low power density. Higher power in city motor is better suited for EV applications.

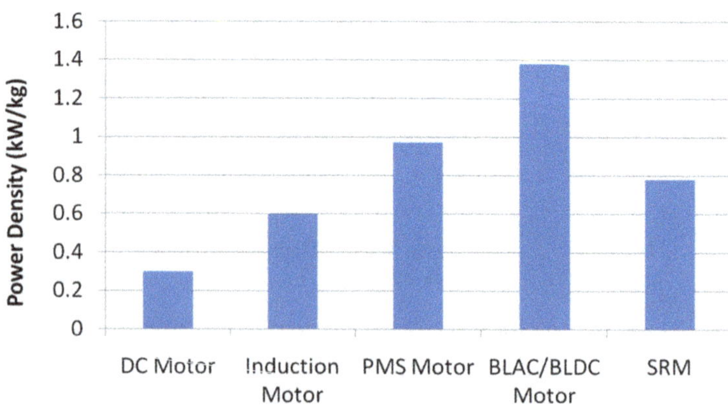

Fig. 3.2 Correlation of DC motor, induction motor, PMS motor, BLDC motor, and SRM with power density

4.2 Energy Efficiency

The efficiency of the electrical motor establishes a connection between the mechanical and electrical yield. Every electric motor manufactured is built to perform with optimal efficiency at the measured output. Based on the experimental analysis, it is observed from Fig. 3.3 that the induction motors show 91% efficiency while the BLDC motors show greater than 95% efficiency. This is mainly due to the lack of rotor losses in the BLDC motor, deeming it to be the most efficient and productive motor. On the other hand, DC motors contribute only 78% efficiency and are not preferred by electric vehicle manufacturers.

4.3 Reliability

As indicated by authors in [32], when comparing the different motors based on fidelity such as support and breakability, the most reliable electric motors would be the SRM and induction motors. The DC motors are found to be slightly dependable while the PM motors pursue reliability as shown in Fig. 3.4. DC motor switches and brushes allow entry of current in the armature and are found to be ill-equipped and are hence not suitable for tasks that are maintenance-free [33].

4.4 Cost Factor

One of the biggest problems faced by electric vehicle manufacturers is providing the customers, electric vehicles which fit their moderate expense and are also

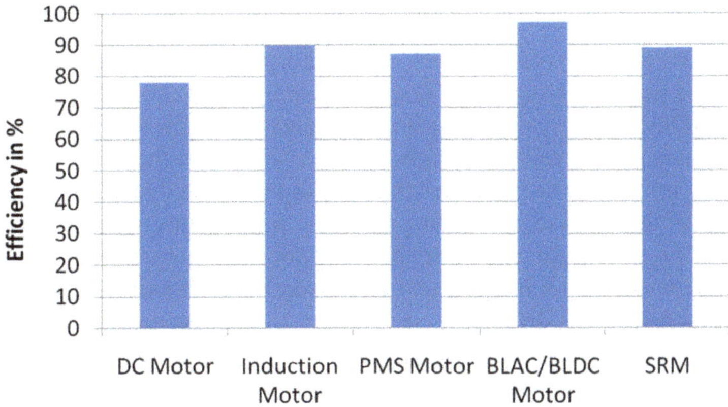

Fig. 3.3 Correlation of DC motor, induction motor, PMS motor, BLDC motor, and SRM with efficiency

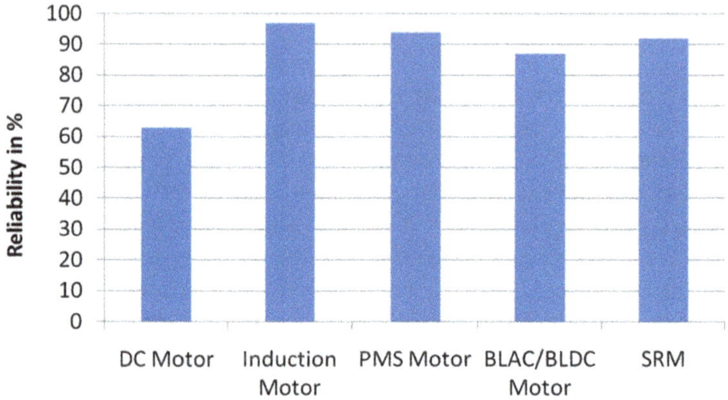

Fig. 3.4 Correlation of DC motor, induction motor, PMS motor, BLDC motor, and SRM with reliability

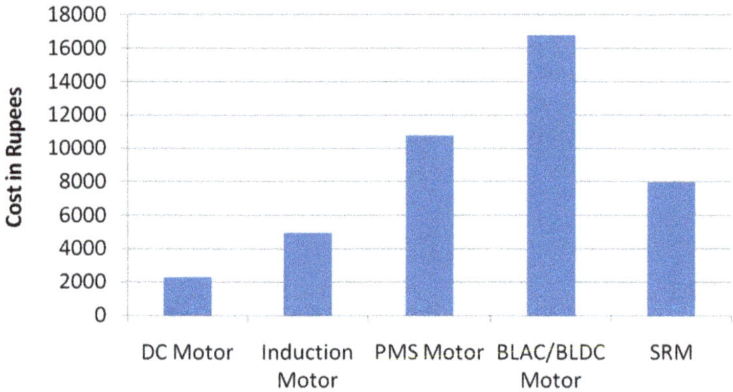

Fig. 3.5 Correlation of DC motor, induction motor, PMS motor, BLDC motor, and SRM with cost

competitive with similar class fuel vehicles. Taking the cost factor into consideration the IM would be the best match, followed closely by SRM and DC motors as shown in Fig. 3.5. Due to their economic rate, the induction engine is commonly used by many manufacturers in electric vehicles. As the capacity of the motor increases, the cost of a DC motor will be more than that of a similar capacity AC motor. If one has to pick one of the two motors holding the same power capacity, a motor with high torque and low speed will cost more than a motor with low torque and high speed. Similarly, a motor with a high current and lower operating voltage will cost more than that of a motor with the low occurrence and higher operating voltage. However, it is the design of the rotor that will play a crucial factor in increasing the cost of a typical DC motor. In fact, analysis indicates that an AC motor that is considered to be less expensive will be of low quality when compared with a high power-rating DC motor. The use of permanent magnets in PMDC motors makes it more expensive and the amount of magnet used is related to the power. On the other hand, iron pole pieces and copper windings are used in induction motors.

5 Conclusion and Future Scope

This chapter attempts at analyzing the diverse electric motors that are used in electric vehicles. Based on the observation, the following conclusions are drawn regarding the different motors examined: The DC motors are not easy to control but at capable of producing large torque at less speed. However, they have deficient efficacy, have an expansive structure, and possess large support costs. The BLDC motor is small in size, can generate high productivity, and has advanced power density. The drawback is the expensive control requirements. An Induction motor is capable of yielding productivity of over 90%. They exhibit substantial area, low power density, fidelity, and average acceleration. At lesser accelerations, synchronous machines are observed to be more proficient and can improve the usage of batter and have a propulsive extent. When steady torque was needed, the synchronous motor is preferred. SRMs are more suitable when the cost is a factor (as it is very economical), adapt to the internal failure capacity, have good efficiency and high reliability. Based on the type of electric vehicle and the characteristics of the vehicle built, the choice of the motor could be either of the five motors surveyed to yield the best results.

References

1. Rimpas, D., & Chalkiadakis, P. (2021). Electric vehicle transmission types and setups: A general review. *International Journal of Electric and Hybrid Vehicles, 13*(1), 38–56.
2. Ren, Q., Crolla, D. A., & Morris, A. (2009, September). Effect of transmission design on electric vehicle (EV) performance. In *2009 IEEE vehicle power and propulsion conference* (pp. 1260–1265). IEEE.
3. Karthikeyan, M., Hasan, B. K., Prakash, S., & Hariprabhu, M. (2021). Implementation of PV based bidirectional Dc To Dc converter using fuzzy logic controller for electric vehicle application. *Turkish Journal of Physiotherapy and Rehabilitation, 32*, 3.
4. Un-Noor, F., Padmanaban, S., Mihet-Popa, L., Mollah, M. N., & Hossain, E. (2017). A comprehensive study of key electric vehicle (EV) components, technologies, challenges, impacts, and future direction of development. *Energies, 10*(8), 1217.
5. Hori, Y., Toyoda, Y., & Tsuruoka, Y. (1998). Traction control of electric vehicle: Basic experimental results using the test EV "UOT electric march". *IEEE Transactions on Industry Applications, 34*(5), 1131–1138.
6. Lulhe, A. M., & Date, T. N. (2015, December). A technology review paper for drives used in electrical vehicle (EV) & hybrid electrical vehicles (HEV). In *2015 international conference on control, instrumentation, communication and computational technologies (ICCICCT)* (pp. 632–636). IEEE.
7. Kumar, M. S., & Revankar, S. T. (2017). Development scheme and key technology of an electric vehicle: An overview. *Renewable and Sustainable Energy Reviews, 70*, 1266–1285.

8. Park, G., Lee, S., Jin, S., & Kwak, S. (2014). Integrated modeling and analysis of dynamics for electric vehicle powertrains. *Expert Systems with Applications, 41*(5), 2595–2607.
9. Xue, X. D., Cheng, K. W. E., & Cheung, N. C. (2008, December). Selection of electric motor drives for electric vehicles. In *2008 Australasian universities power engineering conference* (pp. 1–6). IEEE.
10. Zhang, D., Qi, T., Wang, S., & Ling, Z. (2021). Effect of series/parallel circuits of eccentric switched reluctance motor on vehicle ride comfort. *Journal of the Brazilian Society of Mechanical Sciences and Engineering, 43*(4), 1–12.
11. Zheng, C., Wang, Y., Liu, Z., Sun, T., Kim, N., Jeong, J., & Cha, S. W. (2021). A hybrid energy storage system for an electric vehicle and its effectiveness validation. *International Journal of Precision Engineering and Manufacturing-Green Technology*, 1–16.
12. Banda, G., & Kolli, S. G. (2021). An intelligent adaptive neural network controller for a direct torque controlled eCAR propulsion system. *World Electric Vehicle Journal, 12*(1), 44.
13. Vuddanti, S., Shivanand, M. N., & Salkuti, S. R. (2021). Design of a one kilowatt wireless charging system for electric vehicle in line with Bharath EV standards. *International Journal of Emerging Electric Power Systems, 22*(3).
14. Shirley, D. R. A., Amruthavarshni, R. B., Durainathan, A., & Karthika, M. P. (2021, May). QR-based inventory management system (QR-IMS) of passenger luggage using website. In *2021 5th international conference on intelligent computing and control systems (ICICCS)* (pp. 1180–1185). IEEE.
15. Rivera, S., Kouro, S., Vazquez, S., Goetz, S. M., Lizana, R., & Romero-Cadaval, E. (2021). Electric vehicle charging infrastructure—From grid to battery. *IEEE Industrial Electronics Magazine, 15*, 37–51.
16. Hu, Y., Li, Y., Li, Z., & Zheng, L. (2021). Analysis and suppression of in-wheel motor electromagnetic excitation of IWM-EV. *Proceedings of the Institution of Mechanical Engineers, Part D: Journal of Automobile Engineering, 235*(6), 1552–1572.
17. Tremblay, O., & Dessaint, L. A. (2009). Experimental validation of a battery dynamic model for EV applications. *World electric vehicle journal, 3*(2), 289–298.
18. Senthilkumar, M., Kavitha, V. R., Kumar, M. S., Raj, P. A. C., & Shirley, D. R. A. (2021, March). Routing in a wireless sensor network using a hybrid algorithm to improve the lifetime of the nodes. In *IOP conference series: Materials science and engineering* (Vol. 1084, No. 1, p. 012051). IOP Publishing.
19. Abd Elhafez, A. A., Aldalbehia, M. A., Aldalbehia, N. F., Alotaibi, F. R., Alotaibia, N. A., & Alotaibi, R. S. (2017). Comparative study for machine candidates for high speed traction applications. *International Journal of Electrical Engineering, 10*(1), 71–84.
20. Safi, S. (2010). Alternative motor technologies for traction drives of hybrid and electric vehicles. *Consultant-SDT Drive Technology*.
21. Bălţăţanu, A., & Florea, L. M. (2013, May). Comparison of electric motors used for electric vehicles propulsion. In *Proceeding of International Conference of Scientific Paper AFASES*.
22. De Santiago, J., Bernhoff, H., Ekergård, B., Eriksson, S., Ferhatovic, S., Waters, R., & Leijon, M. (2011). Electrical motor drivelines in commercial all-electric vehicles: A review. *IEEE Transactions on Vehicular Technology, 61*(2), 475–484.
23. Dorrell, D. G., Knight, A. M., Popescu, M., Evans, L., & Staton, D. A. (2010, September). Comparison of different motor design drives for hybrid electric vehicles. In *2010 IEEE energy conversion congress and exposition* (pp. 3352–3359). IEEE.
24. Pellegrino, G., Vagati, A., Boazzo, B., & Guglielmi, P. (2012). Comparison of induction and PM synchronous motor drives for EV application including design examples. *IEEE Transactions on Industry Applications, 48*(6), 2322–2332.
25. Huynh, T. A., & Hsieh, M. F. (2018). Performance analysis of permanent magnet motors for electric vehicles (EV) traction considering driving cycles. *Energies, 11*(6), 1385.
26. Kumar, K., Tiwari, R., Varaprasad, P. V., Babu, C., & Reddy, K. J. (2021). Performance evaluation of fuel cell fed electric vehicle system with reconfigured quadratic boost converter. *International Journal of Hydrogen Energy, 46*(11), 8167–8178.

27. Shirley, D. R. A., Sundari, V. K., Sheeba, T. B., & Rani, S. S. Analysis of IoT-enabled intelligent detection and prevention system for drunken and juvenile drive classification. *Automotive Embedded Systems: Key Technologies, Innovations, and Applications*, 183–200.
28. Emadi, A., Lee, Y. J., & Rajashekara, K. (2008). Power electronics and motor drives in electric, hybrid electric, and plug-in hybrid electric vehicles. *IEEE Transactions on Industrial Electronics, 55*(6), 2237–2245.
29. Plötz, P., Schneider, U., Globisch, J., & Dütschke, E. (2014). Who will buy electric vehicles? Identifying early adopters in Germany. *Transportation Research Part A: Policy and Practice, 67*, 96–109.
30. Tigadi, A. S., Changappa, N., Singhal, S., & Kulkarni, S. (2021). Autonomous vehicles: Present technological traits and scope for future innovation. In M. Kathiresh & R. Neelaveni (Eds.), *Automotive embedded systems. EAI/Springer innovations in communication and computing*. Springer. https://doi.org/10.1007/978-3-030-59897-6_7.
31. Zhou, W., Yang, L., Cai, Y., & Ying, T. (2018). Dynamic programming for new energy vehicles based on their work modes part I: Electric vehicles and hybrid electric vehicles. *Journal of Power Sources, 406*, 151–166.
32. Finken, T., Felden, M., & Hameyer, K. (2008). Comparison and design of different electrical machine types regarding their applicability in hybrid electrical vehicles. *International Conference on Electrical Machines*.
33. Pennycott, A., et al. (2013). Enhancing the energy efficiency of fully electric vehicles via the minimization of motor power losses, *IEEE International Conference on Systems, Man, and Cybernetics*.

Chapter 4
A Comprehensive Study on DC–DC and DC–AC Converters in Electric and Hybrid Electric Vehicles

Sumukh Surya, P. Supriya, and Sheldon S. Williamson

1 Introduction

Air pollution and global warming are some of the major environmental issues caused to the use of conventional fossil fuel-powered vehicles. Tremendous efforts have been made in making the vehicles pollutant free. This can be performed by the use of vehicle electrification technologies, including EVs and HEVs based on the electricity produced from renewable energy sources. Battery EV or in general EV is known as "green vehicle or eco-friendly" vehicles as they have zero emission. The major drawback of such EVs is the range of distance that they can cover. To cover a large distance, larger capacities of batteries are used and those vehicles are called as HEVs. They cover a range of 500 km (TESLA mode) and need to be recharged later.

In modern-day EVs and HEVs, the role of drives and the application of power electronics (DC–DC and DC–AC) in drives is very significant. Different types of motors used in EV/HEV are permanent magnet brushless and switch reluctance motors [1]. The general requirements of the electric machine in the EV sector are (i) high-power density and high-torque density, (ii) low-torque ripple, (iii) high efficiency for wide torque and speed ranges, (iv) wide constant power-operating capability, (v) low cost and high reliability, (vi) low acoustic sound, (vii) high intermittent overload capability, and (viii) efficient working in harsh environment.

Figure 4.1 shows the general configuration of an EV and HEV. The two major power electronic units are DC–DC and DC–AC converters.

S. Surya (✉)
Kirtane and Pandit Information Technology, Bangalore, India

P. Supriya
RNSInstitute of Technology, Bangalore, India

S. S. Williamson
Ontario Tech University, Oshawa, ON, Canada
e-mail: sheldon.williamson@ontariotechu.ca

M. Kathiresh et al. (eds.), *E-Mobility*, EAI/Springer Innovations in Communication and Computing, https://doi.org/10.1007/978-3-030-85424-9_4

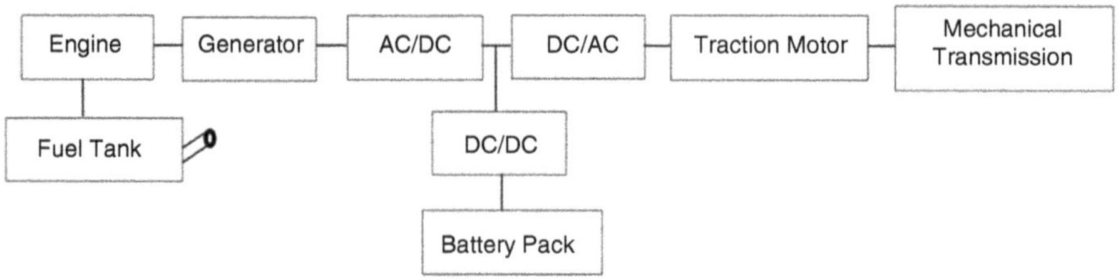

Fig. 4.1 General configuration of EV/HEV

These converters can be classified as (i) unidirectional and (ii) Bidirectional. The loads to the unidirectional converters are sensors, utility, and safety equipment. The bidirectional converters are used for regenerative braking and battery charging and discharging. The input to this converter is fed from a battery source for boosting purposes.

In this chapter, a summary on various DC–DC and DC–AC converters used in EV and HEV is discussed. A mathematical model for constant speed control for a separately excited DC motor is developed and analyzed using MATLAB/Simulink. This simulation showed the importance of DC–DC converters in a drive unit. Various inverters used in EV and HEV were discussed and simulation of a three-level inverter was carried out using built-in blocks of Simulink.

2 Electrical and Thermal Modeling

In [2], the electrical and thermal modeling of an IDU (Integrated Drive Unit) in an electric SUV is presented. The drive unit consists of a permanent magnet synchronous motor (PMSM), inverter, and a transmission gearbox. For cooling purposes, a coolant is circulated through the inverter and PMSM.

The lumped model for the motor and inverter was developed using MATLAB/ Simulink and the input to the model was supplied in the form of heat. The heat generated in the motor and its components was obtained from (Finite Element Analysis) FEA ANSYS Maxwell software tool. The IDC was modeled and simulated for GAC Aion LX SUV EV using the NEDC test cycle.

The drive cycle of an EV refers to the speed as a function of time. Drive cycles are produced by many countries to analyze the performance of EVs (Fuel consumption, CO_2 emission, and Core temperature generated in the battery (T_c)).

The drive cycle of an EV refers to the speed as a function of time. Drive cycles are produced by many countries to analyze the performance of EVs (Fuel consumption, CO_2 emission, and core temperature generated in the battery (T_c)).

In [3], a detailed study on the effect of fast discharge of a battery on T_c for the FTP75 drive cycle was analyzed. A DC motor of 230 V rating was selected to estimate T_c in terms of known surface (T_s) and ambient temperatures (T_{amb}) using

MATLAB/Simulink. The drive cycle data were available in the software in m/s which was converted to power using necessary equations.

Three different IDC models were discussed in [4] viz., (a) linear model, (b) saturated model, and (c) saturated/spatial harmonics model. Among these models, the saturated model is the most accurate result resulting in large computation time [2]. Hence, linear and saturated models were considered for the prediction of torque, no-load speed, and torque-speed characteristics.

It was observed that the thermal resistances lumped parameter thermal network were constants. However, the speed-dependent components were rotor-stator air gap thermal and the rotor end space resistances. The high temperature was observed at the stator teeth, rotor, and winding because of their inner heat generation [2]. Simulations showed that the temperature at the stator teeth reached 120°C for motor speed ~12,000 rpm. This was due to the stator iron loss which is significant in the NEDC cycles [4].

A similar study on the internal temperature of a cell was studied and estimated using a Kalman filter in [2, 5]. A second-order thermal model was considered for estimating the internal temperature. It was shown that with the increase in current, the temperature drastically increased.

3 Power Electronics Intensive Solutions for HEVs

In this section, the role of power electronics in HEV is highlighted and a detailed comparison of the associated advanced power system architectures for HEV as well as electric vehicle (EV) and fuel cell vehicle (FCV) applications is discussed.

During mid-1950s, automotive industries started using 12 V power systems. The battery pack which during this period had six cells set instead of three cells. The demand of electrical power had peaked to 1 kW by the 1990s [6–8].

The conventional electric system of a single 14 V is shown in Fig. 4.1 was able to satisfy (a) vehicular loads like lightings, motor-driven fans, pumps, and compressors. A major drawback of the 14 V system was regarding the control aspect. The switches were manually controlled and hence lacked precision [7, 8]. In addition, the power level did not meet the requirement of 1 kW. The battery voltage varied from 9 to 16 V which created overrating the loads at nominal system voltage (Fig. 4.2).

In view of increasing the electrical load and overcome the disadvantages in the conventional system, mild hybrid EV (MHEV) was introduced with a voltage rating of 300 V. Figure 4.3 shows a block diagram representation of MHEV. This topology takes care of loads including lights, pumps, fans, and electric motors for various functions. In addition, they also include some advanced, electrically assisted vehicular loads, such as power steering, air conditioner/compressor, electromechanical valve control, active suspension/vehicle dynamics, and catalytic converter [7].

In future EVs, power electronics is believed to take measure on three major tasks. (a) Turning ON/OFF loads, which are executed by mechanical switches and relays in conventional cars [8]. (b) A dedicated controller for electrical machinery. (c) For changing system voltage levels and conversion of electrical power from one form to another, using DC–DC, DC–AC, and AC–DC converters.

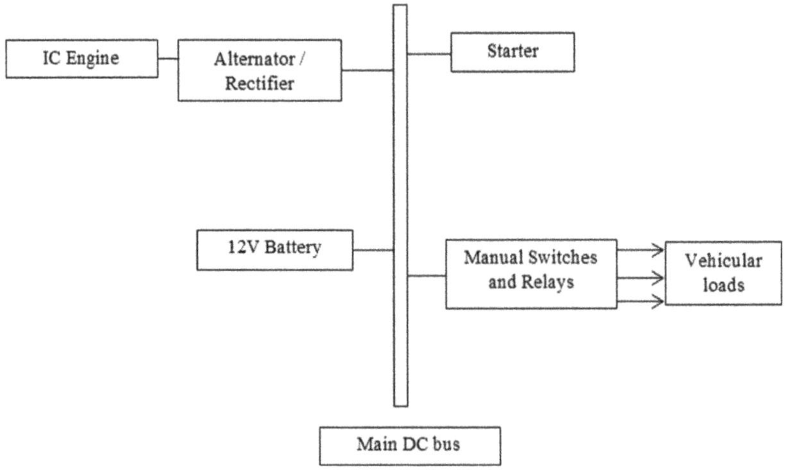

Fig. 4.2 Conventional 14 V DC power system architecture [9]

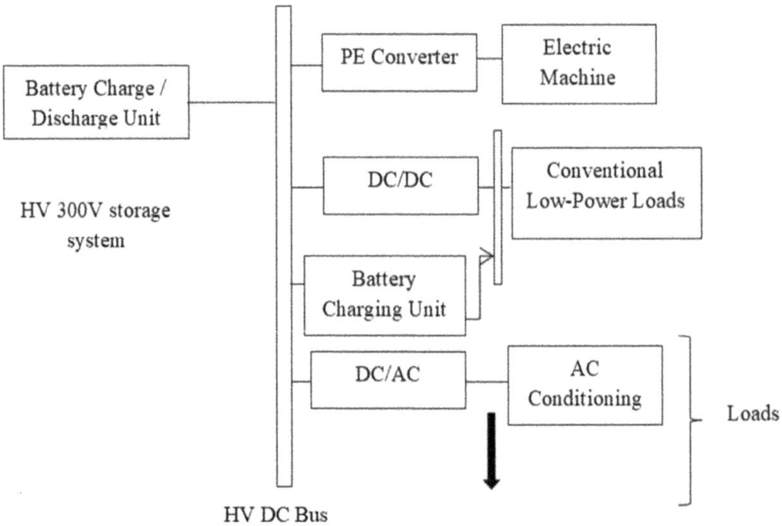

Fig. 4.3 Architecture of Mild HEV [9]

4 Introduction to DC–DC Converters

A DC–DC converter referred to as SMPS (switch mode power supply) is used to change the voltage levels without change in electrical power [10]. The components of a DC–DC converter include (a) electronic switch, (b) inductor, L, (b) capacitor, C, and (c) resistive load. The electronic switch can be a MOSFET, IGBT, or thyristor-based on the operating voltage and frequency.

These DC–DC converters are distinguished as (a) Isolated and (b) Non-Isolated. Few examples of isolated converters are (a) Flyback, (b) Forward, (c) Push–Pull, and so on. These converters possess a transformer at the input side of the converter. This transformer isolates the primary from the secondary and prevents the travel of input voltage spikes to the output secondary side. Based on inductor discharging, continuous conduction mode (CCM) and discontinuous conduction mode (DCM) operations occur [11, 12].

Some examples non-isolated topology are (a) Buck, (b) Boost, (c) Buck–Boost, (d) Cuk, (d) SEPIC, and (e) Zeta converters. Buck and Boost converters are used as a step-down and step-up voltage levels. Buck-boost converters either buck or boost the voltage depending on the duty cycle. Cuk and SEPIC converters are used in EV charging systems as they always produce continuous input current [13, 14].

The modeling of non-isolated and isolated DC–DC converters in open loop is presented in [10, 15]. The converters considered were ideal and operated in CCM. The converters were modeled using volt-sec and amp-sec balance equations using modeled using MATLAB/Simulink. In reality, the switch has a resistor in series, diode contains a forward voltage and resistance, L and C possess equivalent series resistance (ESR).

In [16], the converters were modeled considering the various drops in the converter. Different methods of modeling were presented using MATLAB/ Simulink. It was shown in [16] that the three methods provided similar output voltages. It was concluded from [16] that the built-in models in Simulink under Simscape and Simulink libraries provided only steady-state quantities. However, mathematical modeling of the converter using equations provided information about transient and steady-state quantities.

5 Stability of DC–DC Converters

In any DC–DC converter duty cycle can be considered as one of the inputs. Hence, a DC–DC converter consists of input voltage V_g and duty cycle, d. Hence, there exit two transfer functions viz.; output voltage to input voltage (G_{vg}) and output voltage to duty ratio (G_{vd}). To obtain these transfer functions, there exit numerous techniques like (a) small-signal modeling, (b) state-space averaging, (c) circuit averaging, and so on [17].

It was shown in [17] that the transfer function of average inductor current to duty ratio (G_{id}) for the Boost and Synchronous Boost converter was stable. This is a critical aspect for the converters for power factor correction. In [13, 14], it was shown that G obtained from small-signal model and that of circuit averaging matched. It was shown that G_{vd} for Cuk and SEPIC in CCM and DCM was unstable and stable, respectively.

6 Application of DC–DC Converters in Drives

The mathematical model for a DC separately excited motor is shown below [1]

$$V_a = i_a R_a + L_a \frac{di_a}{dt} + E_a \tag{4.1}$$

$$E_a = k\varphi\omega \tag{4.2}$$

$$T = J\frac{dw}{dt} + \mathrm{B}\omega + \mathrm{T}_L \tag{4.3}$$

$$T_L = k\varphi i_a \tag{4.4}$$

where V_a is the supply voltage (V), i_a (A) is the armature current, R_a and L_a are the armature resistance (Ω), and inductance (H), respectively, E_a is the Back EMF (V), T and T_L are the torque induced in the armature and load torque (Nm), respectively, J is the Moment of Inertia (kgm^2), B is the magnetic co-efficient (T), ω is the speed (rad/s), and ϕ is the flux induced (Wb).

To study the dynamics of the system, this motor was modeled and simulated using MATLAB/Simulink in open-loop and closed-loop configurations. An appropriate step size as per [18] was chosen and the no-load speed and speed—torque relation was analyzed.

The motor constants were defined in MATLAB (3 kW, 220 V). Figure 4.4 shows the Simulink model for the motor.

Figure 4.5 shows the no-load speed of the motor. It is observed that the no-load speed settles in <1 s and the magnitude of the speed was around 180 rad/s.

Since no-load torque was applied, the armature current was reduced to zero under steady-state conditions. Since $T \propto i_a$, T also reaches zero under steady-state conditions (Figs. 4.6, 4.7).

The closed-loop implementation of the motor can be modeled using PWM control. The speed of the machine is controlled using a boost converter. Figure 4.8 shows closed-loop implementation in MATLAB/Simulink.

In order to achieve a constant speed of 110 rad/s, a boost converter operating in CCM was considered. By continuously changing the duty cycle of the converter, the speed is regulated. Figure 4.9 shows a step variation in the speed at $t = 5$ s.

A PID controller is used to control the speed. Manual tuning approach is carried to tune K_p, K_i, and K_d parameters (Fig. 4.10).

Figure 4.11 shows the closed-loop response of the motor for constant speed operation. It is observed that the speed settles to the desired value in ~3 s. The response can be made faster by further tuning the PID parameters. Figure 4.12 shows the variations in the duty cycle for achieving constant speed. The procedure obtaining the actual values of K_p, K_i, and K_d is shown in [19].

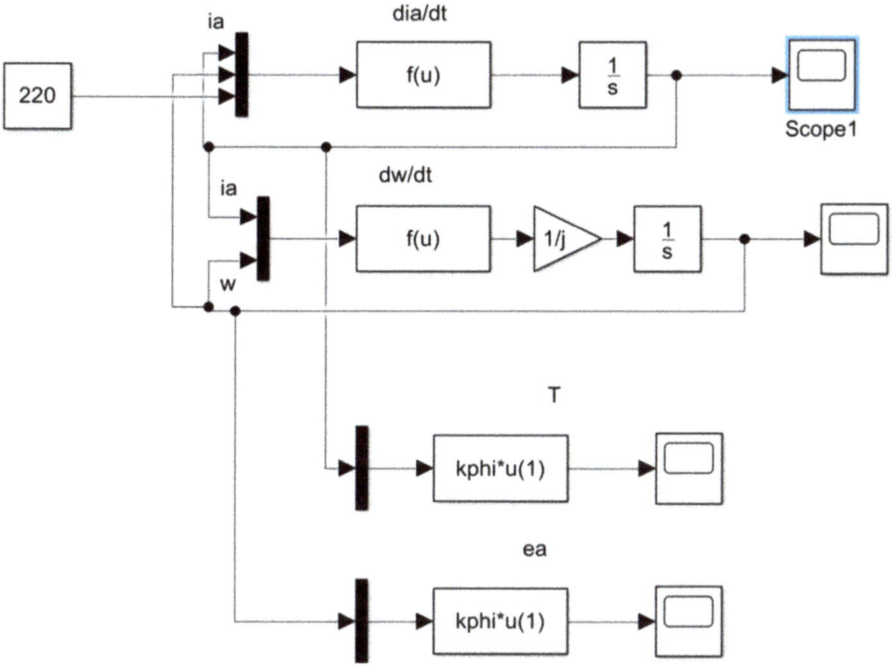

Fig. 4.4 Mathematical model for a separately excited DC motor

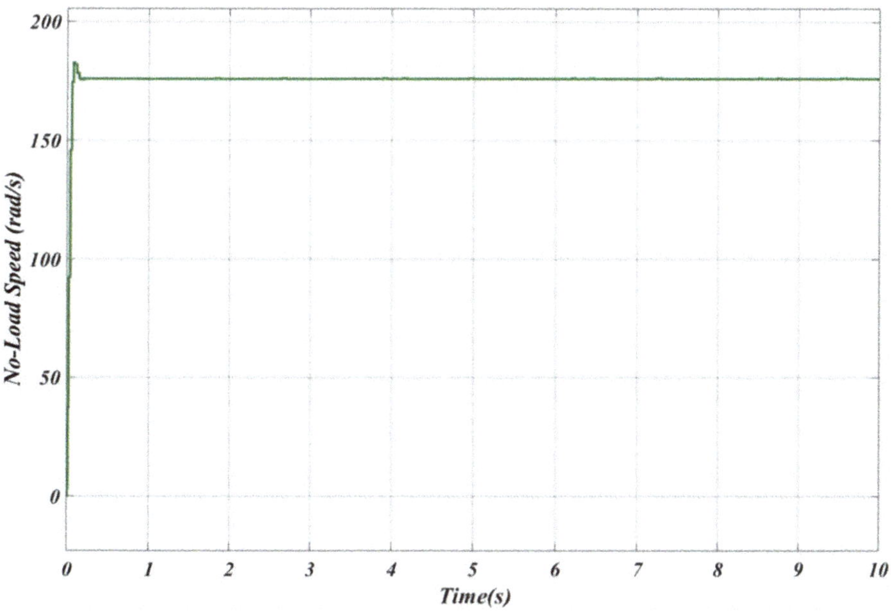

Fig. 4.5 No-load speed versus time

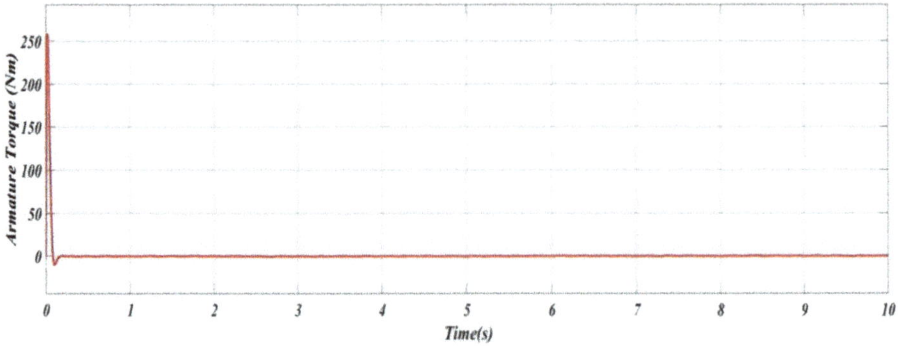

Fig. 4.6 Armature torque versus time

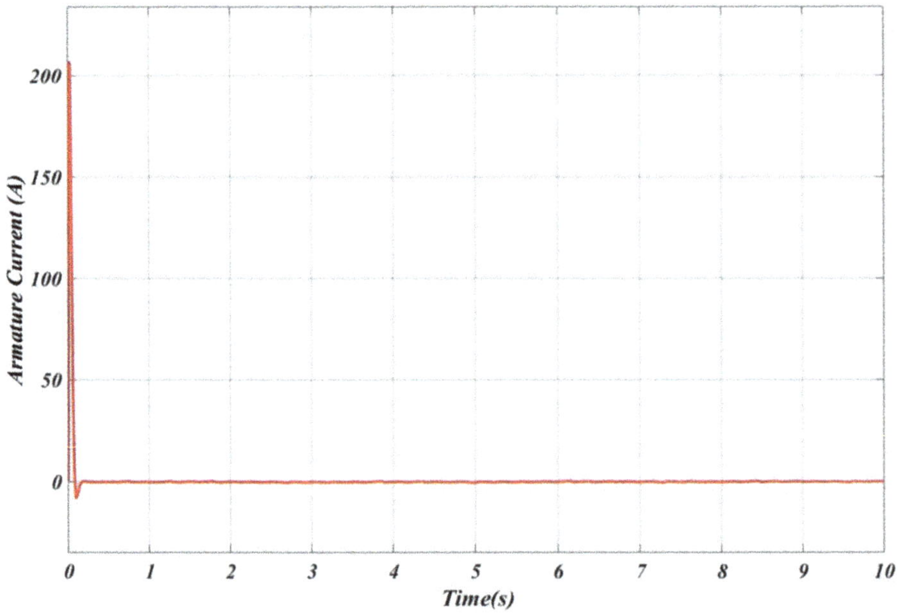

Fig. 4.7 Armature current versus time

7 Control Techniques in DC–DC Converters

In the DC–DC converter the various control techniques are used for

(a) Output voltage
(b) Input voltage
(c) Average current
(d) Peak current mode (PCM)

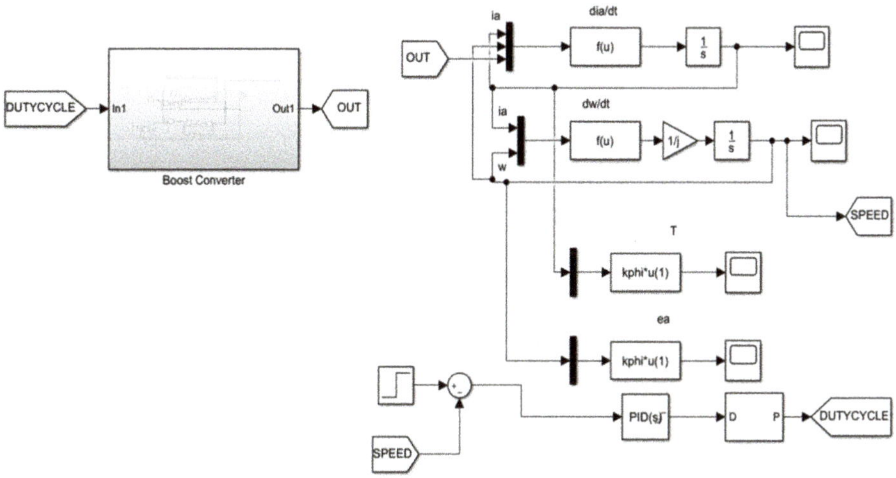

Fig. 4.8 Closed-loop speed control using PWM control

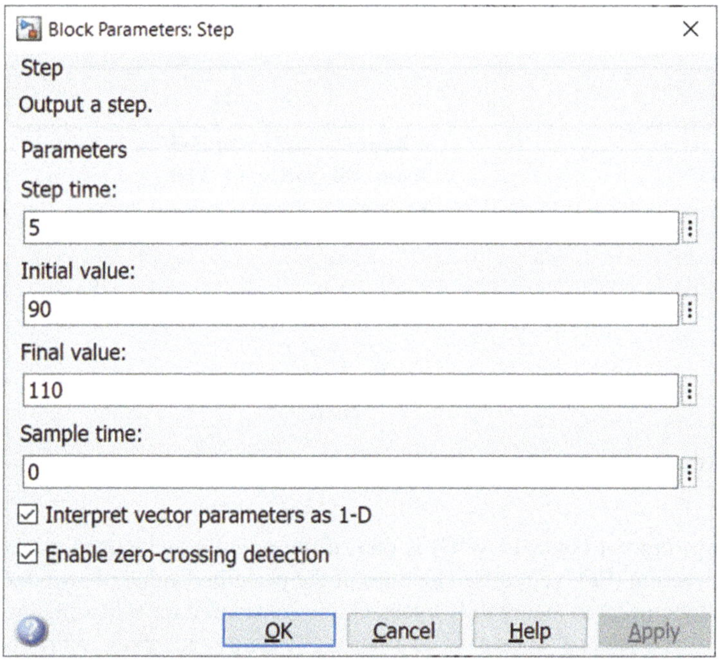

Fig. 4.9 A step variation in the speed at $t = 5$ s.

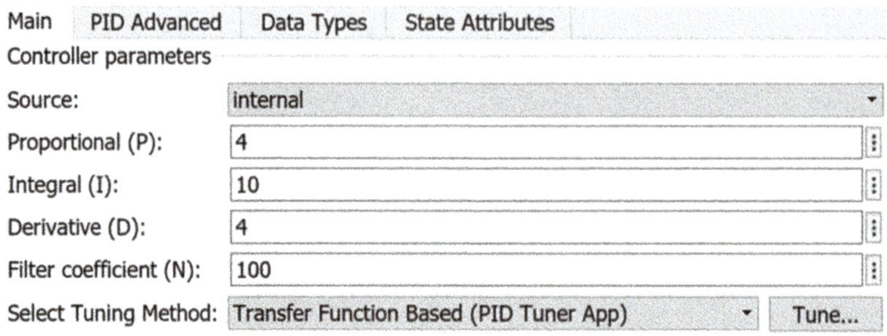

Fig. 4.10 K_p, K_i, and K_d for constant speed operation

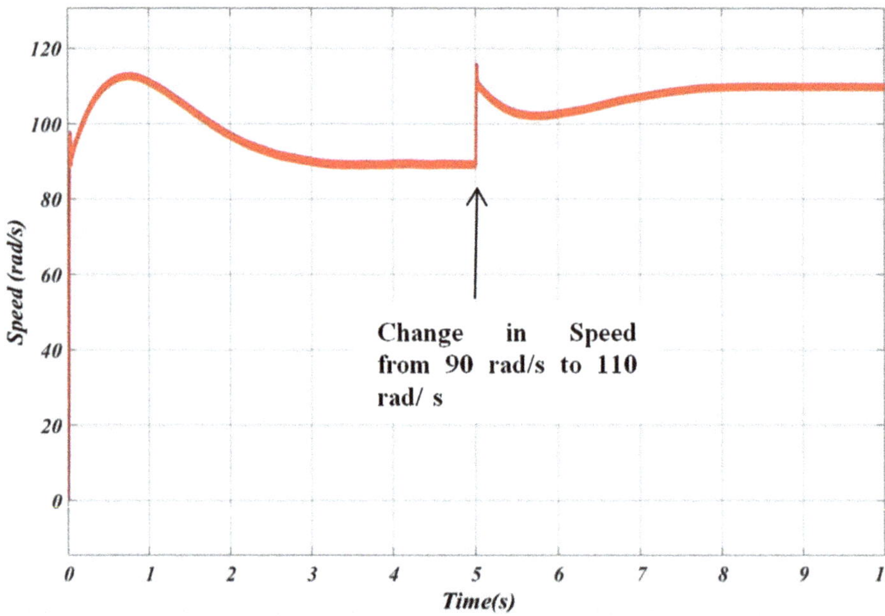

Fig. 4.11 Closed-loop response

Average current control (ACC) is one of the popular techniques used in power factor correction (PFC) circuits. The ratio of the perturbed value of switch current or the inductor current to that of duty cycle, G_{vd} is considered for achieving unity power factor. A controller is necessary to control V_g and i_g in phase with each other. Figure 4.13 shows the block diagram representation of closed-loop ACC in a switched converter. Though ACC cannot provide instantaneous control, some of the advantages of ACC over PCM include (a) better noise immunity and (b) possible control for a wide range of applications.

Where R_f is the sensing resistor, v_c is the reference voltage based on reference current i_c, G_{ci} is the controller, V_g and V_0 are the input and output voltages.

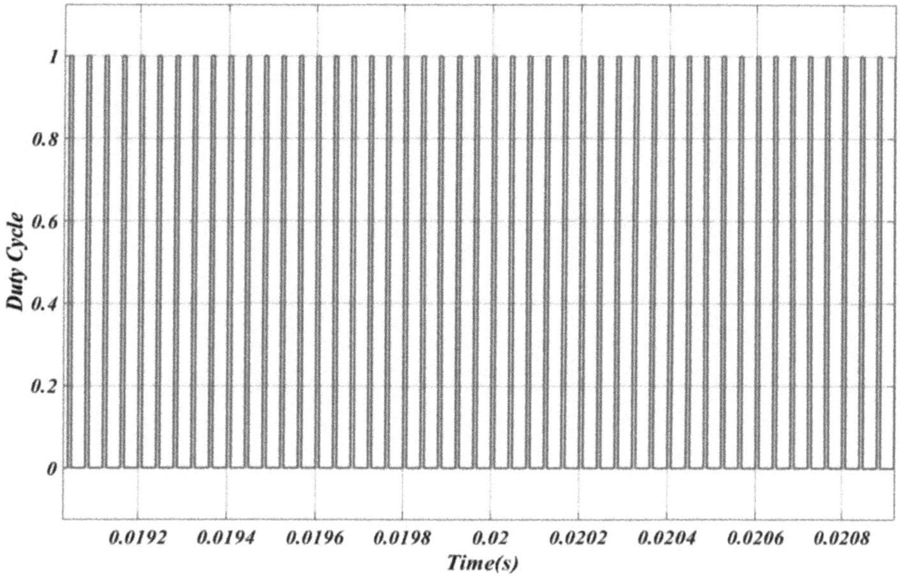

Fig. 4.12 Variation in D versus time

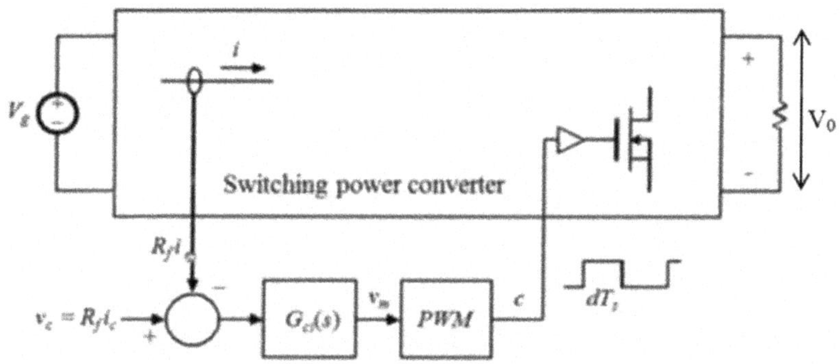

Fig. 4.13 Closed-loop ACC in a switched converter [14]

PCM is mainly used for the protection of switches from current transients. Figure 4.14 shows the closed-loop implementation of PCM applied to a buck converter [12].

The control current $i_c(t)$ is used as a reference and compared with the switch or the inductor current $i_s(t)$. When $i_s(t)$ reaches its peak value, the switch (transistor) is switched off. The analog comparator consists of two inputs viz., (a) sensed switch current and (b) control input—which sets the peak current.

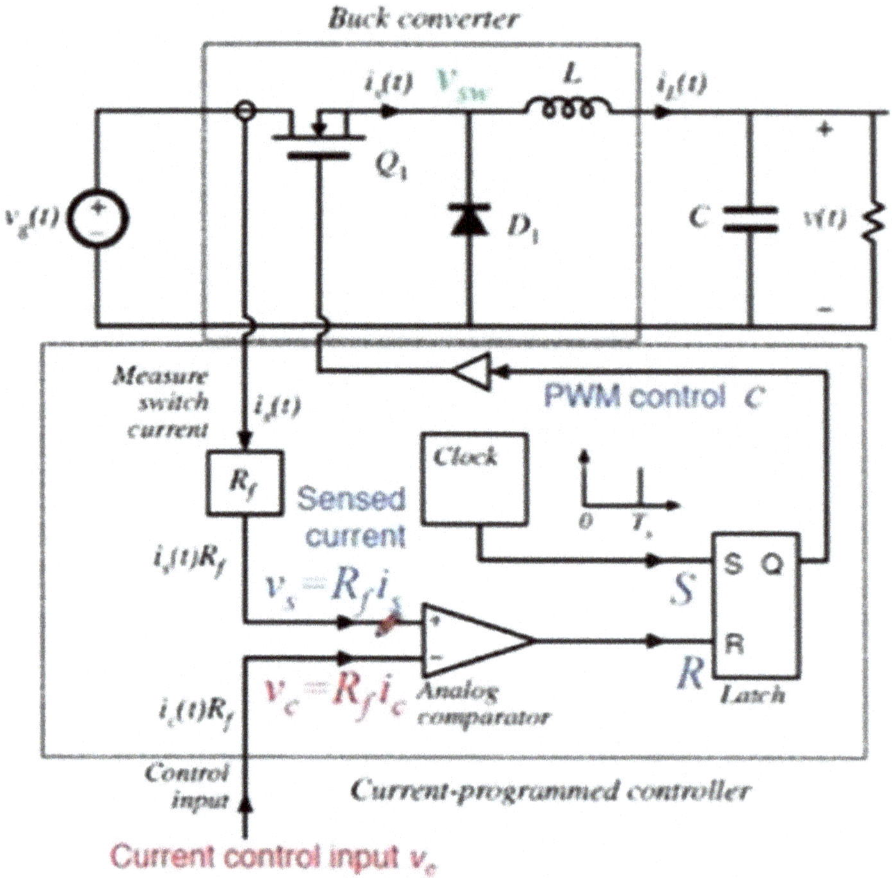

Fig. 4.14 Closed-loop PCM applied to a buck converter [12]

SR latch during the start, sets the signal enabling the switch ON condition. The current $i_s(t)$ starts ramping up. When $i_L = i_c = i_s$, reset signal would be enabled. This cycle repeats and control is ensured. It was clearly shown in [12] that $D > 0.5$ PCM would be unstable in all the converters.

8 DC–DC Converters in EV/HEV

In order to achieve high voltage gain and high efficiency, zero current switching (ZCS) and zero voltage switching (ZVS) switched DC–DC converters are used [20]. The duty cycle plays a very important in selecting the converter. An illustration of the selection of duty cycle is shown below.

Design a converter for the following specifications.

Specification	Value
Output voltage	36 V
Input voltage	5 V
Output current	0.5 A
Current ripple	10%
Voltage ripple	1%
Switching frequency	25 kHz

If selection is a boost converter,

$$D = 0.86 \quad \left(\frac{V_0}{V_g} = \frac{D}{1 - D} \right) \tag{4.5}$$

If selection is either Buck–Boost, Cuk, or SEPIC, $D = 0.87$.
If selection is flyback converter,
Let $D = 0.35$
$(N_2/N_1) = 14$ based on

$$\frac{V_0}{V_g} = \left(\frac{D}{1 - D} \right) \frac{N_2}{N_1} \tag{4.6}$$

Re-calculate duty ratio for $\frac{N_2}{N_1} = 14$

$$\mathrm{D} = 0.33 \tag{4.7}$$

Hence, a flyback converter is best suited for this application as $D = 0.86/0.87$ causes higher losses. Therefore, in general, it is desired to choose a converter having smaller duty ratio.

The conventional converters suffer from the following disadvantages

(a) High ripple current
(b) D increases with the increase in the input voltage
(c) Increased Voltage stress on the diode

These problems can be eliminated by using an interleaved boost converter. These converters generate losses due to hard switching and low efficiency for a higher output voltage. Hence, it is preferable to use zero current transition (ZCT) IBC with an auxiliary circuit [20]. This circuit reduces the reverse recovery time of the output diode and will turn on naturally at zero current. Figures 4.15 and 4.16 show the conventional boost and the Conventional IBC, respectively [20].

In order to achieve higher efficiency, a zero voltage zero current switch interleaved boost converter (ZVZCS IBC) is used. Figure 4.17 shows the circuit diagram of ZVZCS IBC.

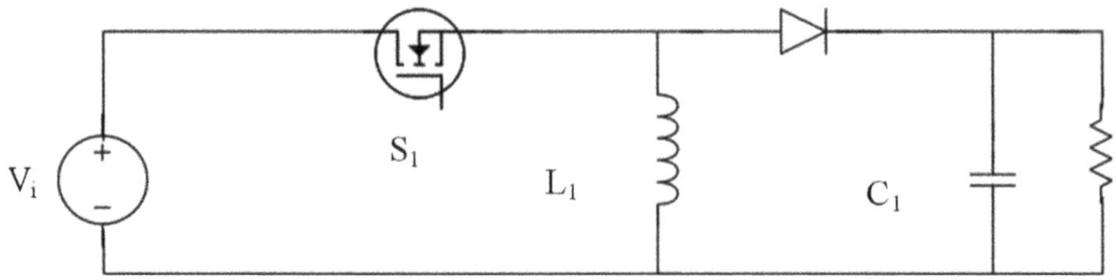

Fig. 4.15 Conventional boost converter [20]

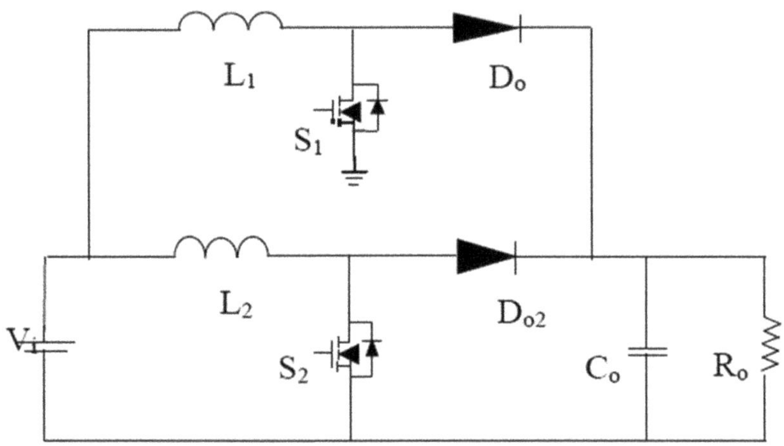

Fig. 4.16 Conventional IBC [20]

In each phase of interleaved boost converter, an auxiliary circuit is placed which helps to accomplish a ZCS turning on of main switch and ZVS turning off of main switch [20].

It was shown in [21] that the four-phase IBC diminished the input current ripple till $D = 75\%$ and can be eliminated for greater than 90% of D.

In order to generate a high gain in converters, KY converters are used [19]. These converters provide a swift transient response. They reduce the stress on the output capacitor and voltage ripple at the output [19]. The gain can be adjusted using various PWM techniques. Figure 4.18 shows a KY converter with gain $M = 1 + D$.

Figure 4.19 shows the circuit diagram of a KY boost converter which is integrated with a KY converter and a traditional boost converter. Voltage transformation ratio, M of the KY boost converters can be expressed as:

$$M = (2 - D)/(1 - D) \tag{4.8}$$

These converters are used in applications where the current ripple is fixed and high efficiency is required.

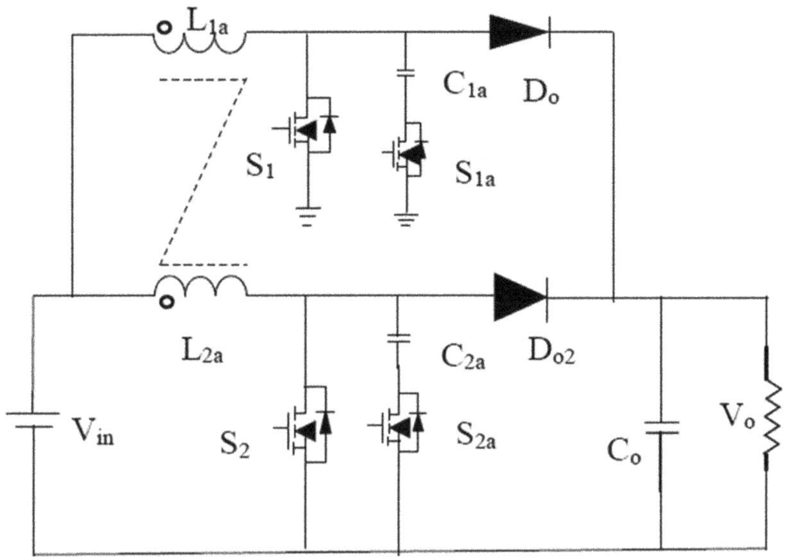

Fig. 4.17 ZVZCS IBC with coupled inductor [20]

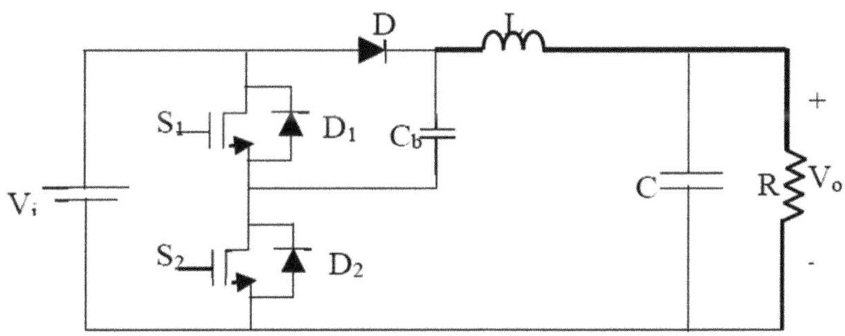

Fig. 4.18 KY converter [19]

9 DC–AC Converters in EV/HEV

9.1 Comparative Analysis of Two- and Three-Level Inverter

A performance comparison study of two and three-level inverters has been carried out in [22]. The use of EV and HEV reduces carbon emission compared to IC engines. The battery source used in EV is supplied to the inverters to drive motors which are coupled with the car wheels.

In order to increase power density in the system, maximum power in the system should be transmitted to the load instead of wasting in the power electronic

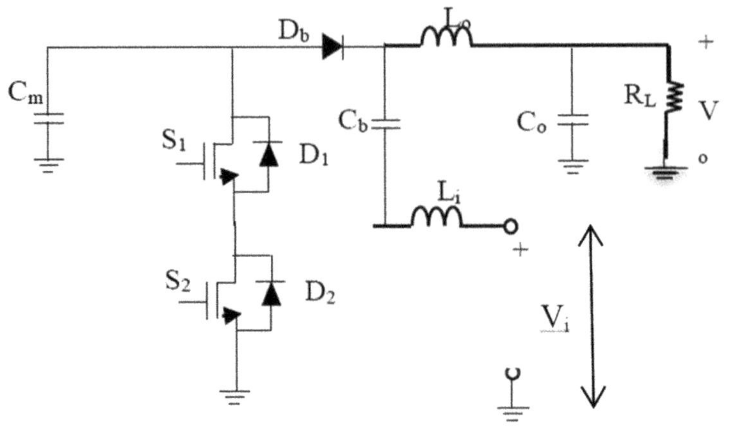

Fig. 4.19 KY converter [20]

converters. DC–AC inverter system provides high-power density for EVs. The function of an inverter is to change a DC input voltage to a symmetrical AC output voltage of desired magnitude and frequency.

9.2 2-Level DC/AC Traction Inverter Drive

Two-level inverters are most widely used DC–AC inverter. This type of inverter has problems with higher switching losses at higher switching frequencies. This problem worsens if the DC link voltage increases as the device voltage needs to be increased. To reduce the torque ripple of the machine, the switching frequency need to be increased. Figure 4.20 shows the two-level inverter-based PMSM traction.

Drive connected with the car wheel. In the above circuit I_{invrms}, I_{invavg}, I_{caprms} are the RMS values of inverter currents supplied to the load, average current supplied by the battery, and ripple current supplied by DC link capacitor, respectively.

In the above model, if the machine speed increases above the base speed, the switching frequency will be kept constant at 12 kHz depending on DC link voltage and torque requirement of the load. Usually, DC link voltage would be fixed at 450 V to keep switch losses minimum. But this introduces more current for the same value of output power when compared to higher DC bus voltage. In order to avoid this higher value of current many electrical companies are replacing their two-level inverter voltage drive with the three-level inverter. The three-level inverter reduces switching losses and for the same DC link voltage, voltage rating also reduces.

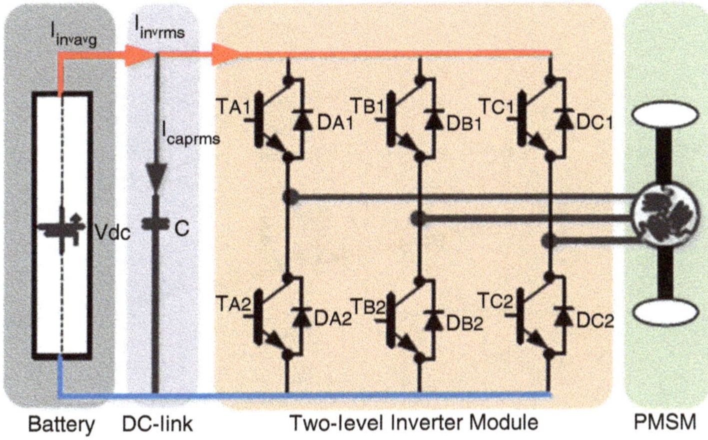

Fig. 4.20 Two-level DC/AC traction inverter [22]

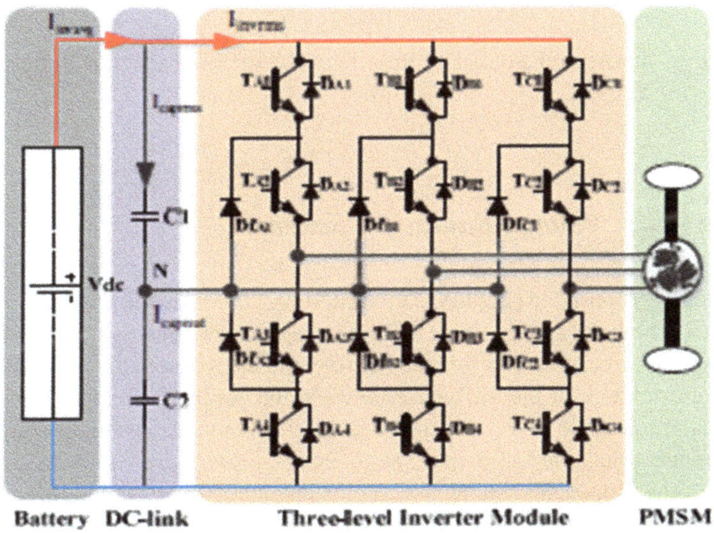

Fig. 4.21 Three-level neutral point clamped DC/AC traction inverter [22]

9.3 Three-Level Neutral Point Clamped (NPC) DC/AC Traction Inverter

Figure 4.21 shows a three-level NPC inverter circuit connected to the vehicle wheel through PMSM. It can be observed that the switch count is doubled when compared to a two-level inverter, due to which the voltage stress on all the switches is reduced by half when compared to a two-level inverter. There are additional six diodes

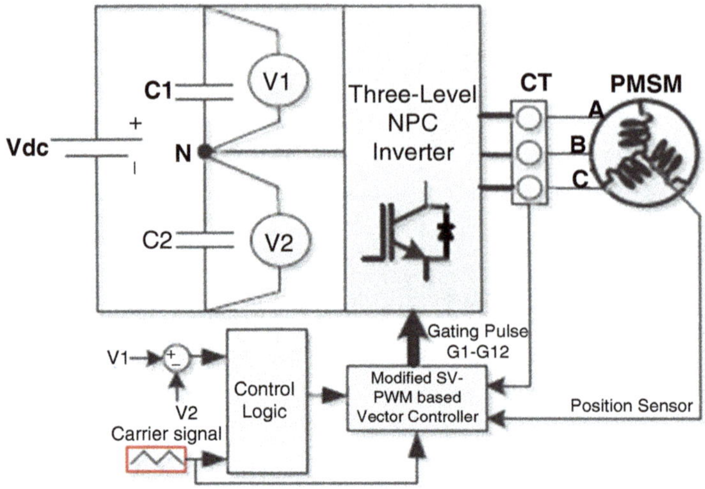

Fig. 4.22 NPC inverter control circuit with DC-bus voltage balancing [22]

connected to the neutral point DC link, because of this capacitor voltage will be unbalanced.

9.4 DC-Link Voltage Balancing Scheme

Figure 4.22 is proposed in [22] to balance DC-link voltage in the three-level inverter. The above scheme proposes two DC-link capacitor voltages that are measured by two voltage sensors and the difference between the measured value is passed through a logic control block. If the difference between the capacitor voltage is greater than the particular tolerance value, the circuit uses an upper capacitor to discharge, and if the difference voltage is lower than the tolerance value it uses a lower capacitor to discharge. The proposed scheme also avoids unsymmetrical switching.

Simulation results in [22] indicate that the total conduction loss for the three-level inverter is higher than the two-level inverter because of the presence of more switches and diodes when compared to the two-level inverter.

In [22] a detailed comparison of two-level and three-level inverters is carried out with a main focus on the switching losses. A low switching loss-based DC link voltage balancing algorithm is used which keeps the two capacitor voltage differences below tolerance level.

9.5 Five-Leg Inverter for an Electric Vehicle in Wheel Motor Drive

A 5-leg single inverter was used for controlling 2 three-phase PMSM motors independently [22]. This method provided an innovative solution to reduce the switch count and hence, decreased the overall cost of the drive. This arrangement also reduces the controller and sensor costs.

Figure 4.23 shows the proposed five-leg inverter-based in-wheel EV motor drive. The five-leg inverter is a single inverter that can drive two motors independently. This inverter consists of five legs, which are primarily a pair of arms that consist of power switching devices and diodes.

Figure 4.24 shows the switching arrangement of a five-leg inverter based in-wheel motor drive. The "C" phase of each motor is connected to one leg in common; the A and B phases of each motor are connected to other legs. As "C" phase of each motor is connected to the common leg of the inverter, it causes a difference in switching pattern.

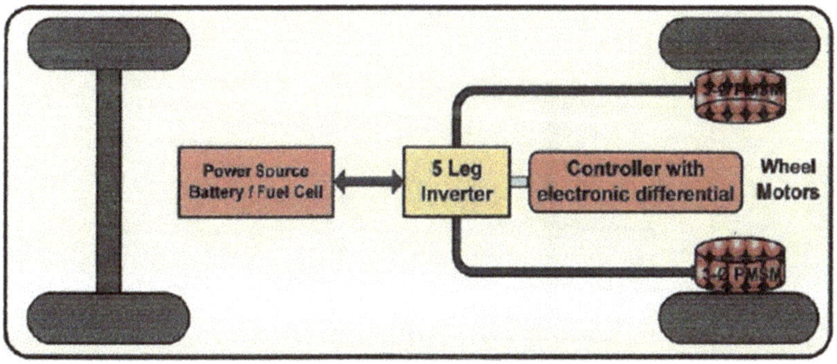

Fig. 4.23 Five-leg inverter-based in-wheel motor drive [23]

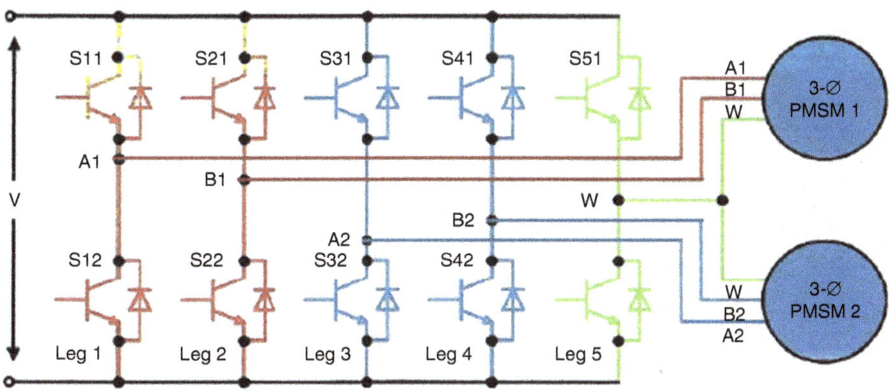

Fig. 4.24 Configuration of a five-leg inverter with 2 three-phase PMSM in-wheel motors [23]

Numerous modulation methods have been proposed for five-leg inverters. However, these methods cannot be used for in-wheel motor drives as each wheel motor requires a dissimilar control torque in order to maintain vehicle stability during cornering.

A modulation method known as modified expanded two-arm modulation (ETAM) can be adopted for in-wheel EV drive applications [23]. By doing so, a voltage utility factor (VUF) of 50% is achieved. The VUF is defined as the ratio of the inverter and DC-link voltage. The in-wheel motor drive methodology was analyzed in terms of converter efficiency, torque ripple, converter rating, vehicle performance, and cost.

10 Simulation of Three-Level Inverter

Simulation of a three-level inverter is carried using sine pulse width modulation (SPWM) method. Figure 4.25 shows the model developed using MATLAB/Simulink.

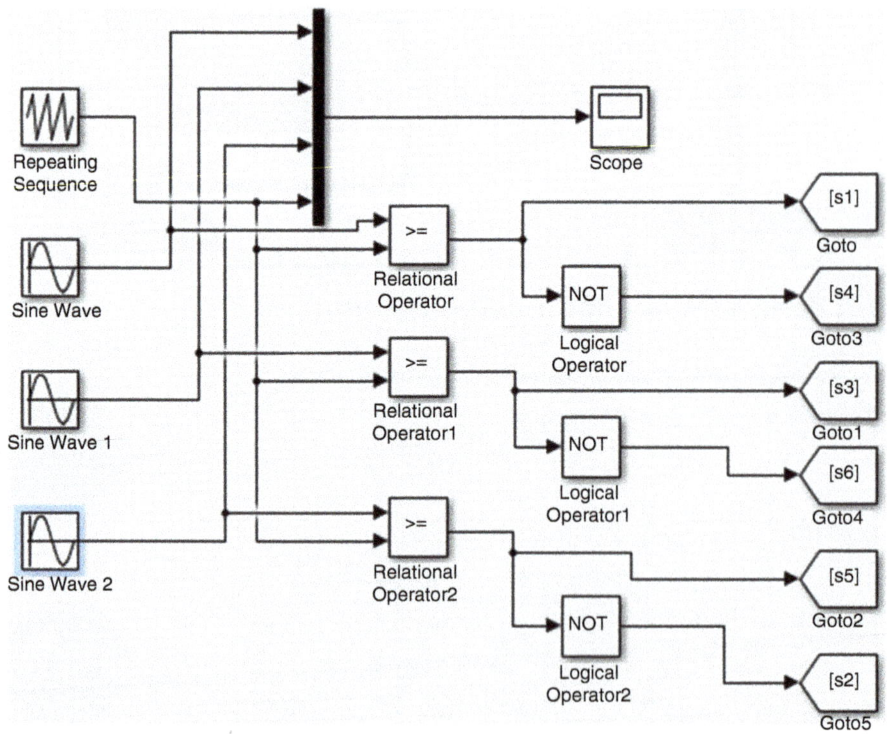

Fig. 4.25 MATLAB/Simulink model of SPWM

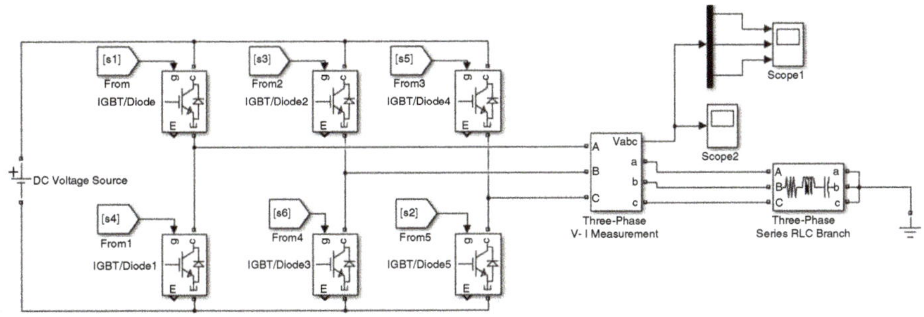

Fig. 4.26 Simulink model for three-level inverter

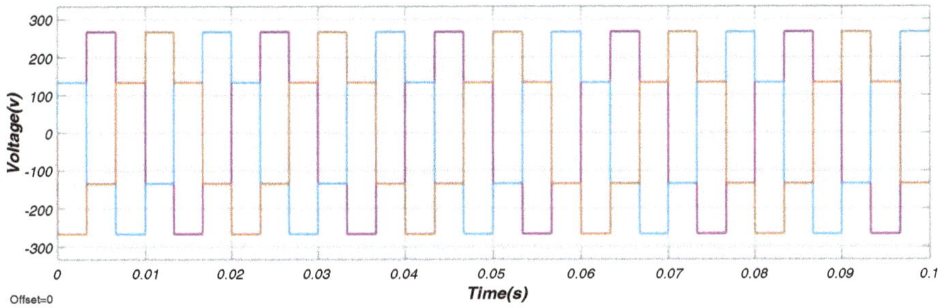

Fig. 4.27 Output voltages of a three-level inverter

Figure 4.25 shows the PWM generation using Simulink. A modulated signal (Sine) is compared with a carrier signal (Triangular). Whenever the amplitude of the sine wave becomes greater than that of the triangular signal, a pulse is generated. This pulse is used to trigger the switches $S1$, $S2$, . . ., $S6$. The switching frequency and the DC-link voltage were 6 kHz and 400 V, respectively (Figs. 4.26, 4.27).

11 Conclusion

In this chapter, a brief introduction on the choice of selection of motors for EV / HEV application is discussed. DC motors are not suited for EV applications as they suffer from commutation failure and need regular maintenance. The mathematical model for a separately excited DC motor was developed and simulated using MATLAB/ Simulink. The results of the simulation clearly showed the dependence of armature torque on the armature current. The emergence of power electronics in HEV and the associated architecture was briefly touched upon. Later, an introduction to DC–DC and DC–AC converters was provided. Constant speed control for a separately excited DC motor was achieved using a simple boost converter. The different types of DC–DC and DC–AC converters used in EV/HEV were shown. The simulation of three-level inverter using built-in Simulink blocks was also discussed.

References

1. Chau, K. T., & Li, W. Overview of electric machines for electric and hybrid vehicles. *International Journal of Vehicle Design, 64*(1), 46–71.
2. Surya, S., Marcis, V., & Williamson, S. (2021). Core temperature estimation for a Lithium ion 18650 cell. *Energies, 14*(1), 87.
3. Surya, S., & Arjun, M. N. (2020). Effect of fast discharge of a battery on its core temperature. In *2020 International conference on futuristic technologies in control systems & renewable energy (ICFCR)* (pp. 1–6). IEEE.
4. Li, X., Yao, M., Yang, Q., Wang, M., & Yang, B. (2020). Modeling of an integrated drive unit in an electric vehicle. In *2020 IEEE transportation electrification conference & expo (ITEC)* (pp. 125–127). IEEE.
5. Surya, S., Bhesaniya, A., Gogate, A., Ankur, R., & Patil, V. (2020). Development of thermal model for estimation of core temperature of batteries. *International Journal of Emerging Electric Power Systems* 1, ahead-of-print.
6. Krein, P. T., Roethemeyer, T. G., White, R. A., & Masterson, B. R. (Oct 1994). Packaging and performance of an IGBT-based hybrid electric vehicle. In *Proc. IEEE workshop power electron. transport.*, Dearborn, MI (pp. 47–52).
7. Kassakian, J. G. (2001). The future of electronics in automobiles. In *Proc. 13th Int. Symp. Power Semicond. Devices ICs, Osaka, Japan* (pp. 4–7).
8. Ehsani, M., Gao, Y., Gay, S. E., & Emadi, A. (Dec. 2004). *Modern electric, hybrid electric, and fuel cell vehicles: Fundamentals, theory, and design*. CRC Press.
9. Emadi, A., Williamson, S. S., & Khaligh, A. (2006). Power electronics intensive solutions for advanced electric, hybrid electric, and fuel cell vehicular power systems. *IEEE Transactions on Power Electronics, 21*(3), 567–577.
10. Surya, S., & Srividya, R. (2021). Isolated converters as LED drivers. Cognitive Informatics and Soft Computing. Springer, Singapore, pp. 167–179.
11. Hart, D. W. (2011). *Power electronics*. Tata McGraw-Hill Education.
12. Erickson, Robert W., and Dragan Maksimovic. Fundamentals of power electronics. Springer Science & Business Media, 2007.
13. Surya, S., & Williamson, S. (2021). Generalized circuit averaging technique for two-switch PWM DC-DC converters in CCM. *Electronics, 10*(4), 392.
14. Surya, S., Channegowda, J., & Naraharisetti, K. (2020). Generalized circuit averaging technique for two switch DC-DC converters. *arXiv preprint arXiv:2012.12724*.
15. Surya, S., & Arjun, M. N. (2021). Mathematical modeling of power electronic converters. *SN Computer Science, 2*(4), 1–9.
16. Surya, S. (2021). Mathematical modeling of DC-DC converters and Li ion battery using MATLAB/Simulink. In *Electric vehicles and the future of energy efficient transportation* (pp. 104–143). IGI Global.
17. Surya, S., & Williamson, S. (2021). Modeling of average current in ideal and non-ideal boost and synchronous boost converters. *Energies, 14*, 5158.
18. Surya, S., & Patil, V. (2019). Cuk converter as an efficient driver for LED. In *2019 4th International conference on electrical, electronics, communication, computer technologies and optimization techniques (ICEECCOT)* (pp. 49–52). IEEE.

19. Surya, S., & Singh, D. B. (2019). Comparative study of P, PI, PD and PID controllers for operation of a pressure regulating valve in a blow-down wind tunnel. In *2019 IEEE International Conference on Distributed Computing, VLSI, Electrical Circuits and Robotics (DISCOVER)* (pp. 1–3). IEEE.
20. Verma, R., & Shimi, S. L. (2018). Analysis and review of DC-DC converter for electric vehicle. In *2018 Second international conference on intelligent computing and control systems (ICICCS)* (pp. 1649–1654). IEEE.
21. Li, W., & He, X. (2010). Review of nonisolated high-step-up DC/DC converters in photovoltaic grid-connected applications. *IEEE Transactions on Industrial Electronics, 58*(4), 1239–1250.
22. Choudhury, A., Pillay, P., & Williamson, S. S. (2014). A performance comparison study of space-vector and carrier-based PWM techniques for a 3-level neutral point clamped (NPC) traction inverter drive. In *2014 IEEE international conference on power electronics, drives and energy systems (PEDES)* (pp. 1–6). IEEE.
23. Jain, M., & Williamson, S. S. (2010). Modeling and analysis of a 5-leg inverter for an electric vehicle in-wheel motor drive. In *CCECE 2010* (pp. 1–5). IEEE.

Chapter 5
Control Systems for Hybrid Electric Vehicle

B. Sathya, R. Neelaveni, and M. Kathiresh

1 Introduction

The word "Hybrid" refers to the combination of two things. A hybrid vehicle refers to the one which uses two methods for propulsion. Hybrid electric vehicle (HEV) is a combination of an internal combustion engine (ICE) and an electric system for propulsion. The main components of HEV are shown in Fig. 5.1 and are explained as follows:

- **Battery (auxiliary)**: It helps in providing electricity to a car at the start and also helps in powering all the accessories of the vehicle.
- **DC/DC converter:** The accessories in the car require low voltage DC so the available high-voltage DC is converted to low-voltage DC by the DC/DC converter. It also helps in recharging the auxiliary battery.
- **Electric generator:** During the process of braking, the generator generates electricity from rotating wheels and also transfers the energy to the traction battery pack.
- **Electric traction motor:** The motor uses power from the traction battery and helps in driving the wheels of the vehicle.
- **Exhaust system:** The emissions generated after combustion are exhausted out from the engine via this system.
- **Fuel tank (gasoline):** This tank stores the fuel required by the vehicle till demanded by the engine for use.
- **Internal combustion engine (spark-ignited):** In this configuration, the spark from the spark plug helps in igniting the air-fuel mixture in the combustion chamber.

B. Sathya (✉) · R. Neelaveni · M. Kathiresh
PSG College of Technology, Coimbatore, Tamil Nadu, India
e-mail: bsa.eee@psgtech.ac.in; rnv.eee@psgtech.ac.in; mkh.eee@psgtech.ac.in

M. Kathiresh et al. (eds.), *E-Mobility*, EAI/Springer Innovations in Communication and Computing, https://doi.org/10.1007/978-3-030-85424-9_5

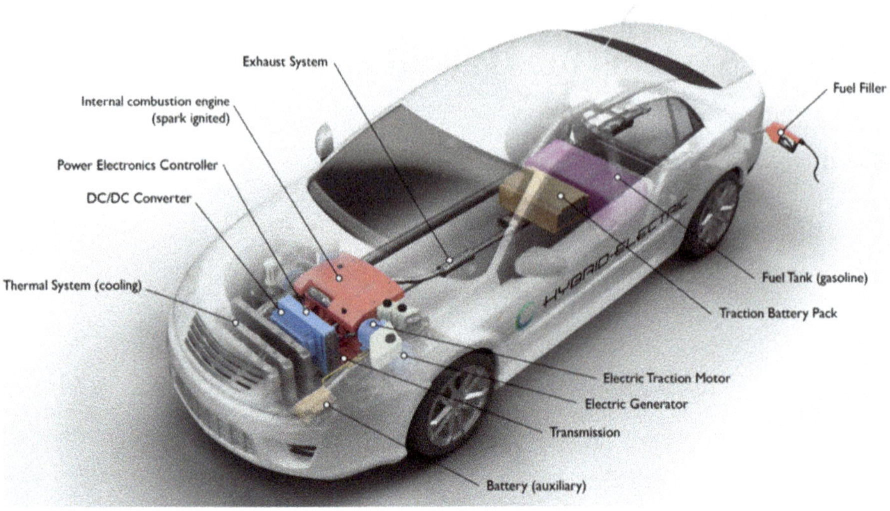

Fig. 5.1 Hybrid electric vehicle[1]

- **Power electronics controller:** This unit helps in managing the power delivered by the traction battery pack, and in turn helps in controlling the speed and torque of the traction motor.
- **Thermal system (cooling):** On continuous operation, the engine, the motor, the power electronics controller unit, and other components get heated up. For proper operation, it is necessary to maintain proper operating temperature for all these units and is done by the thermal system.
- **Traction battery pack:** It helps in storing electricity for the electric traction motor to use when needed.
- **Transmission:** The mechanical power from the engine and the motor is transferred to the wheels by the transmission system.

2 What Is Control?

Control of a system refers to the process of comparing the output of the system with the desired output and take corrective actions based on the error signal. This type of control is called closed-loop control and the block diagram representation of a closed-loop control system is shown in Fig. 5.2.

The input fed to the system is otherwise called the desired output. The error signal is generated by comparing the desired output with the output obtained from the system. The error signal is fed as input to the controller which in turn drives the actual system so as to obtain the desired output. The term “Plant/process” refers to the part of the system which is to be controlled. To better understand the concept of control, let us consider the example of speed control in HEV. Assume that the

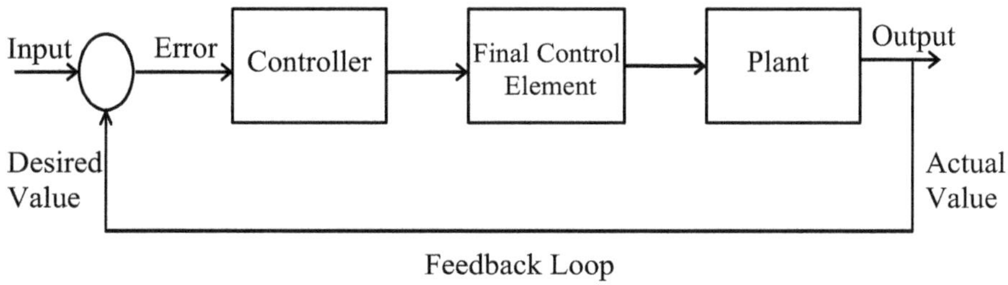

Fig. 5.2 Block diagram representation of closed-loop control system

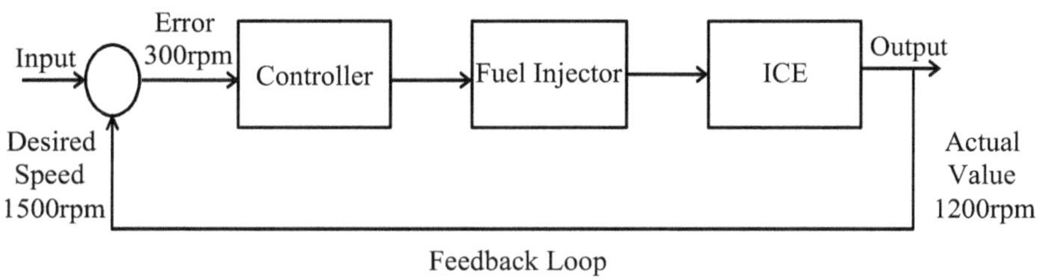

Fig. 5.3 Basic speed control loop for ICE

required speed of the Internal Combustion Engine (ICE) is 1500 rpm, but the actual speed is 1200 rpm as shown in Fig. 5.3. On comparing both, an error signal of 300 rpm is generated which is fed to the controller. Since the error signal is positive, the controller (which is discussed in the latter part of the chapter) commands the fuel injector to inject more fuel, by which the speed of ICE is increased. If the error speed is negative, the fuel injection will be reduced enabling the reduction of the speed of the ICE. In an HEV, there are many variables to be controlled and hence a multivariable control system design is required. The need for control in an HEV and the various ways by which the necessary control could be implemented is discussed in the sections below.

3 Need for Control of HEV

When the ICE extracts more energy, the temperature of the exhaust reduces. At lower temperatures, the combustion of hydrocarbons may not occur and hence the emission of gases is reduced. An increase in compression ratio not only increases the temperature in ICE but also improves fuel efficiency. This increase in temperature also causes an increased level of oxide emissions. This emphasizes the fact that there is always a conflict between maintaining higher fuel efficiency and reducing the emission of gases hazardous to the environment. HEV helps in finding solutions to increase fuel efficiency and reduce exhaust emissions. From the design point of view, there are several factors that decide the performance of HEV. This includes the type of power train, configurations of various components used, and energy management strategy (EMS). The design of an energy-efficient power train makes the EMS the most crucial aspect, as it is required to handle several other conflicting objectives of control such as to provide comfort to drive, maintain fuel economy, reduce exhaust emissions, and preserve battery state of charge (SOC). The main role of Energy Management Strategies is to split the power supply between the engine and the motor so as to increase the fuel efficiency and thereby optimize the performance of the hybrid electric vehicle by reducing the emissions. The minor role of the control includes the preserving battery SOC and maintaining the operating temperature of various components. Thus the EMSs should include the use of controllers in such a way that it achieves global optimization owing to the complex power train structures. The various steps involved in formulating EMS optimization problems are,

1. Choose the power train topology.
2. Model the power train.
3. Formulate EMS and find the optimal solution.

4 Power Train Topologies for HEV

The different power train topologies of HEV include series–HEV(S-HEV), parallel HEV (P-HEV), and series–parallel HEV (SP-HEV), and Plug in type (P) HEV as shown in Fig. 5.4.

In **S-HEV**, the output power from the generator is combined with the power output of the battery and is transferred to the motor, via a dc-bus, which drives the wheels of the vehicle. This topology is mostly preferred in stop-and-go driving situations and not preferred for highways or interurban driving.

In **P-HEV**, the ICE and EM are connected to a torque coupling and the mechanical output (torque) is transmitted to the wheels through a gear arrangement. The losses are lesser compared to S-HEV as they involve mechanical energy rather than electrical energy.

The **SP-HEV** combines the advantages of both S-HEV and P-HEV.

A **plug-in hybrid electric vehicle (PHEV)** possesses the same configuration as an HEV but with an external electric charging plug, bigger electrical components (i.e., electric motor and battery), and a downsized engine.

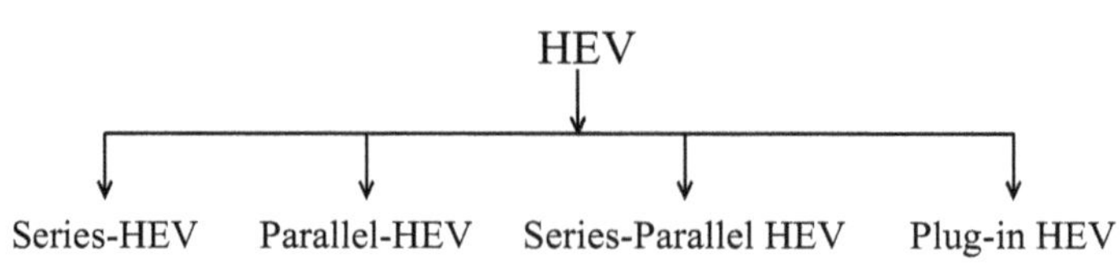

Fig. 5.4 Different power train topologies

5 Operating Modes of HEV

There are various operating modes for an HEV. It includes the battery alone mode, in which the battery alone provides the necessary power for vehicle propulsion, engine alone mode in which the ICE is the only cause for propulsion, combined mode in which both the battery and ICE provides the necessary power to propel the vehicle, the split power mode in which the power from the ICE is split to propel the vehicle and charge the battery. An overview of the various operating modes of HEV is given in Table 5.1.

6 Modeling of Power Train

The second step in formulating an efficient EMS is modeling the power train. The better the model, the better is the results. To model the power train, there are three different approaches and they are forward-facing, backward-facing, and combined forward–backward-facing models. The forward modeling approach uses cause and effect relationships, which implies that the output is always a function of the input. In contrast, the backward modeling approach is non–causal. In the perspective of EMS development, the forward approach is accurate when compared to the backward approach.

7 Energy Management Strategies (EMS)

The selection of powertrain topology and the use of EMS helps in improving fuel efficiency and reducing emissions. So the main objective of an EMS is to model the power train and choose appropriate operating modes to share power which in turn

Table 5.1 Possible operating modes of (P) HEVs

		Powertrain topologies			
Sl. No.	Operation modes	Series	Parallel	Series-Parallel	Plug-in
1.	Battery alone mode	✓	✓	✓	✓
2.	Engine alone mode	✓	✓	✓	✓
3.	Combined mode	✓	✓	✓	✓
4.	Power split mode	✓	✓	✓	✓
5.	Stationary charging	✓	✓	✓	✓
6.	Regenerative braking	✓	✓	✓	✓
7.	Engine-heavy mode	–	–	✓	–
8.	Electric-heavy mode	–	–	✓	–
9.	Charging battery mode	–	–	–	✓
10.	Extended driving	–	–	–	✓

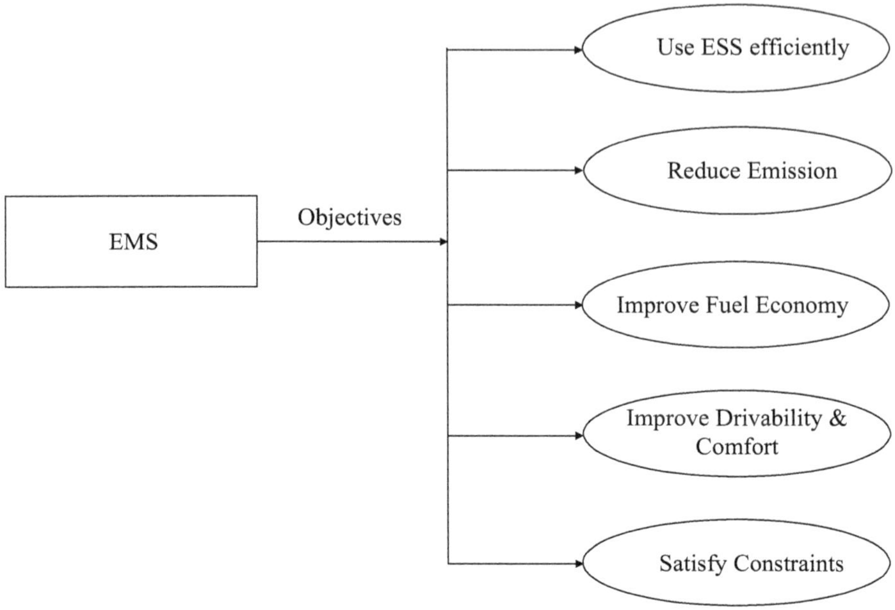

Fig. 5.5 Overview of objectives of EMS for an HEV

reduces the emission of exhaust, improves fuel efficiency, ensures comfortability in driving, and maintains SOC of battery. The general overview of EMS for an HEV is shown in Fig. 5.5.

The three types of EMS include rule-based (RB) EMS, optimization-based (OB) EMS, and learning-based (LB) EMS. The RB- EMS works on a set of predefined rules which does not include any prior knowledge of the trip. It is further sub-classified into deterministic and Fuzzy- logic EMS. The OB-EMS works based on certain prior information about the driving conditions. Based on the apriori, the OB-EMS can be classified into online and offline optimization. The dynamic programming (DP) approach, metaheuristic search methods like genetic algorithm (GA), particle swarm optimization (PSO), and several other techniques are offline methods that help in achieving the global optimization whereas approaches like model predictive control (MPC) falls under the category of online OB-EMS. LB-EMS learns from the previously available training data, which is the driving data in the case of HEV. With the advent of machine learning and deep learning approaches the LB-EMS has proven to show promising potential. The combination of all the three types can help in forming a versatile EMS called as the Integrated EMS. The iEMS along with the different types of EMS is shown in Fig. 5.6.

Generally, the information from the GPS or the Intelligent Transportation System (ITS) helps in updating the parameters of EMS and is called adaptive EMS.

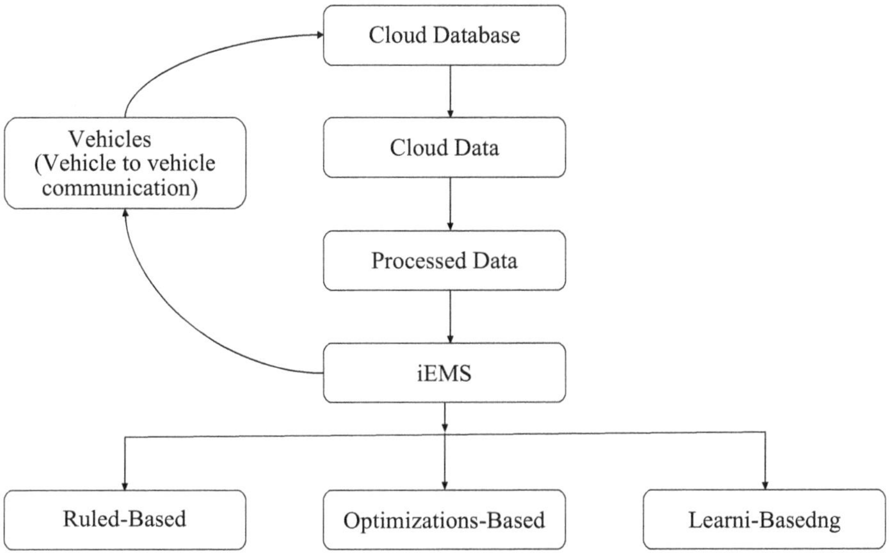

Fig. 5.6 Classification of EMS and the concept of Integrated EMS

7.1 Rule-Based Energy Management Strategies (RB-EMSs)

Without prior knowledge of the drive cycle, RB-EMS uses heuristic approaches, human expertise, or assumptions to solve the problems. The main advantage of the RB-EMS is that the real-time implementation using look-up tables or state–machine is simple and feasible. Besides this, the main drawback of this method is that it may not yield an optimal solution. Hence, to improve the performance, RB-EMS is integrated with several other optimization techniques. Few such integrated approaches include a combination of ECMS with rule-based approach [2], state machine control by using ECMS [3], and a multi-mode EMS based on driving pattern identification using learning vector quantization and a neural network [4]. There are two types of RB-EMS. They are deterministic and fuzzy–logic-based EMSs.

7.1.1 Deterministic Strategies

Based on heuristics and human expertise, the rules for the deterministic approaches are designed. These rules are framed in such a way that the ICE and other sources of energy in HEV are controlled to work under optimal working conditions as shown in Fig. 5.7, in such a way that they yield high fuel efficiency and reduce transmission loss of energy. The optimal working conditions refer to optimal working point or optimal efficiency region or optimal operating line as quoted in the literature [6–8]. These rules are also designed to split power between ICE and EM and this

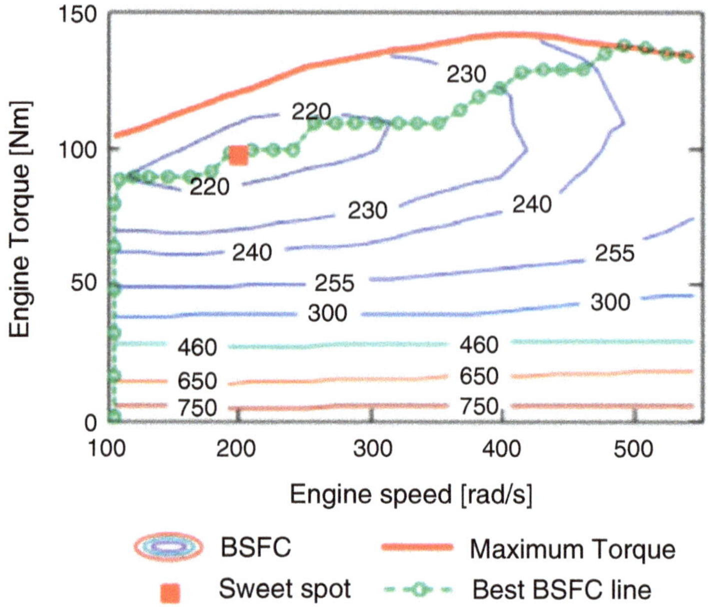

Fig. 5.7 Efficiency map and operating points for ICE based HEV[5]

approach is called the frequency-decoupling approach. In this approach, the low-frequency power is provided by the sources with slow dynamics such as ICE in HEV, and the required high-frequency power is compensated with sources with faster dynamics.

Optimal Working Condition-Based Strategies

Thermostat (On/Off) Strategy

In this strategy, providing a constant torque and speed the ICE is made to work at optimal efficiency points to maintain the battery SOC within the limits. This is done by ICE being turned ON/OFF as when required. The difference in power at the sweet point and the demand is either used to charge the battery (battery charging mode) or used to assist the battery to supply the required load (hybrid traction mode). In this strategy, besides the overall HEV efficiency being low, the efficiency of the engine-generator set is high. This type of strategy is mostly used for series-type HEVs in city drive applications.

Power Follower (Baseline) Strategy

In this, the ICE is used as the main source of power and the battery is used as the auxiliary source of power. If the speed of the vehicle is less than the minimum required value, the battery supplies the required power. If the power demand of the vehicle is greater than the power generated by the engine, the battery assists the engine with its power and gets charged during regenerative braking. The engine charges the battery when the battery SOC is lesser than the minimum limit. This strategy not only helps to maintain the SOC of batteries but also improves the overall efficiency of the system. By combining these two, hybrid approaches were developed to improve fuel efficiency in both series [9] and parallel HEV [10].

State Machine Strategy (Multi-Mode Strategy)

This approach is also referred as a multi-mode approach. The specific operation of a vehicle is considered as a state and the relationships between the states depend upon certain conditions imposed and the input values at each state. Basically, HEV operates in different modes such as engine mode in which the ICE propels the vehicle, boosting mode in which both ICE and batteries propel the vehicle, and charging mode in which ICE helps in propelling the vehicle and charging the battery. The transition between these modes occurs in several situations like a change in the operating condition of the vehicle or when there is a change in driver command or when there is a failure in the system or subsystem. This approach ensures that there is a smooth transition from one state to another and stable operation of the entire system.

Frequency-Decoupling Strategies

In this strategy, the rules are designed to split power between ICE and EM. In this approach, the low-frequency power is provided by the sources with slow dynamics such as ICE in HEV, and the required high-frequency power is compensated with sources with faster dynamics. The different ways of realizing frequency-decoupling are through a simple low-pass filter (LPF), a gliding average strategy [11] (known as a Phlegmatising strategy), or a time-frequency representation tool such as a wavelet transform (WT).

7.1.2 Fuzzy Logic (FL) Strategies

Unlike the deterministic approach, this approach incorporates the decision-making property of fuzzy logic. This strategy frames rules based on the human experience. This process involves the following stages: input quantization, fuzziness, fuzzy reasoning, inverse fuzziness, and output quantization. The membership function and the FL rules determine the performance of this strategy. The main advantage is that the rules can be tuned easily and the others include its robustness, flexibility, and adaptability. Besides these advantages, optimal performance cannot be obtained using this strategy.

Optimized fuzzy-rules control: This strategy incorporates optimization techniques to tune the controller so as to achieve the objectives. Several works in the literature have used evolutionary algorithms such as PSO [12], GA [13], and Bee algorithm [14].

Adaptive fuzzy logic control: Fuzzy logic rule-based strategy is integrated with adaptive algorithms to develop adaptive control systems. Few such developments include the development of a neuro-fuzzy interference system to reduce fuel consumption and increase vehicle torque [15] and the development of a Compensation Fuzzy Neural Network (CFNN) to enhance vehicle performance.

Predictive fuzzy logic control: This approach aids in predicting the future state of the vehicle. A predictive FL-RB is designed to predict the future states of traffic using the information from GPS in [16].

7.2 Optimization-Based (OB) EMSs

The objective of OB-EMSs involves either minimization or maximization of the cost function subject to constraints in case of constraint optimization or without any constraints in case of unconstrained optimization. The cost function could be the maximization of fuel efficiency, minimization of exhaust emissions subject to constraints maintaining battery SOC or maintaining torque, speed, or power within the limits. Based on the prior information about the driving conditions, the OB-EMSs can be classified into offline strategy or online strategy.

7.2.1 Offline Strategies

This strategy requires prior information about the drive cycles to attain the noncausal and global optimal solution. Consider an example of parallel HEV, in which the optimization problem could be formulated as minimization of fuel consumption and exhaust emission. Let the constraints be maintaining battery SOC or comfortability in driving. Once the problem is formulated with the constraints, a suitable algorithm like the power split approach could be incorporated to solve the problem. Based on the method incorporated to solve the problem, the offline OB-EMSs could be classified into direct, indirect, gradient, and derivative-free types and are shown in Fig. 5.8.

Direct Algorithms

Dynamic programming (DP) is the most widely used algorithm in this approach to solve the optimization problem. In DP, the nonlinear dynamic optimization problem is subdivided into several subproblems. For each subproblem, an individual cost function is defined and the solution for each subproblem is arrived by using the

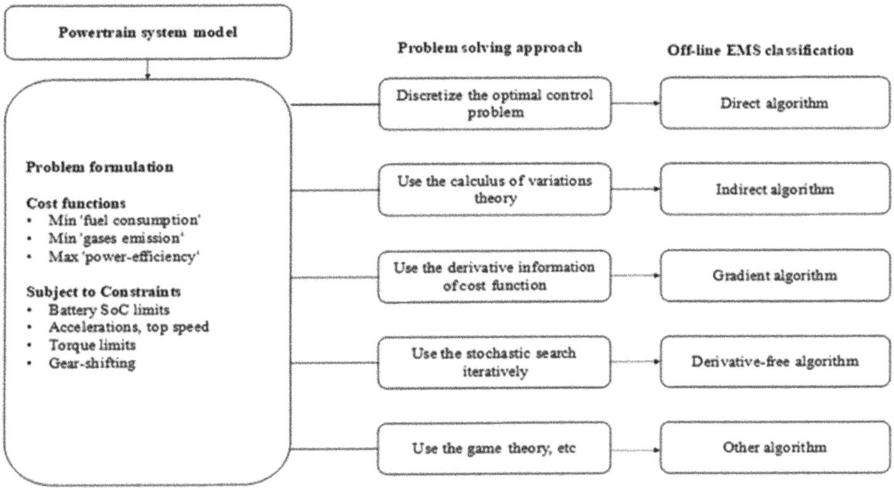

Fig. 5.8 Classification of offline OB-EMSs

backward/forward programming technique. The major disadvantage of this method is high computation time and because of this, real-time implementation becomes infeasible.

Indirect Algorithms

This approach uses the calculus of variations to solve the optimization problem. An extension of the calculus of variation namely Pontryagin's minimum principle (PMP) is the most preferred algorithm and is used in this approach. The main idea being the reduction of constrained global optimization problems to local Hamiltonian minimization problems. Since this is an iterative process, it involves high computational time which makes it difficult for real-time implementation.

Gradient Algorithms

This approach uses derivative information of the objective function to solve the optimization problem. Linear programming (LP) and quadratic programming (QP) types of approaches are used to arrive at the optimal solutions. Once the optimization problem is formulated with constraints, software packages like MATLAB can be used to find solutions provided a vehicle model is available. Since an approximate model of the vehicle is used, these approaches provide only near-optimal solutions.

To avoid a dependency on the derivatives, derivative-free algorithms are used to find the global optimal solution. Other algorithms include the application of game theory to solve optimization problems.

7.2.2 Online Strategies

In contrast to the offline strategy which involves noncausal and global optimization, the online strategy provides causal and local optimal solutions. Equivalent consumption minimization strategies and model predictive control are the most widely algorithms for real-time implementations.

Equivalent Consumption Minimization Strategies (ECMS)

The ECMS, as a realization of offline PMP for parallel HEVs, was developed in [17]. In this, a global optimization problem of PMP is modified as local optimization to minimize fuel consumption. The ECMS calculates the fuel consumption required to recharge the batteries during regenerative braking.

Model Predictive Control Based Strategies (MPC)

To overcome the drawback of DP, MPC was introduced. The major drawback of using DP is that it requires future information on the state of the vehicle. But in practice, it is impossible to obtain such information. Hence, MPC calculates the optimal points over a prediction horizon so as to minimize the constrained objective function, applies it to the process/plant, and moves forward the prediction horizon. These steps are repeated until an optimal solution has arrived.

Robust Control (RC)

The main objective of RC is to find a feedback controller such that it minimizes the consumption of fuel. In this approach, the nonlinear time-varying system is approximated to a linear time-in varying system. Because of this approximation, RC offers sub-optimal solution. RC is insensitive to parameter uncertainties and hence stable and robust. An example of RC is usage of H-infinity controller [18].

Sliding Mode Control (SMC)

SMC provides a robust control against the highly nonlinear and time-varying vehicle dynamics. Typical applications of SMC include speed control of the engine [19], smooth, and steady operation of the vehicle in all three operating modes [20].

7.3 *Learning-Based EMSs*

This strategy employs data mining techniques to arrive at the optimal solution. The major advantage of this approach is that there is no need for precise information about the model. But the process of creating the database is highly difficult and time-consuming. This approach is highly efficient as it is adaptive and deals with a large dataset. The efficiency of the algorithm can be further improved by integrating this approach with other approaches. Based on the type of learning this approach is further classified into reinforcement learning, supervised/unsupervised learning, neural network learning, and classification learning approach.

7.3.1 Reinforcement Learning

In this approach, the learning agent continuously interacts with the environment. At every instant of time, the agent observes the state of the environment, and based on the state action is chosen, and the current state is updated. Based on the updation the environment transits into a new state. A reward is given back to the learning agent based on the transition. This process of training continues till an optimal solution is obtained. The major advantage of this process is that without any learning rule, the system learns the optimal law by itself. An illustration of this process is shown in Fig. 5.9.

7.3.2 Supervised Learning

In this approach, a set of data named training data with labels is fed to the model and predictions are made. If there is an error in the prediction the training is continued.

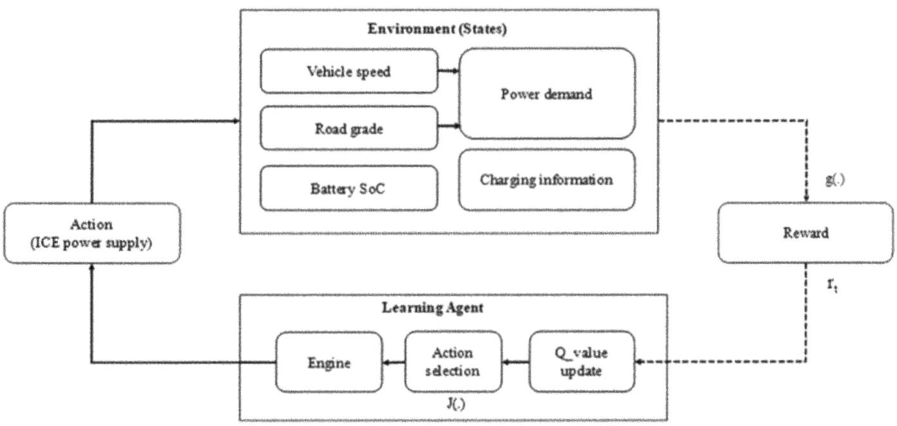

Fig. 5.9 Illustration of reinforcement learning

This process is repeated till the model output has a higher level of accuracy. This is similar to a classification problem. Chin et al. [21] used this to assess the performance of the selection algorithm.

7.3.3 Unsupervised Learning

In this method, model creation is done based on certain rules and mathematical process. Clustering is a type of unsupervised learning algorithm in which similar elements are grouped together into a single set.

7.3.4 Neural Network Learning

Neural network is the one that mimics the human brain neurons. These neurons are connected in a certain manner to form layers which in turn form the network. The network is trained to achieve a certain objective. The training is repeated till the error output is tolerable. One such use of NNL to maintain SOC of battery is discussed in [22].

An overall comparison of all the three strategies is given in Table 5.2.

Table 5.2 Comparison of different EMSs

Rule-based algorithm		
Strategy	Advantages	Challenges
Deterministic	• Simple	• Low fuel economy
Fuzzy-logic	• Robust and adaptive	• Requires calibration of control parameters for different drive cycles
Offline strategy: Optimization based		
DP	• Global optimality • Benchmark for other EMSs	• Driving cycle information needed apriori • High computational cost
PMP	• Global trajectory optimal control	• Computationally complex
Gradient	• Fast computation	• Derivative information of objective function is needed • Computationally complex
Derivative-free	• Stochastic solution can be used to get rid of local optima	• Requires more number of iterations to ensure the optimal solution
Game theory	• Comprehensive tradeoff of conflicting objectives	• Computationally complex
Online strategy: Optimization based		
ECMS	• Online implementation	• Problem of local optima
MPC	• Solutions close to global optima can be obtained with less computational effort	• Future driving information is required

8 Summary

To improve fuel efficiency and to minimize pollution control of HEV is important. Power train and Energy Management Strategies are the most important factors which help in deciding the performance of HEV. The various power train topologies such as Series, Parallel, Series-parallel, and.

Plug-in types are discussed in detail. The various operating mode of HEV for each of these power train topologies are discussed. The various objectives of EMS and its types are outlined. The three types of EMS include rule-based (RB) EMS, optimization-based (OB) EMS, and learning based (LB) EMS. The various approaches or strategies which help in building these EMS are explained. The various advantages and challenges in each of these approaches are discussed. These help upcoming researchers to design control systems for HEV in such a way that it tackles the contradictory objectives of improving efficiency and reducing emissions.

References

1. https://afdc.energy.gov/vehicles/how-do-hybrid-electric-cars-work
2. Hemi, H., Ghouili, J., & Cheriti, A. (2013). A real time energy management for electrical vehicle using combination of rule-based and ECMS. In *2013 IEEE electr. Power energy conf. IEEE* (p. 1–6). https://doi.org/10.1109/EPEC.2013.6802927.
3. Li, Q., Su, B., Pu, Y., Han, Y., Wang, T., Yin, L., et al. (2019). A state machine control based on equivalent consumption minimization for fuel cell/supercapacitor hybrid tramway. *IEEE Transactions on Transportation Electrification, 5*, 552–564. https://doi.org/10.1109/TTE.2019.2915689
4. Song, K., Li, F., Hu, X., He, L., Niu, W., Lu, S., et al. (2018). Multi-mode energy management strategy for fuel cell electric vehicles based on driving pattern identification using learning vector quantization neural network algorithm. *Journal of Power Sources, 389*, 230–239. https://doi.org/10.1016/j.jpowsour.2018.04.024
5. Tran, D.-D.. *Renewable and sustainable energy reviews*. https://doi.org/10.1016/j.rser.2019.109596.
6. Park, J., Oh, J., Park, Y., & Lee, K.. (2006). Optimal power distribution strategy for series-parallel hybrid electric vehicles. In *2006 Int. Forum strateg. Technol. IEEE* (p. 37–42). https://doi.org/10.1109/IFOST.2006.312240.
7. Sun, D., Lin, X., Qin, D., & Deng, T. (2012). Power-balancing instantaneous optimization energy management for a novel series-parallel hybrid electric bus. *Chinese Journal of Mechanical Engineering, 25*, 1161–1170. https://doi.org/10.3901/CJME.2012.06.1161
8. Ahn, K., & Papalambros, P. Y. (2009). Engine optimal operation lines for power-split hybrid electric vehicles. *Proceedings of the Institution of Mechanical Engineering Part D Journal of Automobile Engineering, 223*, 1149–1162. https://doi.org/10.1243/09544070JAUTO1124
9. Kim, M., Jung, D., & Min, K. (2014). Hybrid thermostat strategy for enhancing fuel economy of series hybrid intracity bus. *IEEE Transactions on Vehicular Technology, 63*, 3569–3579. https://doi.org/10.1109/TVT.2013.2290700
10. Zhiguo, Z., Yu, Z., Minglu, Y., & Yang, Z. (2009). Torque distribution strategy for single driveshaft parallel hybrid electric vehicle. In *2009 IEEE intell. veh. symp. IEEE* (p. 1350–1353). https://doi.org/10.1109/IVS.2009.5164481.

11. Tritschler, P. J., Bacha, S., Rulliere, E., & Husson, G. (2010). Energy management strategies for an embedded fuel cell system on agricultural vehicles. In *XIX int. Conf. Electr. Mach. - ICEM 2010*. IEEE (p. 1–6). https://doi.org/10.1109/ICELMACH.2010.5608314.
12. Wu, J., Zhang, C.-H., & Cui, N.-X. (2012). Fuzzy energy management strategy for a hybrid electric vehicle based on driving cycle recognition. *International Journal of Automotive Technology, 13*, 1159–1167. https://doi.org/10.1007/s12239-012-0119-z
13. Poursamad, A., & Montazeri, M. (2008). Design of genetic-fuzzy control strategy for parallel hybrid electric vehicles. *Control Engineering Practice, 16*, 861–873. https://doi.org/10.1016/j.conengprac.2007.10.003
14. Derakhshan, M., & Shirazi, K. H. Optimized fuzzy controller for a power-torque distribution in a hybrid vehicle with a parallel configuration. *Proceedings of the Institution of Mechanical Engineering Part D Journal of Automobile Engineering, 228*, 1654–1674. https://doi.org/10.1177/0954407013496183
15. Mohebbi, M., Charkhgard, M., & Farrokhi, M. (2005). Optimal neuro-fuzzy control of parallel hybrid electric vehicles. In *2005 IEEE veh. Power propuls. Conf. IEEE* (pp. 252–256). https://doi.org/10.1109/VPPC.2005.1554566.
16. Hajimiri, M. H., & Salmasi, F. R. (2006). A fuzzy energy management strategy for series hybrid electric vehicle with predictive control and durability extension of the battery. In *2006 IEEE Conf. Electr. Hybrid Veh.* (pp. 1–5). https://doi.org/10.1109/ICEHV.2006.352279. IEEE.
17. Paganelli, G., Guerra, T. M., Delprat, S., Santin, J.-J., Delhom, M., & Combes, E. (2000). Simulation and assessment of power control strategies for a parallel hybrid car. *Proceedings of the Institution of Mechanical Engineering Part D Journal of Automobile Engineering, 214*, 705–717. https://doi.org/10.1243/0954407001527583
18. Reyss, O., Duc, G., Pognant-Gros, P., & Sandou, G. (2009). Multivariable torque tracking control for E-IVT hybrid powertrain. *International Journal of Systems Science, 40*, 1181–1195. https://doi.org/10.1080/00207720902985427
19. Gokasan, M., Bogosyan, S., & Goering, D. J. (2006). Sliding mode based powertrain control for efficiency improvement in series hybrid-electric vehicles. *IEEE Transactions on Power Electronics, 21*, 779–790. https://doi.org/10.1109/TPEL.2006.872373
20. Ayad, M. Y., Becherif, M., & Henni, A. (2011). Vehicle hybridization with fuel cell, supercapacitors and batteries by sliding mode control. *Renewable Energy, 36*, 2627–2634. https://doi.org/10.1016/j.renene.2010.06.012
21. Chin, H. H., & Jafari, A. A. (2011). A selection algorithm for power controller unit of hybrid vehicles. In *2011 14th int. IEEE conf. Intell. Transp. syst. IEEE* (pp. 324–328). https://doi.org/10.1109/ITSC.2011.6082910.
22. Liu, R., Shi, D., & Ma, C. (2014). Real-time control strategy of Elman neural network for the parallel hybrid electric vehicle. *Journal of Applied Mathematics, 2014*, 1–11. https://doi.org/10.1155/2014/596326

Chapter 6
Power Flow in Hybrid Electric Vehicles and Battery Electric Vehicles

S. Madhu, A. Ashwini, and Karanam Vasudha

1 Introduction

Increasing levels of pollution, global warming and population are important concerns as we move towards the current century [1]. To overcome the effects of these problems on the environment and also for the human community, a shift must be required to sustainable transportation [2, 3]. The solutions are – to use a low or zero-emission vehicle, promote the use of public transport, use more and more renewable energy sources for charging, thus becoming less dependent on the depleting fossil fuels. With the interest to reduce the impact of emissions given out by the internal combustion engine vehicles and with the tremendous technological developments happening in the field of battery-operated vehicles, the interests of many researchers are shifting towards Electric vehicles [4]. Various steady developments and cutting-edge technologies are recorded in the past few decades which has led to the recent development of electric vehicles such as Autonomous driverless cars. Electric vehicles have greater advantages in all dimensions. Hence, wide use of electric vehicles (EV) is encouraged for transportation.

Electric vehicles implement modern electric propulsion consisting of electric machines, power electronic converters, electric energy sources, electronic controllers and storage devices. Based on the group of subsystems involved in the process and based on whether they transmit energy to wheels or energy to charge the battery, there are various modes of power flow operation possible. In Hybrid EV (HEV) and some of the Battery Electric Vehicles (BEVs) two power flow paths are possible. Different sets of components are involved in the transmission of energy from the sources to the wheels. These sets of components which operate together in the process are called Powertrain or drivetrains.

S. Madhu (✉) · A. Ashwini · K. Vasudha
BNM Institute of Technology, Bengaluru, Karnataka, India
e-mail: madhus@bnmit.in; ashwinia@bnmit.in; kvasudha@bnmit.in

M. Kathiresh et al. (eds.), *E-Mobility*, EAI/Springer Innovations in Communication and Computing, https://doi.org/10.1007/978-3-030-85424-9_6

2 Types of Electric Vehicles

The vehicles can be classified in various ways, based on propulsion devices, energy sources and energy carriers. The classification based on the propulsion systems is internal combustion engine vehicles, hybrid EVs and all-electric vehicles [5]. The conventional IC engine vehicle has an internal combustion engine for propulsion. The energy sources used here are exhaustible fossil fuels, whereas EV runs on electrical energy derived from batteries, fuel cells and renewable energy sources. The important class of this classification is the Hybrid EVs which are further classified as Series HEV, parallel HEV, Series-parallel HEV and complex HEV. Plug-in EV is another upcoming type of HEVs.

The HEV is classified based on different aspects like the percentage of battery and ICE involvement for driving, power flow path during various modes of operation, etc. The different types of conventional HEV are Micro HEV, Mild HEV and Full HEV. In micro HEV, the battery is used for the start–stop and regenerative braking only. In mild HEV, the battery is used for start, stop, regenerative braking and power assistance. In full HEV, the battery is used for start, stop, regenerative braking, power assistance and electric launch. A more versatile operation is possible in the case of HEV [6].

All EV uses batteries and ultracapacitors as sources. Six types of configurations are existing for power transfer, among them three are widely used viz. battery EVs, Fuel cell EVs and Ultracapacitor EVs.

The design aspects depend on the energy requirement at the wheels and also on the power flow path. In the power flow path, a major amount of energy is required for all the components for its operation and also there is some amount of power loss that is unavoidable. In this regard, the power flow in the vehicle between the energy sources and the wheels is very important to be understood. In this chapter, the main focus is on power flow in HEV and BEV. The power flow in various modes of operation for all types of HEVs and BEVs is explained in the preceding sections.

3 Power Flow in HEVs and BEVs

The energy required for propulsion in HEVs is provided by the IC Engine, fuel tank and battery while in BEVs battery and ultracapacitors are the energy sources. Along with the energy source/s other key components of these EVs are electric motors & generators, power converters and controllers.

The various typical modes of operation are Acceleration of the vehicle, normal driving of the vehicle, light load condition, regenerative braking (deceleration), battery charging during driving, battery charging during standstill. There can be multiple energy sources to deliver the power as per the requirements. Let us consider the sources as 1 and 2 and with 2 as a battery, there can be various modes of operation possible based on power flow path as mentioned below:

- Sources 1 and 2 both supply power to the wheels
- Source 1 or source 2 alone supply power to the wheels
- Source 2 (here Battery) receives power from source 1
- Source 2 receives power from wheels and stores it
- Source 1 supplies power to the wheels and also to source 2
- Source 1 supplies power to the supply 2 and supply 2 to the wheels
- Source 1 supplies power to the wheels and the wheels deliver power to source 2

In the above modes of power flow operation, the power flow path may or may not include the power converters depending on the type of power and the voltage levels at different subsystems. The power requirements to propel the vehicle due to frequent changes in the system requirements as per the drive cycle needs different power flow paths. As per the need during acceleration, deceleration, normal drive, etc. load power may be steady power or dynamic power. The operation of the vehicle with the two supplies operating in coordination favours all the conditions as per the need. The subsystems in any configuration have to be decided based on the types of sources available, the need for power converters and electric motors. In the preceding section, the power flow modes of operation for various configurations are presented. The architecture considered here is typical structures that may vary as per the power levels required, AC or DC power, voltage levels, etc.

Based on the component integration, its location and working we obtain various vehicle designs. The various types of HEVs and BEVs and their architecture are presented in this section.

3.1 Power Flow in HEV

A full hybrid electric vehicle can operate with IC engine mode alone, an electric mode alone or in a combination of both the IC engine and motor together. It requires a complicated transmission system known as electronic variable transmission which is generally a high power device of 50–60 kW with a system voltage of 500–600 V [7, 8]. All the parts of HEV that work together with the IC engine to move the wheels and other parts of the vehicle to move are together referred to as a drivetrain. This includes the transmission, differential, driveshaft, axles and wheels. HEV configuration also increases its performance and an increase in the speed is provided when needed by the vehicle. The ICE can provide energy for charging the batteries and also batteries can regain energy through regenerative braking. Hence, HEVs are predominantly ICE-driven vehicles that can use an electric drivetrain for increasing their performance.

In HEV, the traction power required for the load is delivered together by the IC engine and the electrical motor. The amount of power and direction of power flow depends upon the load demand. Depending upon the architecture, HEV has different modes of operation viz. start-up, acceleration, normal driving, Braking, battery charging, etc. In this section, power flow for various modes of operation for each configuration is presented [7, 9–12].

3.1.1 Series HEV Powertrain and Modes of Operation

Figure 6.1 shows the architecture of a Series Hybrid drivetrain. The mechanical energy from the IC engine is converted into electrical energy by a generator. This power will charge the battery or it gets added with the battery power through the power converter for driving the electric motor. The function of the power converter is to act as a power coupler and control the flow of power from the generator and battery to the motor during acceleration and from the motor to the battery during braking. The IC engine, fuel tank and generator are the primary energy sources while the batteries are energy storage devices.

Based on the power flow there are four modes of operation in series HEV.

1. **Start-up/normal driving/acceleration mode:** Fig. 6.2a shows the power flow diagram during starting or normal driving or acceleration. In this mode, the electrical energy to the drive motor is supplied by both the battery and IC engine through the converter. The output of the motor is then delivered to the wheels via transmission.
2. **Light load mode:** Fig. 6.2b shows the power flow diagram at light load conditions. In this mode, the output of the IC engine is more than the power required for the propulsion of the vehicle. The additional electrical energy generated is hence used for battery charging. This continues till the battery is charged to its full capacity.
3. **Braking/deceleration mode:** Fig. 6.2c shows the power flow diagram when the vehicle is braking or decelerating. In this mode, the motor behaves like a generator to convert the kinetic energy of the wheels into electrical energy. This electrical output from the generator is used to charge the energy storage device through the power converter.
4. **Battery charging/standstill mode:** Fig. 6.2d shows the power flow diagram when the vehicle is at rest. In this mode, the IC engine continues charging the battery through the generator even if the vehicle stops or comes to a standstill position.

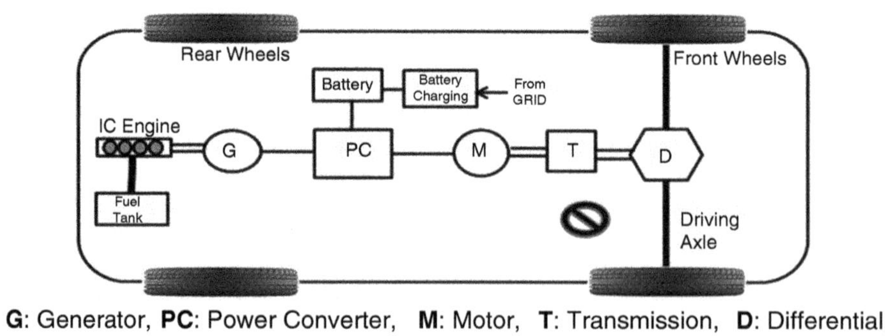

Fig. 6.1 Architecture of Series HEV configurations. *G* Generator, *PC* Power converter, *M* Motor, *T* Transmission, *D* Differential

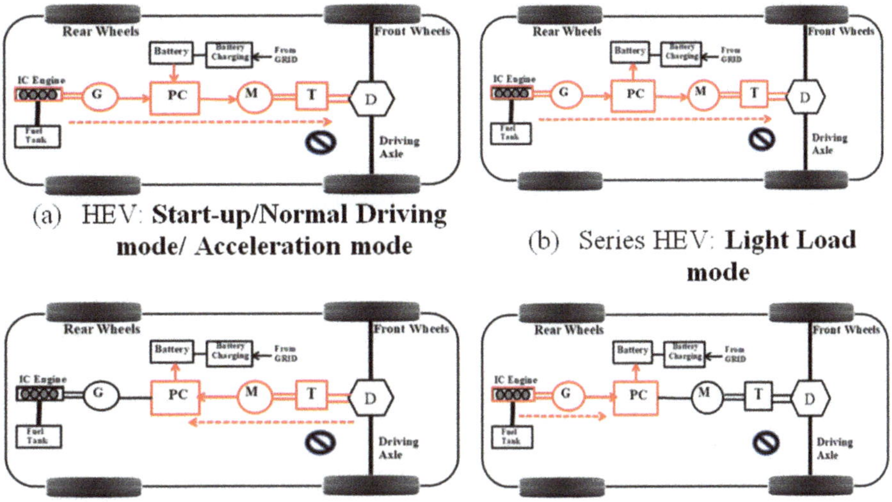

(a) HEV: **Start-up/Normal Driving mode/ Acceleration mode**

(b) Series HEV: **Light Load mode**

(c) HEV: **Deceleration/Braking Mode**

(d) Series HEV: **Battery charging (Standstill)**

Fig. 6.2 Different power flow modes in series HEV

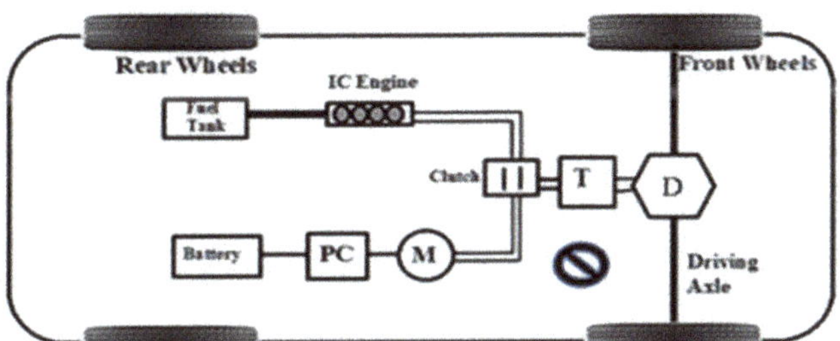

G: Generator, **PC**: Power Converter, **M**: Motor, **T**: Transmission, **D**: Differential

Fig. 6.3 Architecture of parallel HEV configurations. *G* Generator, *PC* Power converter, *M* Motor, *T* Transmission, *D* Differential

3.1.2 Parallel HEV Powertrain and Modes of Operation

Figure 6.3 shows the architecture of the Parallel Hybrid drivetrain. The IC engine and the electrical motor are coupled to drive the wheels through a mechanical coupler or a clutch. The power to drive the wheels is delivered either by the motor or IC engine or both. The IC engine is the primary power source while the motor can also work as a generator for charging the battery during braking/deceleration and absorbs power from the ICE when its power output is more than the power needed for traction.

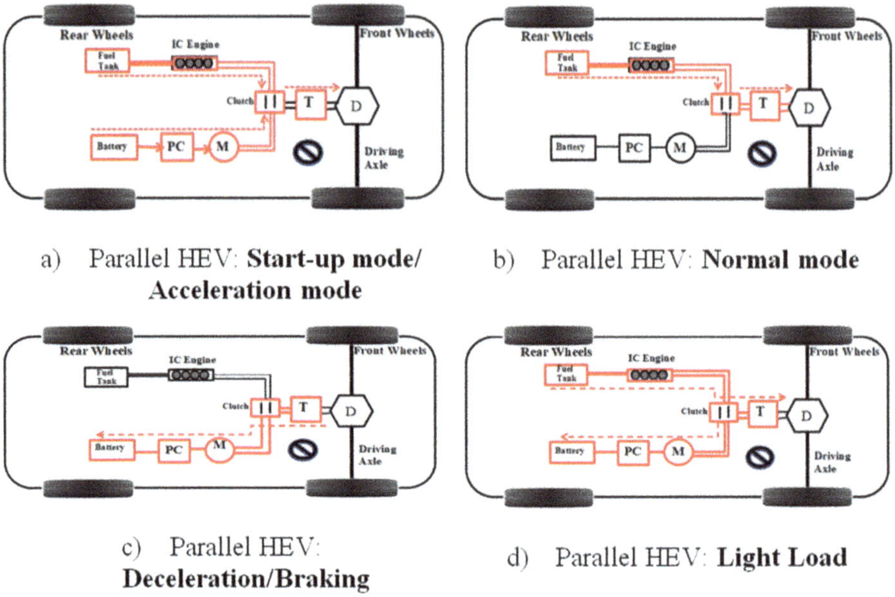

a) Parallel HEV: **Start-up mode/ Acceleration mode**

b) Parallel HEV: **Normal mode**

c) Parallel HEV: **Deceleration/Braking**

d) Parallel HEV: **Light Load**

Fig. 6.4 Power flow modes in parallel HEV

Based on the power flow there are four modes of operation in Parallel HEV.

1. **Start-up/acceleration:** Fig. 6.4a shows the power flow diagram during starting or acceleration. In this operating mode, both the motor and the ICE supply power to drive the wheels via transmission. Typically, the energy distribution for propulsion is 80–20% for IC engines and electric motor.
2. **Normal driving:** Fig. 6.4b shows the power flow diagram during normal driving conditions. In this mode, the power for driving the wheels is delivered only by the ICE and the motor will be in standstill condition.
3. **Deceleration/braking:** Fig. 6.4c shows the power flow diagram during braking/deceleration. In this mode, the generator action is provided by the motor itself and the available kinetic energy at the wheels is converted into electrical energy. This energy is then used to supply charge to the battery through the power converter.
4. **Light load and battery charging:** Fig. 6.4d shows the power flow diagram at light load condition. In this mode, IC Engine delivers the power required to drive the wheels as well as charges the battery through the electric motor and power converter.

3.1.3 Series-Parallel HEV Powertrain and Modes of Operation

Figure 6.5 shows the architecture of the Series-Parallel Hybrid drivetrain. This configuration employs both mechanical and electrical power couplers. It has the

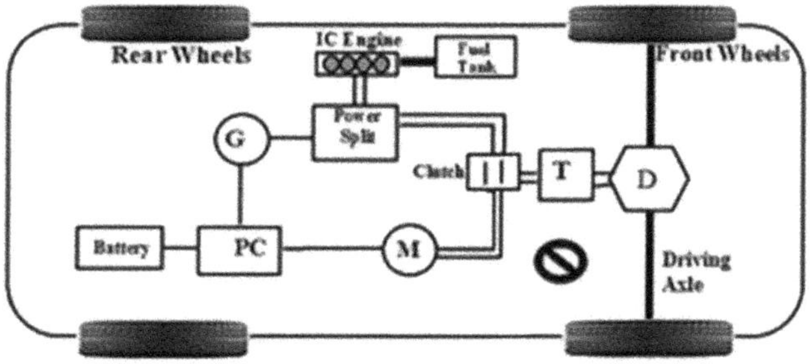

Fig. 6.5 Architecture of series-parallel HEV configurations. *G* Generator, *PC* Power converter, *M* Motor, *T* Transmission, *D* Differential

characteristics of series and parallel HEVs. It allows more operating modes than that of series or parallel drive alone.

Based on the power flow there are 12 modes of operation in Series-parallel HEV. As this system involves the characteristics of both series and parallel hybrid systems, several operating modes are possible which can be classified under two categories viz. ICE dominated and EM dominated. This mode is also called Dual-mode HEV [13].

The operating modes under the ICE dominant system are:

1. **Start-up:** Fig. 6.6a shows the power flow diagram at the start. In this mode, the battery alone offers the power required for driving the vehicle and the IC engine is in off condition.
2. **Acceleration:** Fig. 6.6b shows the power flow diagram during acceleration. In this operating mode, the electric motor and the IC engine share the traction power to drive the wheels.
3. **Normal driving:** Fig. 6.6c shows the power flow diagram during the normal driving condition. In this operating mode, the power required for propulsion is delivered only by the ICE and the electric motor in the off condition.
4. **Deceleration/braking:** Fig. 6.6d shows the power flow diagram during deceleration or braking. In this mode, kinetic energy at the wheels is transformed into electrical energy by the motor working like a generator and charges the battery through a power converter.
5. **Battery charging while driving:** Fig. 6.6e shows the power flow diagram of the battery charging while driving. In this mode, the IC engine delivers the power required for both traction and battery charging.
6. **Battery charging during standstill:** Fig. 6.6f shows the power flow diagram of the battery charging when the vehicle is at rest. In this mode, when the vehicle halts, the ICE charges the battery through the power converter.

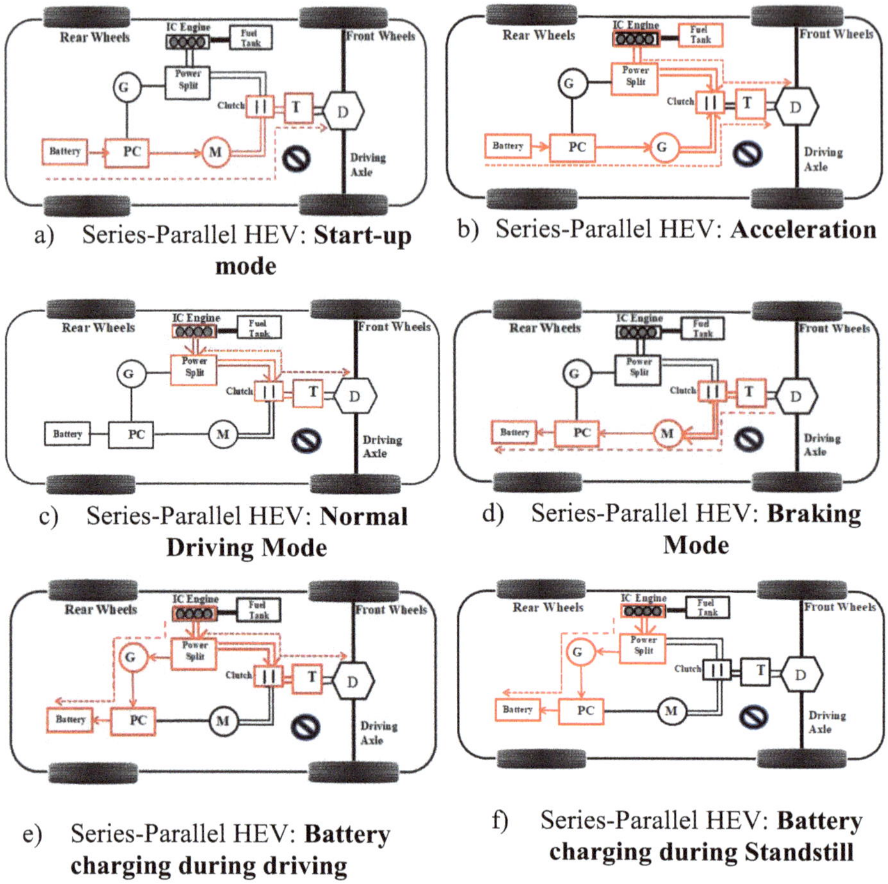

a) Series-Parallel HEV: **Start-up mode**

b) Series-Parallel HEV: **Acceleration**

c) Series-Parallel HEV: **Normal Driving Mode**

d) Series-Parallel HEV: **Braking Mode**

e) Series-Parallel HEV: **Battery charging during driving**

f) Series-Parallel HEV: **Battery charging during Standstill**

Fig. 6.6 Power flow modes in series-parallel HEV (ICE dominant)

The operating modes under EM dominated system are:

1. **Start-up:** Fig. 6.7a shows the power flow diagram at the start. In this mode, the battery alone supplies the required power for driving the vehicle whereas the ICE is in the off state.
2. **Acceleration:** Fig. 6.7b shows the power flow diagram during acceleration. In this mode, mechanical energy from the IC engine is transferred to the wheels along with a part being transmitted through the generator and motor. Also, energy from the battery is transmitted through the electric motor.
3. **Normal driving:** Fig. 6.7c shows the power flow diagram during the normal driving condition. In this mode, the IC engine delivers the required power to the wheels directly and also through the generator. So, in this operation, the battery is not used.

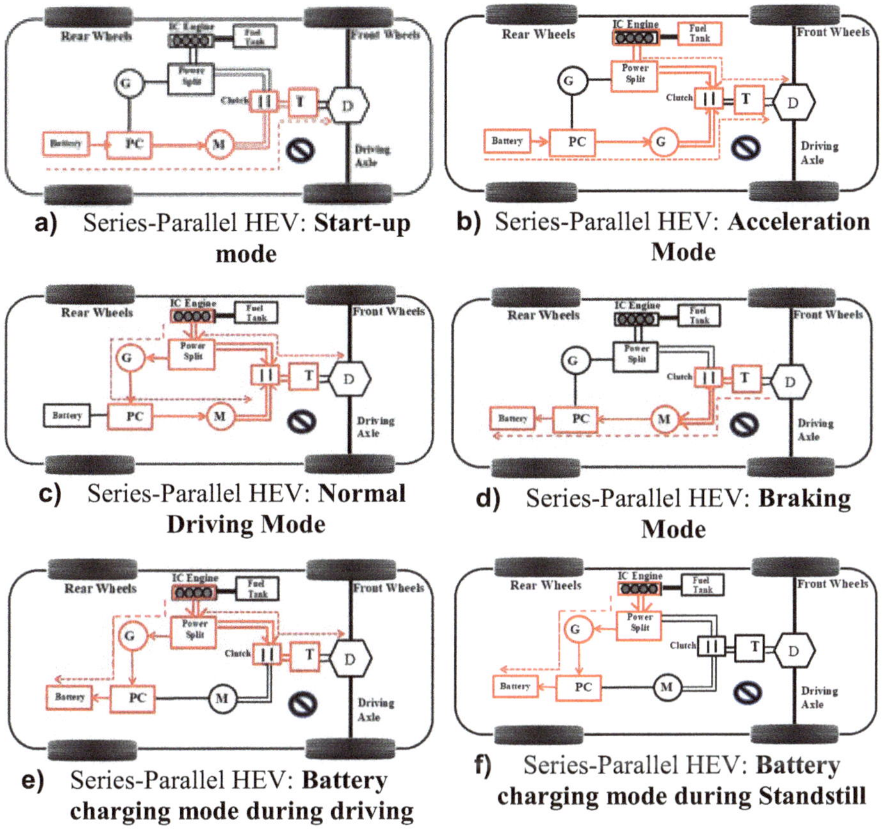

a) Series-Parallel HEV: **Start-up mode**

b) Series-Parallel HEV: **Acceleration Mode**

c) Series-Parallel HEV: **Normal Driving Mode**

d) Series-Parallel HEV: **Braking Mode**

e) Series-Parallel HEV: **Battery charging mode during driving**

f) Series-Parallel HEV: **Battery charging mode during Standstill**

Fig. 6.7 Power flow modes in series-parallel HEV (electric motor dominant)

4. **Braking/deceleration:** Fig. 6.7d shows the power flow diagram during deceleration or braking. In this mode of operation, the kinetic energy at the wheels is transformed into electrical energy by the motor working like a generator and charges the battery via a power converter.
5. **Battery charging while driving:** Fig. 6.7e shows the power flow diagram of the battery charging while driving. In this mode, the IC engine delivers the power required for both traction and battery charging.
6. **Battery charging during standstill:** Fig. 6.7f shows the power flow diagram of charging the battery during standstill. In this operating mode, when the vehicle comes to rest, the ICE charges the battery through the power converter.

3.1.4 Complex HEV Powertrain and Modes of Operation

Figure 6.8 shows the architecture of the Complex Hybrid drivetrain. This configuration is similar to a series-parallel hybrid electric vehicle as it has both generator and motor. Generally, dual axle transmission is employed by which both the front and the rear wheels are driven. A power converter is connected between the battery and motor/generator to provide electrical coupling and facilitate bi-directional power flow.

The configuration in Complex HEV is of two types viz. Front hybrid Rear electric and Front electric Rear hybrid which is based on the placing of propulsion devices. Here both the front wheels and the rear wheels are driven by two subsystems. Based on the power flow, both have six modes of operation.

In **Front Hybrid Rear Electric (F-H-R-E) dual axle complex HEV**, there is a hybrid system of IC engine & motor connected to the front wheel transmission and a pure electric motor connected to rear wheel transmission. The modes of power flow operation are:

1. **Start-up:** Fig. 6.9a shows the power flow diagram during starting. In this mode, the required traction power for starting is supplied by the battery to both the motors connected to the front and rear transmission through the power converters whereas the ICE is in the off condition.
2. **Full-throttle Acceleration:** Fig. 6.9b shows the power flow diagram during acceleration. In this mode, the ICE and front-wheel motor drive the front wheel while the second motor drives the rear wheel.
3. **Normal driving/battery charging:** Fig. 6.9c shows the power flow diagram during normal driving conditions. In this operating mode, the IC engine supplies the driving power to the front wheels and part of the energy is used to charge the battery. So, in normal driving, only front wheels are used to drive the system while the rear transmission is inactive.
4. **Light load:** Fig. 6.9d shows the power flow diagram during the light load operation. In this mode, the power is delivered to the front wheels by the front motor while the rear motor and IC engine are in the off state.

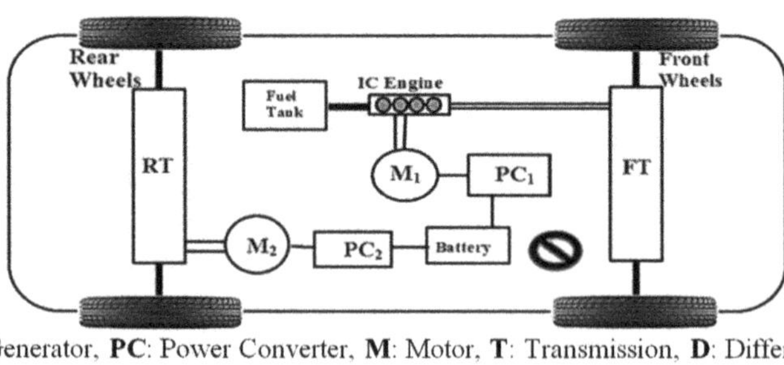

Fig. 6.8 Architecture of different HEV configurations. *G* Generator, *PC* Power converter, *M* Motor, *T* Transmission, *D* Differential

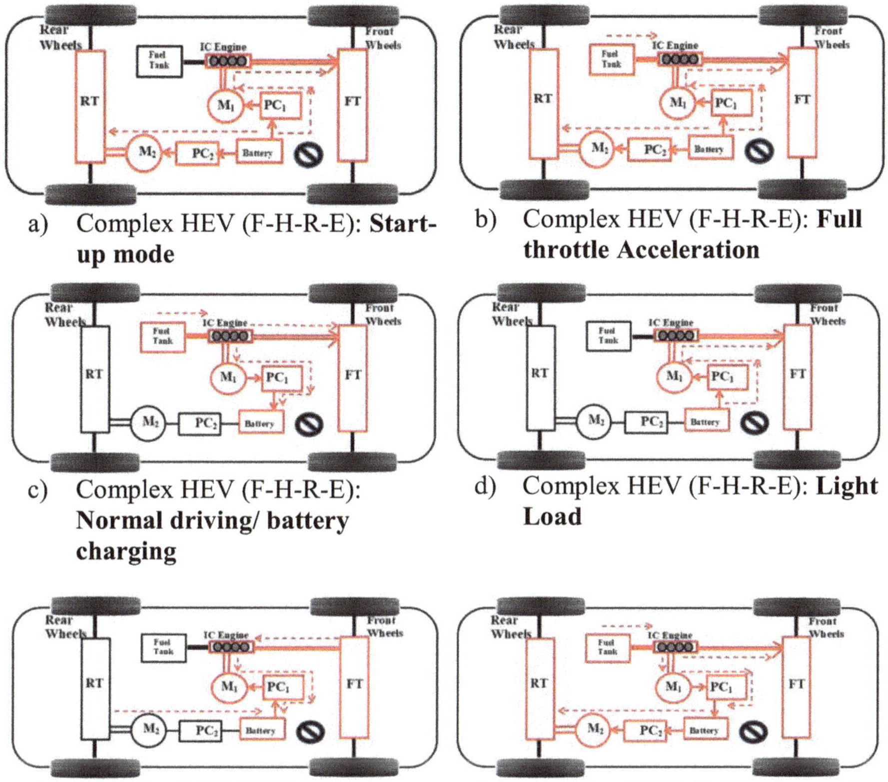

a) Complex HEV (F-H-R-E): **Start-up mode**

b) Complex HEV (F-H-R-E): **Full throttle Acceleration**

c) Complex HEV (F-H-R-E): **Normal driving/ battery charging**

d) Complex HEV (F-H-R-E): **Light Load**

e) Complex HEV (F-H-R-E): **Braking/deceleration**

f) Complex HEV (F-H-R-E): **Axle Balancing**

Fig. 6.9 Power flow modes in front hybrid rear electric (F-H-R-E) dual axle complex HEV

5. **Braking/deceleration:** Fig. 6.9e shows the power flow diagram during deceleration or braking. In this mode, motors at the front wheel transmission and rear-wheel transmission act as generators and charge the battery simultaneously.
6. **Axle balancing:** Fig. 6.9f shows the power flow diagram during axle balancing. This is a unique operating model. If the front wheel is undergoing sliding or due to bad road, there is a sliding force in the front wheels. To overcome this an equal and opposite force can be extracted or put in the rear wheel thereby balancing the sliding force. Such system can balance itself and can move in difficult terrains as well.

In the **Front Electric Rear Hybrid (F-E-R-H) dual axle complex HEV**, a pure electric motor is coupled to the front wheel transmission while there is a hybrid system of IC engine and a motor connected to rear wheel transmission. The modes of power flow operation are:

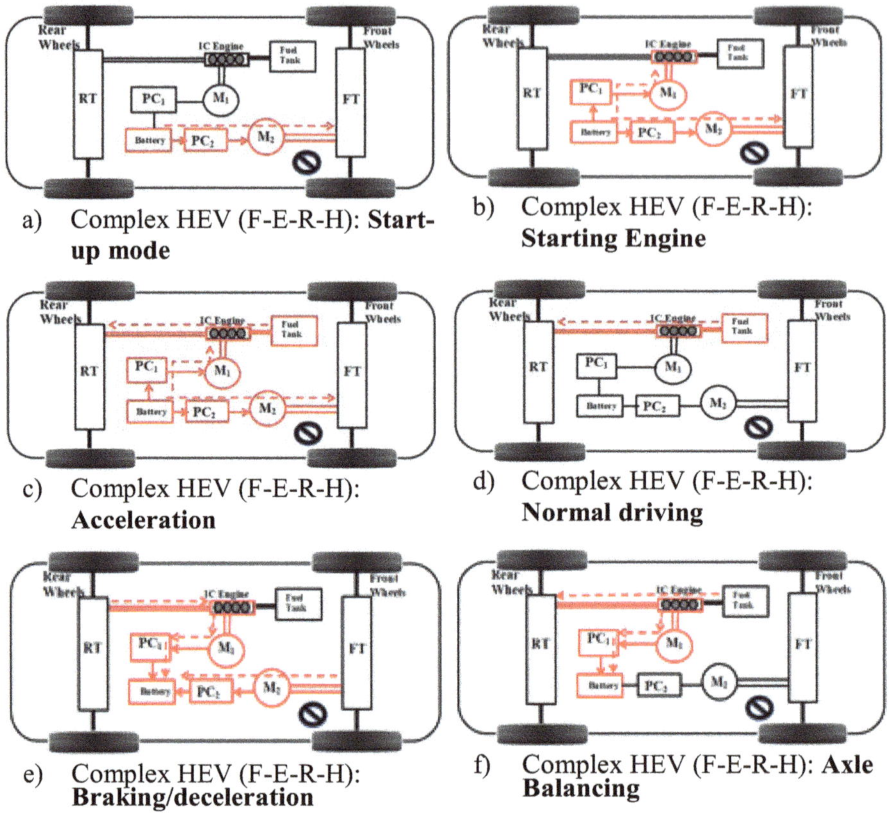

a) Complex HEV (F-E-R-H): **Start-up mode**

b) Complex HEV (F-E-R-H): **Starting Engine**

c) Complex HEV (F-E-R-H): **Acceleration**

d) Complex HEV (F-E-R-H): **Normal driving**

e) Complex HEV (F-E-R-H): **Braking/deceleration**

f) Complex HEV (F-E-R-H): **Axle Balancing**

Fig. 6.10 Power Flow modes in Front Electric Rear Hybrid (F-E-R-H) dual axle complex HEV

1. **Start-up:** Fig. 6.10a shows the power flow diagram during starting. In this mode, the required traction power for starting is supplied only to the front transmission by the front motor connected to the battery via power converter while the IC engine and rear motor are in the off condition.
2. **Starting engine:** Fig. 6.10b shows the power flow diagram during the start of the IC engine. The second motor works as an integrated starter generator to start the IC engine as cranking.
3. **Acceleration:** Fig. 6.10c shows the power flow diagram during acceleration. In this mode, the front-wheel motor propels the front wheel whereas the rear-wheel motor and ICE drive the rear wheel.
4. **Normal driving:** Fig. 6.10d shows the power flow diagram during normal driving. In this mode, the IC engine delivers power to the rear transmission to drive the rear wheels. So, in normal driving, only rear wheels are used to drive the system while the front transmission is inactive.

5. **Braking/deceleration:** Fig. 6.10e shows the power flow diagram during deceleration or braking. In this mode, the motors at the front wheels and rear wheels act as generators, extracting the power from the wheels and charging the battery simultaneously.
6. **Battery charging:** Fig. 6.10f shows the power flow diagram during normal driving. In this mode, the IC engine delivers power to the rear transmission to drive the rear wheels and some amount of energy is used to charge the battery. So, in normal driving, only rear wheels are used to drive the system while the front transmission is inactive.

3.2 Power Flow in Battery Electric Vehicles

A BEV is a system integrating three types of subsystems viz. energy source, traction and support device. The energy subsystem comprises of a battery source, energy management and refuelling system. The traction system comprises a power electronics converter, motor and its controller, propulsion system and driving wheels. The auxiliary system consists of a power supply, power and temperature controllers [11]. The BEV is relatively flexible because of the absence of many mechanical components which are present in ICE vehicles. The only moving part in BEV is the motor which requires a power supply to operate and it can be placed at various locations in the BEV. The motor will run with the supply given to it through the electrical wires. Due to such flexibility, various configurations are possible to be implemented.

BEV uses batteries and ultracapacitors as sources. Six types of configurations are existing for power transfer in BEV's among them three are widely used viz. BEV with longitudinal front-wheel drive, Fixed gear and No clutch, Transverse front-wheel drive, dual motor drive, In-wheel motor drive, Outer rotor motor drive [11, 14]. In this chapter, BEV with longitudinal front-wheel drive architecture is considered to explain the modes of operation. The power flow in Battery Electric Vehicles can be explained with seven modes of operation indicating the power flow. Among these modes, G2V and V2G are popular, along with these two new modes H2V and V4G are also discussed. EVs can store energy in their batteries and Bidirectional DC–DC converters play a major role in the charging and discharging process of Batteries. However, here charge controller operation is not focused and it is assumed that the charge controller is an integral part of the Energy Storage System (ESS) [15–18].

1. **Grid-to-Vehicle (G2V):** In the G2V mode, power flow will be from the grid to the charger and then to the EV battery. This mode is used to charge batteries and is shown in Fig. 6.11a.
2. **Vehicle-to-Grid (V2G):** When there is a requirement for power in the grid, the battery has to supply power to the grid. In this mode, power flows in the reverse

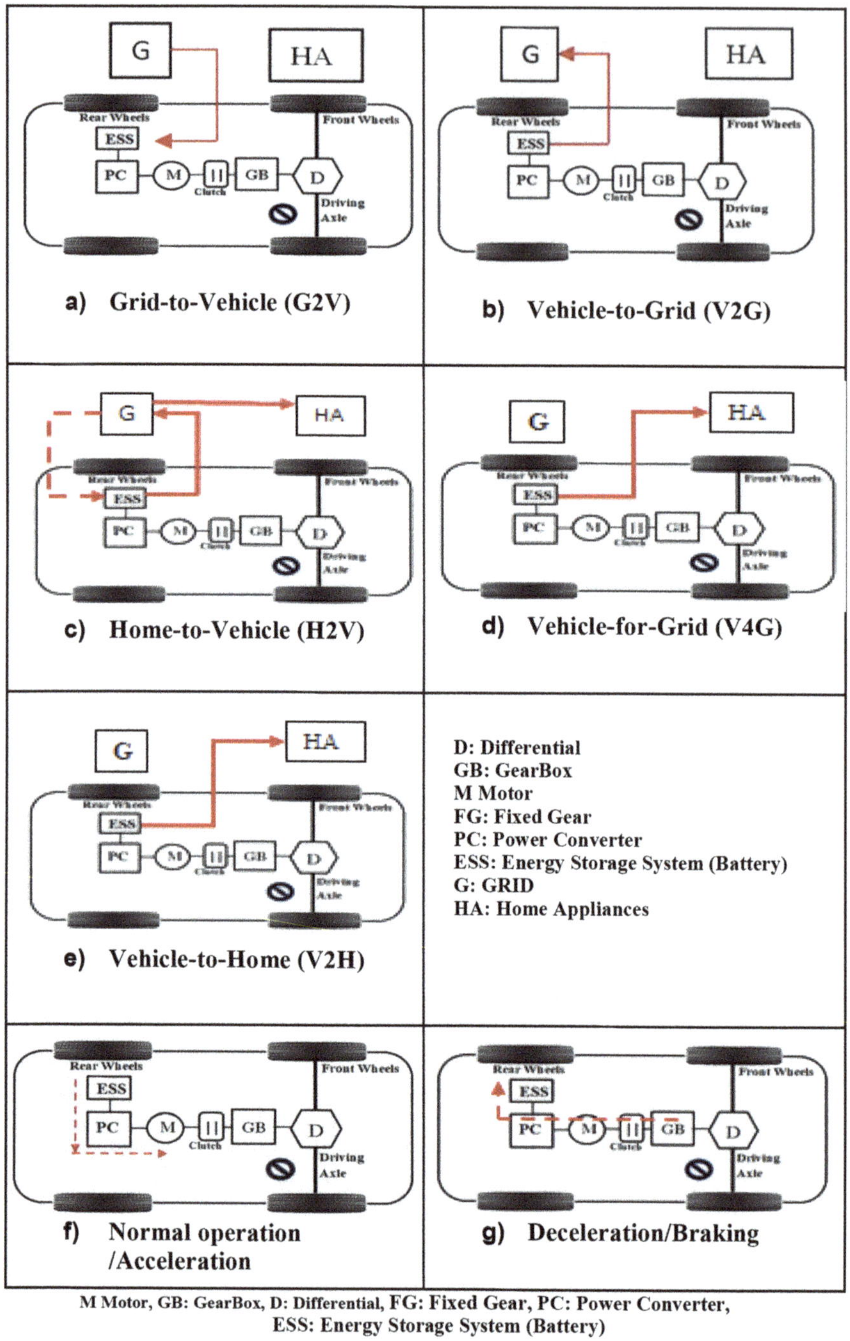

Fig. 6.11 Power flow modes in battery electric vehicles. *M* Motor, *GB* Gear box, *D* Differential, *FG* Fixed gear, *PC* Power converter, *ESS* Energy storage system (battery)

direction from the battery to the grid via the controller and the control during the mode will be taken care of by the power grid manager. Figure 6.11b shows the power flow path for this operating mode.

3. **Home-to-Vehicle (H2V):** In this mode based on the power consumed by home appliances two ways of power flow are possible. H2V is a modified form of G2V and V2G.

   (a) H2V combined with G2V
   (b) H2V combined with V2G

   These two modes help in avoiding overcurrent tripping in the circuit breaker which is installed in the house The power flow diagrams for the modes are shown in Fig. 6.11c.

4. **Vehicle-for-Grid (V4G):** When EV is non-operating in G2V mode and V2G modes it can be used for reactive power compensation acting as an active power filter for the appliances connected in-home as shown in Fig. 6.11d. However, this mode can also be combined with V2G or G2V modes, in that case, power flow to or from the battery will be limited.
5. **Vehicle-to-Home (V2H):** In case of power outages when emergency power is required then EV batteries can be used as offline UPS to deliver power by operating in this mode as shown in Fig. 6.11e.
6. **Normal operation/Acceleration:** When the EV is in motion during starting, normal operation and during acceleration, this mode comes into the picture. The power flow will be from battery to power converter then to motor as shown in Fig. 6.11f.
7. **Deceleration/Braking:** When the motor vehicle slows down by applying brakes or during stopping, Electric Vehicles are to be operated in this mode and the power flow path is as shown in Fig. 6.11g.

3.3 Power Flow in Fuel Cell HEV (FCHEV)

The fuel cells are the prime supplier of energy in fuel cell EV (FCEV). These cells use their chemical reactions to generate electricity and the electrical motor for propelling the vehicle. The fuel cells cannot be considered as the main source of energy for BEV since fuel cells cannot have regenerative energy. Therefore, batteries are usually adopted as the other energy source.

The fuel cell generates the electric power and this power is given to the motor to drive the wheels. Excess energy, if any, can be stored in the batteries. The fuel cells have higher efficiency with low emissions of carbon. The power density of the fuel cells is suitable for EV applications. A typical powertrain configuration of Fuel cell EV is as shown in Fig. 6.12. There are various energy management techniques that help to monitor the system and focus on important aspects such as energy losses, fuel economy and efficiency [5, 19].

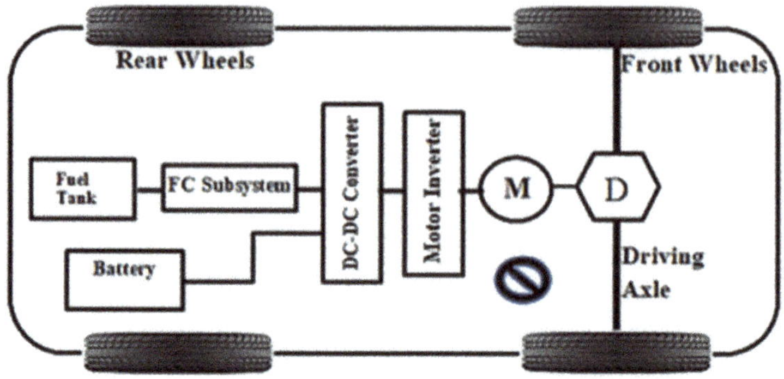

Fig. 6.12 Configuration of FCEV

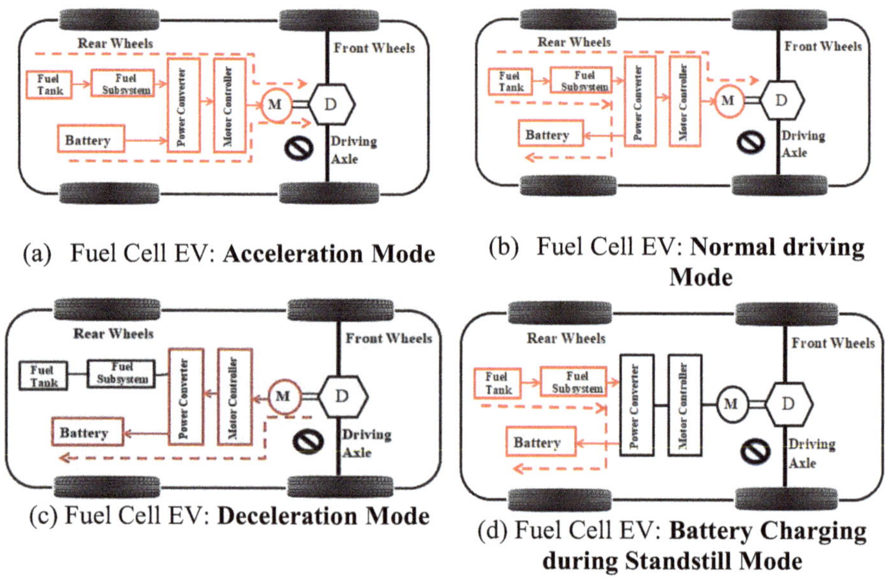

Fig. 6.13 Power flow modes in Fuel Cell HEV

Fuel cell hybrid electric vehicles operate majorly in 3 operating modes as explained below. Other than these modes there is one more least popular method called battery charging mode. Power flow in all these four modes is discussed with diagrams [20, 21].

1. **Acceleration mode:** During acceleration our starting motor requires more power hence both fuel cell and battery will provide power. The fuel cell will be operating with its full rating remaining power will be supplied by the battery as shown in Fig. 6.13a.

2. **Normal operation:** In this mode of operation the motor will demand power depending on the speed operated by the user. If the power demand by the motor is the same as the rated power of the fuel cell, then the fuel cell supplies power to the motor alone. If the demand from the motor is less than the rated power of the fuel cell then the fuel cell provides the power to the motor and the remaining power will charge the battery as the fuel cell has to operate at its rated power always. Power flow for both the cases is shown in Fig. 6.13b.
3. **Deceleration:** During deceleration, the motor operates as a generator and supplies power for the battery. Hence this power charges the battery as shown in Fig. 6.13c.
4. **Battery charging mode:** In this mode of operation the electric motor does not receive power from the fuel cell as well as battery and the vehicle remain in a halt position. However, the battery can be charged through the fuel cell in this mode as shown in Fig. 6.13d. Hence it is called battery charging mode.

3.4 Ultra Capacitor Electric Vehicles (UCEV)

One of the major concerns is supplying for the peak power needs in EV. Ultracapacitors (UC) are used in two-wheeler along with other energy sources in such systems which need both continuous supply, peak power demands and also for low power discharges. In the EVs which implement Ultracapacitors, it will have ICE and/or Battery which acts as the primary source of energy and the Ultracapacitors act as the secondary power source [7, 21]. In the life cycle of the vehicle drive, it has to satisfy the needs of short duration events many times for which such combination is suitable. Even the UC will have an increased life span without any constraint on the depth of discharge. Such combinations also help in the overall reduction in size. The UC-based vehicle offers greater advantages such as operation during peak power, has a longer life cycle, tolerates surges during operation of the battery and provides high magnitude power. The combination of UC with any other primary source has high stability. Figure 6.14 shows the configuration of UC-based electric vehicles. The power flow modes in UCEV are shown in Fig. 6.14. The UC can also be implemented along with both the ICE and the battery. As per the power conversion needs, various power converters can be implemented at different points in the power flow path. A typical structure with UC and battery is shown in Fig. 6.14. Also, the various modes of operations are indicated in the figure such as acceleration, normal driving mode, deceleration and battery charging under a standstill.

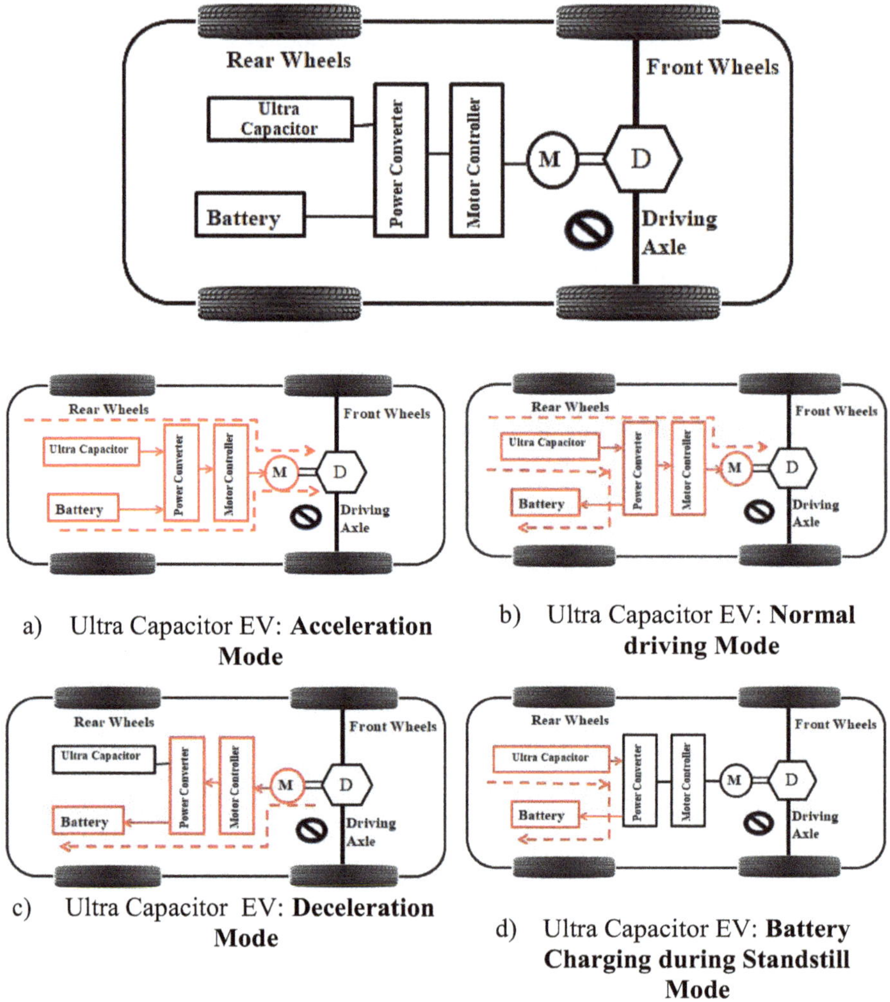

Fig. 6.14 Power flow modes in Ultra Capacitor electric vehicles

4 Current Trends and Technology

The various configurations under HEV and BEV have their advantages and challenges. Wide R&D is being done to improvise the configuration for efficient energy utilisation and reduce the power losses during transmission. A constant and a common trend is still not observed due to the regional variations and road conditions. Previous surveys indicate that the widely used HEV architecture is Parallel configuration and then combined HEV (series-parallel & complex). The reason behind this trend is in the parallel configuration it is possible to reduce the size and rating of both the propelling devices and in complex HEV configuration it is possible to operate in

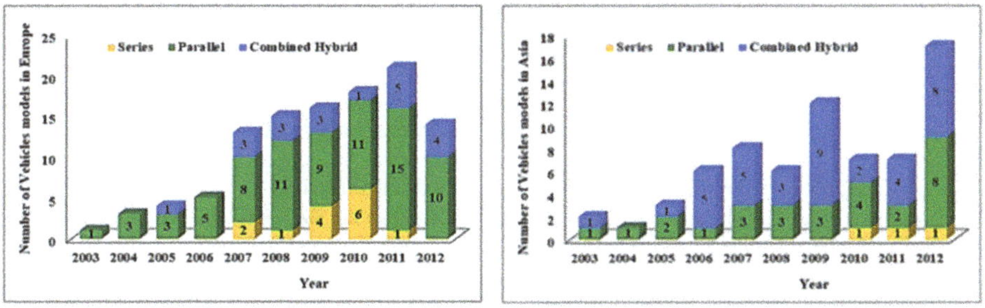

Number of HEV vehicle models with various powertrain configuration in Europe between 2003 and 2012

Number of HEV vehicle models with various powertrain configuration in Asia between 2003 and 2012

Fig. 6.15 Different HEV powertrain architectures in Europe and Asia

both series and parallel modes. The graph in Fig. 6.15 shows the implementation of various HEV configurations in Europe and Asia in various electric vehicle models [12, 22].

Based on the chemical composition, different types of fuel cells are emerging resulting in higher energy efficiency. Due to advances in research towards different types of fuels used in fuel cells, a fuel cell-powered vehicle with a longer driving range and less battery chagrin time can be obtained.

5 Conclusion

In this chapter, the architecture of different configurations of BEV's and HEV's are briefly presented. The various power flow modes of series, parallel, series-parallel, complex HEVs, fuel cell HEV and also for BEV are discussed with the support of power flow diagrams for each mode in all the types. BEVs use battery, fuel cell or ultracapacitor as energy storage systems and the HEV uses IC Engine along with the batteries. HEV's provide high efficiency compared to BEVs. It has been observed that the widely implemented configuration is parallel HEV structure because of its strengths and flexibility. FCHEV and UCEV are also gaining focus due to the developments happening in fuel cell technology and energy supply during peak requirements.

References

1. Boretti, A., & Rosa, L. (2019). Reassessing the projections of the World Water Development Report. *npj Clean Water, 2*(1).
2. Dargay, J., Gately, D., & Sommer, M. (2007). Vehicle ownership and income growth, worldwide: 1960–2030. *The Energy Journal, 28*, 143–170.

3. Chakraborty, S., Vu, H., Hasan, M. M., Tran, D., El Baghdadi, M., & Hegazy, O. (2019). DC-DC converter topologies for electric vehicles, plug-in hybrid electric vehicles and fast charging stations: State of the art and future trends. *Energies, 12*(8), 1569.
4. Kawamoto, R., Mochizuki, H., Moriguchi, Y., & Nakano, T. (2019). Estimation of CO 2 emissions of internal combustion engine vehicle and battery electric vehicle using LCA. *Sustainability, 11*(9), 2690.
5. Das, H. S., Tan, C. W., & Yatim, A. H. M. (2017). Fuel cell hybrid electric vehicles: A review on power conditioning units and topologies. *Renewable and Sustainable Energy Reviews, 76*, 268–291.
6. Polini, C., Crescimbini, F., Di Napoli, A., Lidozzi, A., & Solero, L. (2010). *Power flow analysis for plug-in hybrid electric vehicles*. IEEE.
7. Veer, K., Hari, S., Bansal, O., & Singh, D. (2019). A comprehensive review on hybrid electric vehicles: Architectures and components. *Journal of Modern Transportation, 27*(2), 77–107.
8. Emadi, A., Williamson, S. S., & Khaligh, A. (2006). Power electronics intensive solutions for advanced vehicular power systems. *IEEE Transactions on Power Electronics, 21*(3), 567–577.
9. Chan, C. C. (2002). The state of the art of electric and hybrid vehicles. *Proceedings of the IEEE, 90*, 247–275.
10. Ehsani, M., Gao, Y., & Emadi, A. *Modern electric, hybrid-electric and fuel cell vehicles* (2nd ed.). CRC Press.
11. Un-noor, F., Padmanaban, S., & Mihet-popa, L. (2017). A comprehensive study of key electric vehicle (EV) components, technologies, challenges, impacts, and future direction of development. *Energies, 10*(8), 1–82.
12. Kumar, D. (2019). A comparative review on power conversion topologies and energy storage system for electric vehicles. *International Journal of Energy Research, 44*, 1–23.
13. Lulhe, A. M. (2015). *A technology review paper for drives used in electrical vehicle (EV) & hybrid electrical vehicles (HEV)* (pp. 632–636). IEEE.
14. Guirong, Z., Henghai, Z., & Houyu, L. (2011). The driving control of pure electric vehicle. *Procedia Environmental Sciences, 10*, 433–438.
15. Monteiro, V., & Member, S. (2015). *Operation modes for the electric vehicle in smart grids and smart homes: Present and proposed modes* (Vol. 9545). IEEE. (for BEV-operating modes).
16. Chau, K. T. (2014). *"Pure electric vehicles" alternative fuels and advanced vehicle technologies*. Woodhead Publishing Limited. https://doi.org/10.1533/9780857097422.3.655
17. El-Sharkawy, M. R., & Ghazaly, N. M. (2014). Power management strategies for hybrid electric vehicles. *European Journal of Academic Essays, 1*(5), 63–67.
18. Tie, S. F., & Wei, C. (2013). A review of energy sources and energy management systems in electric vehicles. *Renewable and Sustainable Energy Reviews, 20*, 82–102.
19. Emadi, A., Rajashekara, K., Williamson, S. S., & Lukic, S. M. (2005). Topological overview of hybrid electric and fuel cell vehicular power system architectures and configurations. *IEEE Transactions on Vehicular Technology, 54*(3), 763–770.
20. Feroldi, D., Basualdo, M., Outbib, R., Basualdo, M. S., Feroldi, D., & Outbib, R. (2011). *PEM fuel cells with bio-ethanol processor systems: A multidisciplinary study of modelling, simulation, fault diagnosis and advanced control.* Springer-Verlag. 978-1-84996-184-4.
21. Erjavec, J. *Hybrid, electric & fuel cell vehicles* (2nd ed.). DELMAR Cengage Learning. ISBN-13: 978-0-8400-2395-7.
22. Frieske, B., Kloetzke, M., & Mauser, F. (2013). Trends in vehicle concept and key technology development for hybrid and battery electric vehicles. *2013 World Electric Vehicle Symposium and Exhibition, 6*, 9–20.

Chapter 7
Energy Storage Devices and Front-End Converter Topologies for Electric Vehicle Applications

Sumukh Surya and Sheldon S. Williamson

1 Introduction

Conventional transport system uses fossil fuels to drive vehicles. These vehicles produce a large amount of green house gas (GHG), which are leading causes of climate change and air pollution. According to a report from Intergovernmental Panel on Climate Change (IPCC), the transportation sector is responsible for 14% of all green house gas emissions. Hence, electrification of the existing transportation sector is a much-needed solution to tackle these problems.

EV, HEV, and PHEV have gained utmost importance during recent years. However, the idea of EV arose in the early twentieth century. EVs are preferred as they reduce fuel usage and greenhouse emissions. The main source of power to drive them is derived from the rechargeable battery packs. In order to use the power effectively and enable user safety, power electronic converters are used. The main converters include Front End AC–DC converters, DC–DC converters, and DC–AC inverters.

An EV operates on multiple electric motors for propulsion. It uses a battery, fuel cell, or ultracapacitors as the main power plant. Lithium-ion batteries are preferred over other batteries because of their power density being highest, lightweight, long life and thermal stability, and high terminal voltage. However, it is expensive.

S. Surya (✉)
Kirtane and Pandit Information Technologies, Bangalore, India

S. S. Williamson
Ontario Tech University, Oshawa, ON, Canada
e-mail: sheldon.williamson@ontariotechu.ca

M. Kathiresh et al. (eds.), *E-Mobility*, EAI/Springer Innovations in Communication and Computing, https://doi.org/10.1007/978-3-030-85424-9_7

2 Energy Storage Devices

Fossil fuels are depleting at a massive rate. Hence, prominence is on the usage of power-storing devices. Some of the examples are (a) battery, (b) FC (fuel cell), (c) UC (ultra-apacitors).

Few advantages are mentioned below. (a) Serve power during peak demand (b) Backup power during emergency (c) Dependence on-grid reduces

The choice of selection depends on the

- Operational and maintenance cost
- Capacity self-discharge rates and
- Sensitivity toward environmental changes

2.1 Battery

Battery converts chemical energy contained within its active materials directly into electric energy by means of electrochemical reactions. Different types of batteries are (a) primary (irreversible chemical reaction) and (b) secondary (reversible chemical reaction). Alkaline batteries, Zinc batteries are examples of the primary batteries. Nickel cadmium and lithium-ion batteries are examples of the secondary batteries. Most of the EVs use lithium-ion batteries for powering them. One of the important problems associated with Li-ion batteries is that they are expensive and most often would cost more than the EV itself as the large number of cells is required to form a battery pack in addition to the protection circuits. In order to provide a cost-effective EV, the battery cost has to be lowered. A 10 year 1,50,000 mile warranty on the components for HEV in regard to regulation in California. In [1], it was reported that the second generation of Honda Civic battery packs deteriorated after 5 years. The cost of those battery packs was about $2000 which excluded the shipping and installation costs and tax. Another important drawback is that they are sensitive to high temperatures and catch fire easily [2].

In 2013, all Boeing 787 Dreamliner aircraft was grounded due to the failure of Li-ion battery packs causing large number of deaths. Upon investigation which included X-ray computer tomography scans and digital radiography, it was found that a thermal runway in a single cell got spread to the rest of the battery which leads to the catastrophic incident [2]. Hence, an efficient BMS (Battery Management System) for protecting the battery is required. The function of BMS includes (a) Overcurrent protection (b) Over Voltage protection (c) Protection against high temperature (d) Performance Management (State of Charge and the State of Health) and (e) Diagnostics (Estimation of State of Life). The BMS functionality is shown in Fig. 7.1. It can be inferred from [3] that for the efficient working of the battery, the temperature has to be $<60°C$.

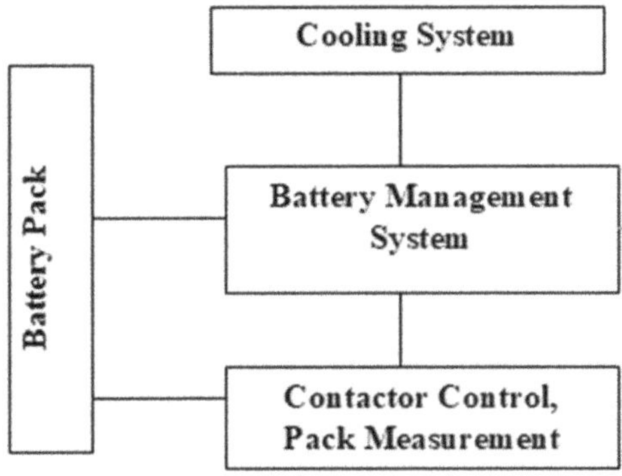

Fig. 7.1 Functionality of battery management system [3]

Fig. 7.2 Poor voltage method [3]

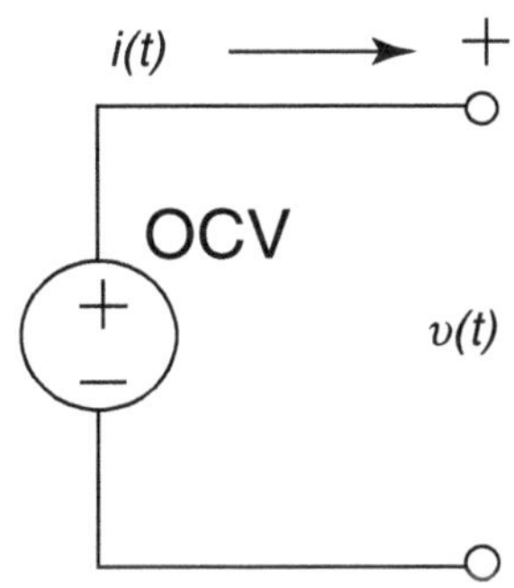

In [3, 4], an ideal model for a battery is shown. Figure 7.2 shows the poor voltage battery model. It does not contain any losses components. It assumes the battery to be a constant voltage source.

The next approach to battery modeling was shown [5] was termed as “Simple Battery Model” which contained a series resistance named ESR (Equivalent Series Resistance) with the battery source and represented the drop in OCV due to flow of current. The limitation with this model is that the internal impedance characteristics and the dependence on SOC are not considered. Hence, a modified version viz. Thevenin’s Battery Model was proposed.

Thevenin’s model can be described as a pair of RC in series with the internal resistance R_0. When a discharge pulse is provided, the terminal voltage (V_T) decays and falls to a lower value. Upon providing sufficient settling time, it diffuses to its OCV. This phenomenon can be modeled as a delay circuit by adding a pair of RC in series with R_0 (Fig. 7.3).

In [6], battery modeling considering five RC pairs is shown. The advantage of such modeling is that the current with high-frequency transients can be easily are captured during pulse charging. It was observed that the battery parameters were independent of the discharge current and ambient temperature (T_{amb}) and showed dependence on SOC.

Fig. 7.3 Thevenin's model [5]

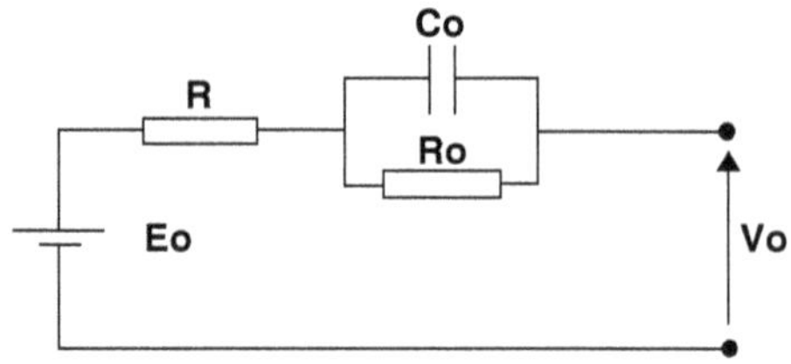

Fig. 7.4 Enhanced self-correcting model

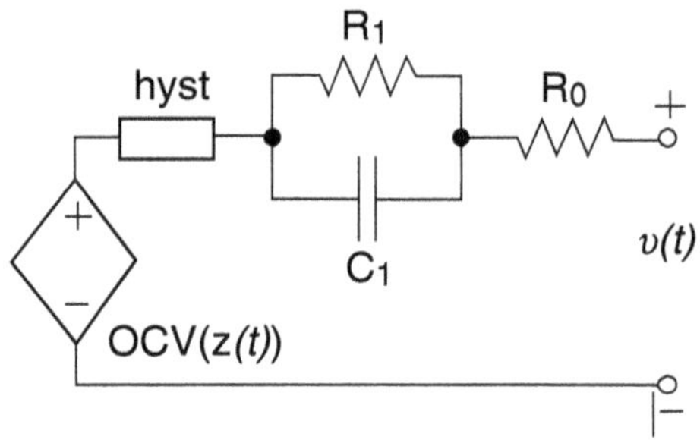

The modified model is based on operation over a range of load combinations in [7]. Figure 7.4 shows an ESC (Enhanced Self Correcting) Model for a battery [3]. This is model captures all the dynamics of the battery and hence ESC model closely matches the real behavior of the battery.

2.2 *Ultra-Capacitor (UC)*

Ultracapacitors possess two metal plates coated with a soft material that is porous and is known as activated carbon. The plates are immersed in a liquid containing positive and negative charged particles. They are also known as EDLC (electric double-layer capacitor). The energy density of UC is much higher than the electrolytic capacitor, No chemical reactions take place in a UC and hence they can run efficiently for many cycles and hence are used in HEV's. Some of the disadvantages of UC are (a) low voltage per individual cell, (b) self-discharge is high in comparison with battery, (c) issues during balancing when more than three UCs are connected in series.

In [8], UCs for vehicular application were modeled using MATLAB/Simulink. Changes in

- V_T under various charge/discharge rates and
- Ambient condition changes were carried out and later, the results obtained during simulation matched with the experimental validation.

"Fast Branch" actions are governed by R_s and C_1. They represent the transient behavior in seconds. R_2 and C_2 are called the "Slow Branch." They help in estimating the internal energy distribution at the end of cycle's charge or discharge. The self-discharge rate of UC is represented by Parallel resistance R_p, which plays a pivotal role in estimating the stored energy for a particular duration of time under no-load conditions. The value of L is usually small [9] and applicable for rapidly changing demands at the load side.

2.3 Mathematical Modeling of UC

The output voltage of a fuel cell as observed from Fig. 7.5 is shown below

$$V_{UC} = V_{C1} - V_{RS} - V_L \tag{7.1}$$

where V_{UC} the output of FC is, V_{C1} the voltage at the terminals of C_1, V_{Rs} the voltage at the terminals of R_S, V_L the voltage at the terminals of L

$$V_{RS} = R_s * (i - i_L) \tag{7.2}$$

$$V_L = Ldi/dt \tag{7.3}$$

where i charge/discharge

The voltage across the UC capacitances C_1 and C_2

$$V_{c1} = \mathrm{V}_c^0 - (1/C_1)\int i_1 dt \tag{7.4}$$

$$V_{c2} = V_c^0 - (1/C_2)\int i_2 dt \tag{7.5}$$

From (7.2), (7.3), (7.4) and (7.5)

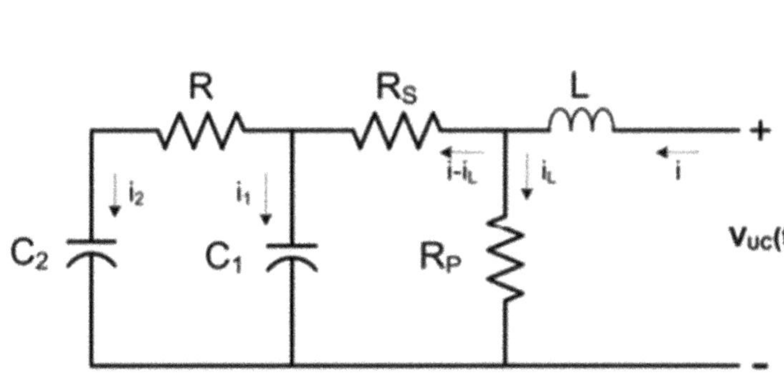

Fig. 7.5 Equivalent electrical circuit for UC [8]

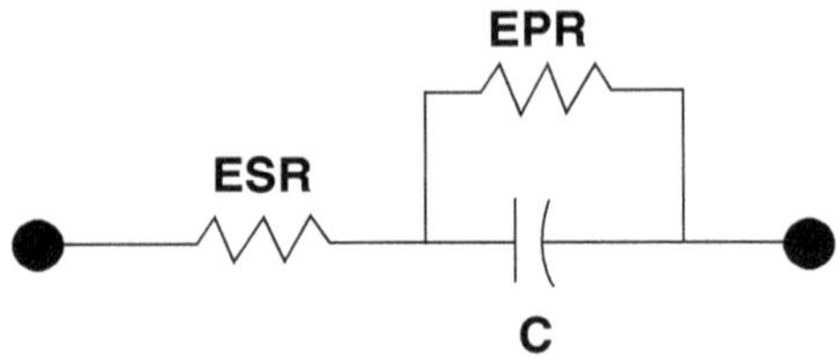

Fig. 7.6 Simplified equivalent circuit of the UC [11]

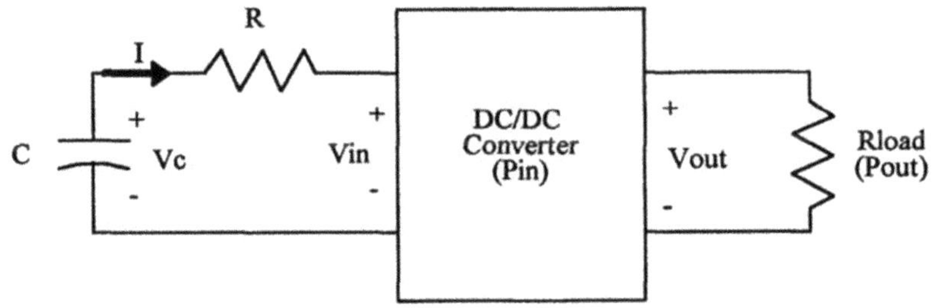

Fig. 7.7 UC feeding a DC–DC converter [11]

$$i_L = V_{Rp}/R_p \tag{7.6}$$

i_L is charge/discharge, i_1 is currently passing through C_1 and i_2 is the current passing through C_2

$$i_1 = i - i_L - i_2 \tag{7.7}$$

$$i_2 = (V_{c1} - V_{c2})/R \tag{7.8}$$

It was shown that the capacity of the UC remains unaffected by the changes in temperature. However, R_s changes with temperature.

In order to mathematical model R_s for temperature variations, Temperature Correction Factor (TCF) is applied [10].

$$TCF = -3.5 * 10^{-7}T^3 + 7.7 * 10^{-5}T^2 - 0.0054T + 1.1 \tag{7.9}$$

$$R_s(T) = TCF * ESR \tag{7.10}$$

In [11], a double layer capacitance's simplified equivalent circuit is shown. In reality, EDLC is a capacitance made up of complex networks, associated with a time constant. ESR is essential at charging and discharging intervals as it causes the internal heating of the capacitor (Figs. 7.6, 7.7, 7.8).

The Leakage effect is contributed by EPR (Equivalent Parallel Resistance) and hence, only storage for large durations is affected. EPR was not considered for simplicity in calculations [11] as eliminating it does not cause any major impact on the results. A capacitor (ideal) bank with a resistance in series is a simple circuit used for analysis and the considered load. Hence, the resistance can be written as

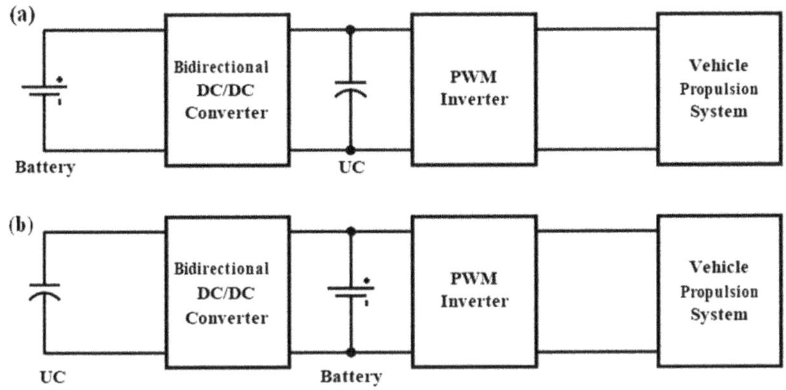

Fig. 7.8 Topologies of battery and UC in EV drive systems

$$R = n_s ESR / n_p \quad (7.11)$$

where n_s represent each string's total capacitor count in series and n_p represents each string's total capacitor count in parallel.

Amongst the two topologies, (b) is preferred for HEV applications. In configuration (a) since, the battery supplies DC/DC converter which leads to decreased energy efficiency. A UC bank of high voltage is required and is high priced. In an EV drive system containing a DC BLDC motor, to enable a larger drive range, extreme acceleration/deceleration accomplishment at affordable rates, a UC bank can be used.

2.4 *Mathematical Modeling of Ultra-Capacitor (UC)*

Electrical energy is obtained from redox reactions due to the chemical energy conversion in an electrochemical cell. Fuel cells work like batteries, but they do not run down or need recharging. They produce electricity and heat as long as fuel is supplied. A fuel cell comprises two electrodes (a) a negative electrode and (b) a positive electrode surrounded by an electrolyte. Hydrogen fuel is fed to the anode and air is let inside the cathode. A catalyst present at the anode separates hydrogen molecules into protons and electrons. These reach the cathode taking different paths. The electrons reach an external circuit and when the load is connected, current flows. The protons move through the electrolyte to the cathode where oxygen and the electrons combine to produce water and heat.

Different types of fuel cells are

1. Molten carbonate
2. Solid oxide
3. Reversible

4. Polymer electrolyte membrane
5. Direct methanol

Any FC consists of two electrodes which are usually porous and conducting materials to tap the electrical power from the electrolyte. Hydrogen (H_2) in its pure form is supplied at the negative electrode and at the positive electrode, oxygen (O_2) is supplied. Since the electrodes are porous, there exists an opportunity for the gas to come in contact with electrochemical reaction. The reaction is usually slow and needs a catalyst to fasten the process. Platinum acts as a catalyst for the electrodes.

Anodic Reaction at the negative electrode

$$H_2 \rightarrow 2H^+ + 2e^-$$

Cathode reaction

The H^+ ions migrate from the anode and reach the cathode. Interacting with O_2, water (H_2O) is produced

$$\frac{1}{2} O_2 + 2H^+ + 2e^- \rightarrow H_2O$$

Hence, the overall reaction is

$$H_2 + \frac{1}{2} O_2 \rightarrow H_2O$$

Advantages of FCs are (1) high efficiency as there are no moving parts, (2) less pollution as outputs are only water and heat, (3) operating time and endurance of FCs are much higher than that of conventional sources, and (4) low-temperature FCs can be used in indoor places

Some of the disadvantages of FCs are (1) FC's cannot be used in critical applications due to sluggish response, (2) fuel reforming technology is expensive, and (3) FC storage is critical.

Figure 7.9 shows the FC stack I–V and power curve. The cell voltage is dominated by the electrochemical kinetics of the anodic methanol oxidation (Region A) at low currents. The cell voltage is further decreased due to limitations of methanol mass transport to the anode and the electric resistance of the membrane (Region B) due to the increase in the load. In Region C, the cell voltage breaks down, and the system approaches unity current density.

Since FC provides a sluggish response; a battery pack is used for avoiding sluggish response and improves the behavior during transients. In addition, for the purpose of regenerative braking, battery packs can also be used. Amongst all the FC's PEM fuel cell is regarded as the best option for automotive application as it has high power density, low operating temperature (80 °C), and high efficiency [13].

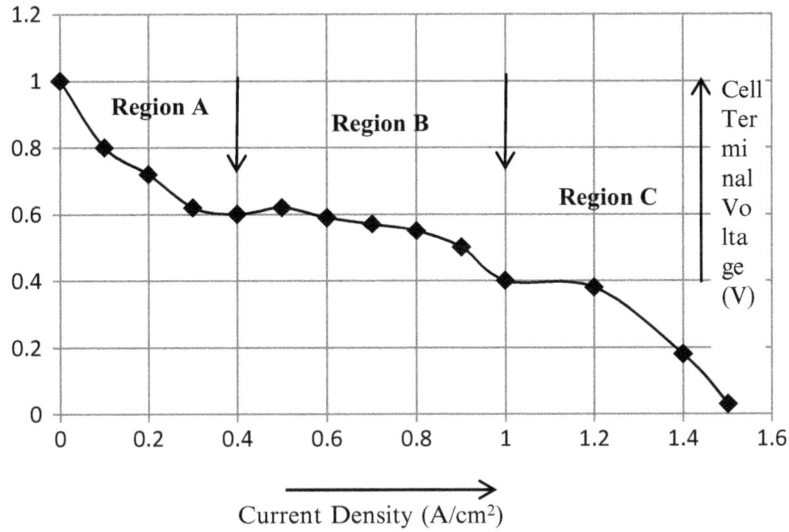

Fig. 7.9 FC stack I–V and Power Curve [12]

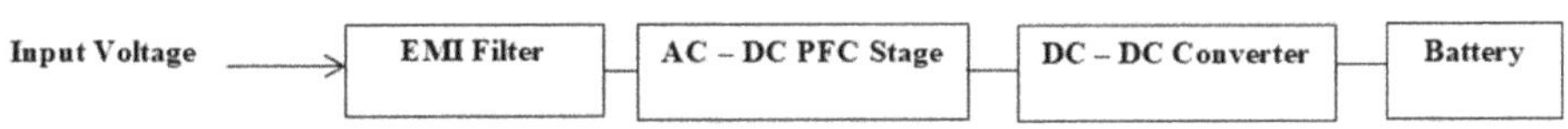

Fig. 7.10 Block diagram of a PFC front-end converter [14]

3 Front-End Converters in On-Board Charging Systems

In [14], comparative analyses on various power factor correction (PFC) converters (Rectifiers) for EV, HEV, and PHEV charging systems and associated problems are presented. Figure 7.10 shows a block diagram of a front-end converter in an EV using the PFC technique.

The AC power from the grid is fed to the AC–DC converter, Rectifier. A capacitor is added at the end of the rectifier to smoothen the current. The DC power from the rectifier feeds the DC–DC converter and later feeds the battery. Lithium-ion batteries require low voltage ripple for charging [15–17]. Hence, a PFC is required to maintain minimum current ripples.

The main idea of using a PFC is to force the DC–DC converter's input voltage in phase with the input current. Hence, a high value of power factor is created due to which smooth charging of Li-ion batteries occurs.

For enabling PFC operation, some of the control techniques used are average current mode control, peak current mode control, hysteresis control, and charge control. Amongst all the control techniques, average current mode control is the most popular control technique [18] because of its stable operation, improved noise rejection, and lower input current ripple.

4 Topologies in PFC

4.1 Conventional Boost PFC Converter

Figure 7.11 shows an uncontrolled rectifier coupled to a conventional boost converter. A Boost converter is a type of non-isolated DC–DC converter used for stepping up the voltage levels.

This configuration is simple and easy to build. However, some of the major setbacks are

1. Inefficient operation for power level >1 kW
2. Turning on the MOSFET creates a spike at C_1 causes higher EMI in the circuit due to diode current having negative spikes [19]
3. Traditional boost devices cause high voltage stress

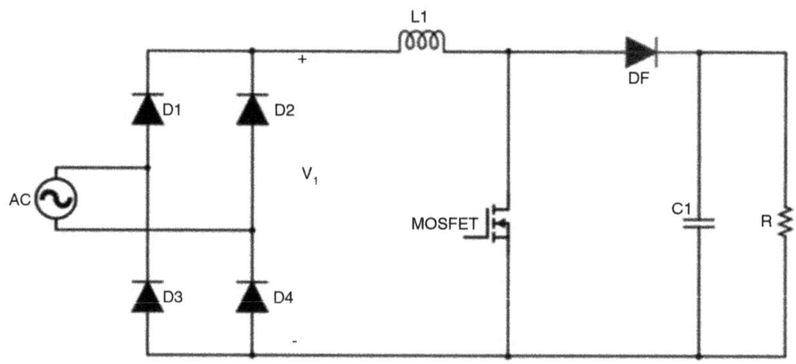

Fig. 7.11 Conventional boost PFC converter [18]

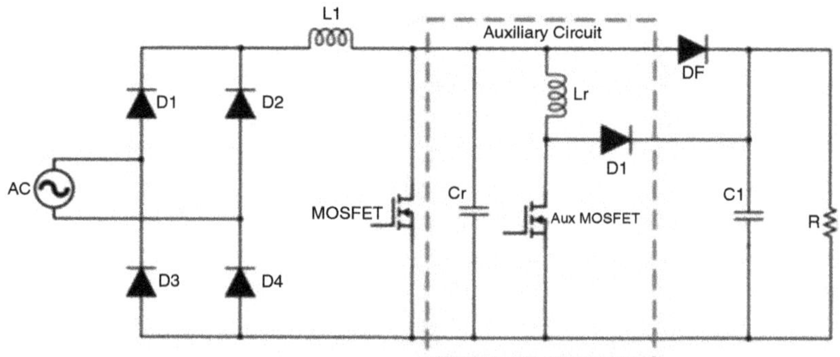

Fig. 7.12 ZVT PFC converter [19]

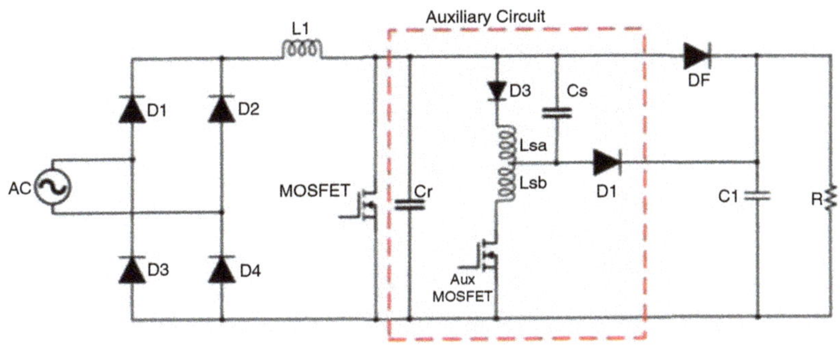

Fig. 7.13 ZCT-ZVT Boost PFC

In addition to the issues shown above, the switching and the conduction losses in the boost converter are high. Hence, a resonant converter is proposed for the PFC converter. In Fig. 7.12, a ZVT (Zero Voltage Transition) PFC converter is shown.

In resonant converters, turning ON or OFF the switches is done at when either the current or voltage is zero. By adopting this technique, the switching losses can be minimized. The soft switching is enabled due to the presence of an auxiliary circuit in the circuit shown in Fig. 7.13. In addition, the auxiliary circuit eliminates the current effect during diode reverse recovery on the circuit. The main problem associated with this circuit is that hard switching is incorporated in an auxiliary circuit. To overcome this problem, the ZCT PFC converter was proposed. Hybrid resonant converters are introduced [20–22] as these converters offer high voltage stress on the DF. In addition, during the resonant time period voltage swings of high magnitude are observed.

4.2 *Interleaved Boost Converter*

The purpose of interleaving is to save energy and to achieve higher efficiency [23]. The performance of a boost DC–DC converter was compared with an interleaved boost converter. Reduction in output voltage ripples, hence reduction in switching losses were the benefits of an interleaved converter.

Two boost converters are connected in parallel combinations, facilitating the ripple cancellation. Hence, the life of the capacitor present at the load increases. One of the main issues with this type of converter is the management of heat at the input side of the diode, which can be avoided using Bridgeless Boost PFC (Fig. 7.14).

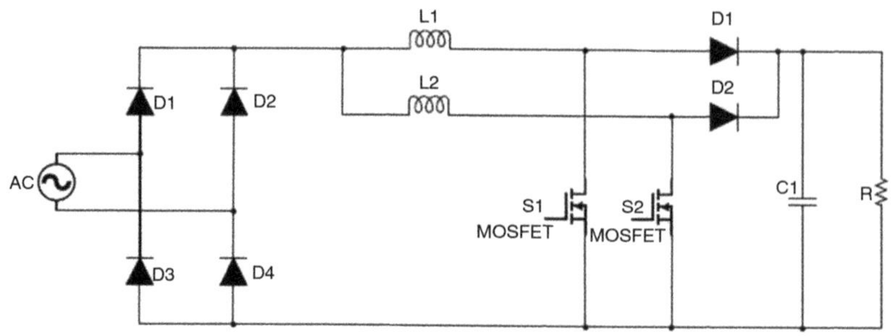

Fig. 7.14 Interleaved boost PFC [14]

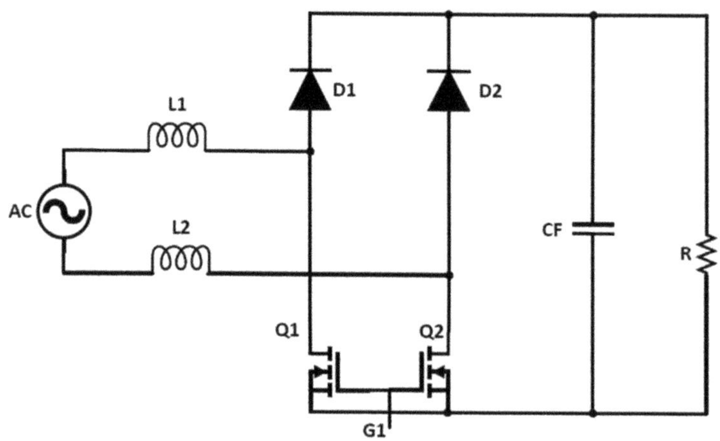

Fig. 7.15 Bridgeless boost PFC [14]

4.3 Bridgeless PFC

By eliminating the diodes, the problem of heat management is solved [14]. Different types of bridgeless boost PFCs are

4.3.1 Bridgeless Boost PFC (Fig. 7.15)

The diodes D3 and D4 are replaced by MOSFET's Q1 and Q2. Common gate signals are provided to Q1 and Q2. Few important aspects of this converter are

(a) This topology is suitable for applications higher than 1 kW
(b) In such topologies, current sensing at the input side is strenuous and hence, CT (current transformer) or a differential amplifier is used.

4.3.2 Dual Boost PFC

Figure 7.16 depicts a dual boost PFC. Separate gating signals to the MOSFETs Q1 and Q2 (Synchronous operation). Some important points about these converters are

(a) Conduction losses are reduced as MOSFET drop < Diode drop
(b) Reduction in gate switching losses
(c) The bus output voltage varies with output voltage half of its magnitude, as two inductors are present on the input side of the converter.

4.3.3 Bridgeless Interleaved Boost PFC

Figure 7.17 shows a bridgeless interleaved boost PFC. It consists of four inductors and four MOSFET.

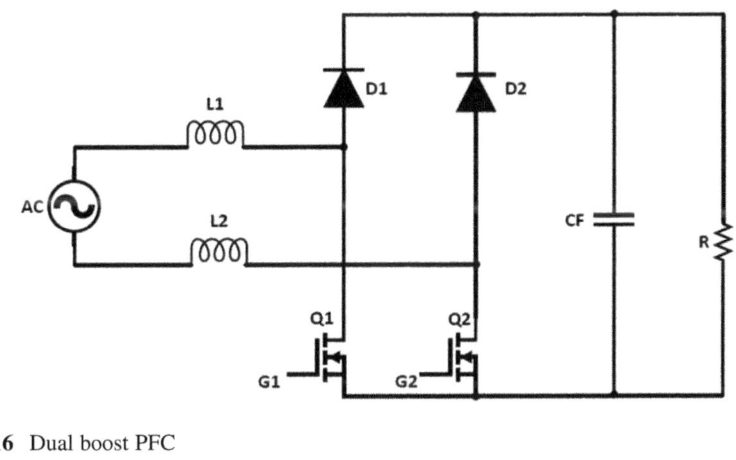

Fig. 7.16 Dual boost PFC

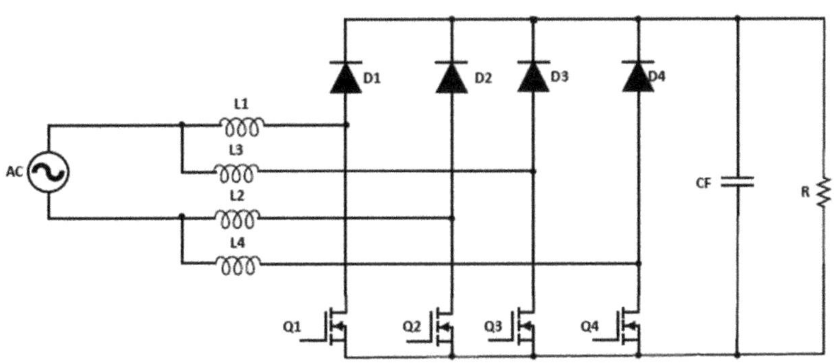

Fig. 7.17 Bridgeless interleaved boost PFC

(a) This topology is best suited for low capacitor output
(b) High-efficiency levels for power levels above 4 kW
(c) High input power factor, low input current harmonics, and suitable for Level II battery charging applications.

In [14], simulation for loss analyses in the case of PFC rectifiers for 1 kW power level was carried out using PLECS 4.1.1 software package. The Bridgeless Interleaved Boost Converter showed a maximum efficiency of 95.5% and loss of 50 W. The conventional boost and interleaved boost converters showed efficiencies of 90% and 94.5%.

5 Battery Charging Techniques

The charging of batteries is based on constant current (CC) and constant voltage (CV). In addition, few standards based on the charging include (a) Level I, (b) Level II, and (c) Level III. In the charging phase, the energy is transferred to the battery and in the final stage, the battery is conditioned and balanced. The plot of CC and CV is shown in [24].

In addition to CC and CV, CT was considered while charging a Li-ion battery [25]. A major setback using Li-ion batteries is that they fade in their capacity due to high temperatures. A closed-loop approach for charging the battery that uses cell voltage and surface temperature (T_s) of the battery is used. The charging current is continuously increased, till the battery T_s meets the threshold value. The proposed method provided 20% faster charging than the traditional CC-CV charging. In addition, a 20% lesser temperature rise than CC–CV was achieved.

A CT-CV-based charging technique was implemented on Nickel Manganese Cobalt (NMC) 18650 and Li-ion cells [26]. The proposed is compared with existing constant CC–CV under various ambient temperatures and C rates. The health degradation of a battery is shown in Fig. 7.18.

An increase in surface temperature (T_S) causes heating of the battery due to the internal resistance, R_i, and in turn the battery life decreases [27–29]. Hence, to maintain T_s under control, the battery current, I_{bat} has to be limited.

An extensive review on various types of SOH estimation techniques is shown in [30]. This plays an important factor during fast discharging. Shows the modeling of different DC – DC converters which can be used as chargers [31–34]. In order to

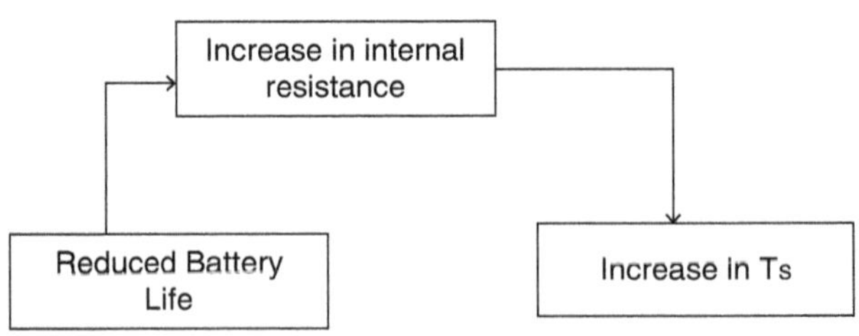

Fig. 7.18 Battery health degeration due to temperature [26]

achieve, closed loop operation of the converter has to be established. In [35], in depth analysis on designing P, PI, PD and PID controllers using MATLAB / Simulink is shown.

6 Conclusion

In this chapter, modeling of energy storage devices like batteries, ultracapacitors, and fuelcell is discussed. The best topologies in front-end AC–DC converters for EV/HEV/PHEV application and their range of power levels are presented. Bridgeless interleaved boost PFC converter was best suited for PFC for power levels >4 kW. It was shown that a battery is best suited for EV application due to its small size and low self-discharge rates. The best AC–DC converter for vehicular application was bridgeless interleaved boost PFC converter for power levels >4 kW.

References

1. IEEE spectrum, Software fix extends failing batteries in 2006-2008 Honda Civic Hybrids: Is Cost Acceptable? (2011). Retrieved August 23, 2011, from http://spectrum.ieee.org/riskfactor/green-tech/advanced-cars/ software-fix-extends-failing-batteries-in-20062008-honda-civic-hybrids-is-cost-acceptable.
2. Williard, N., et al. (2013). Lessons learned from the 787 Dreamliner issue on lithium-ion battery reliability. *Energies, 6*(9), 4682–4695.
3. Plett, G. L. (2015). *Battery management systems, Volume I: Battery modeling*. Artech House.
4. Kim, H., & Ha, H.-D. (1997). Design of interface circuits with electrical battery models. *IEEE Transactions on Industrial Electronics, 44*(1), 81–86.
5. Chan, H. L. (2001). A new battery model for use with battery energy storage systems and electric vehicles power systems. In *2000 IEEE Power Engineering Society Winter Meeting. Conference Proceedings (Cat. No. 00CH37077)* (Vol. 1). IEEE
6. Surya, S., Channegowda, J., Datar, S. D., Jha, A. S., & Victor, A.. (2020). Accurate battery modeling based on pulse charging using MATLAB/Simulink. In *2020 IEEE International Conference on Power Electronics, Drives and Energy Systems (PEDES)* (pp. 1–3). IEEE.
7. Marcos, J., et al. (2001). An approach to real behaviour modeling for traction lead-acid batteries. In *2001 IEEE 32nd annual power electronics specialists conference (IEEE Cat. No. 01CH37230)* (Vol. 2). IEEE.
8. Vural, B., et al. (2009). A dynamic ultra-Capacitor model for vehicular applications. In *2009 International conference on clean electrical power*. IEEE.
9. Zubieta, L., & Bonert, R. (2000). Characterization of double-layer capacitors for power electronics applications. *IEEE Transactions on Industry Applications, 36*(1), 199–205.
10. Miller, J. M., Nebrigic, D., & Everett, M. (2006). *Ultracapacitor distributed model equivalent circuit for power electronic circuit simulation*, Ansoft Leading Insights Workshop, Los Angeles.
11. Spyker, R. L., & Nelms, R. M. (1996). Double layer capacitor/dc-dc converter system applied to constant power loads. In *IECEC 96. Proceedings of the 31st intersociety energy conversion engineering conference* (Vol. 1). IEEE.
12. Boettner, D. D., et al. (2002). Proton exchange membrane fuel cell system model for automotive vehicle simulation and control. *J. Energy Resour. Technol, 124*(1), 20–27.

13. Boettner, D., & Moran, M. (2004). Proton exchange membrane (PEM) fuel cell-powered vehicle performance using direct-hydrogen fueling and on-board methanol reforming. *Energy, 29*(12–15), 2317–2330.
14. Praneeth, A. V. J. S., & Williamson, S. S. (2018). A review of front end ac-dc topologies in universal battery charger for electric transportation. In *2018 IEEE transportation electrification conference and expo (ITEC)*. IEEE.
15. Petersen, L., & Andersen, M. (2002). Two-stage power factor corrected power supplies: the low component-stress approach, In *APEC. seventeenth annual IEEE applied power electronics conference and exposition (Cat. No.02CH37335)* (Vol. 2, pp. 1195–1201).
16. Lu, B., Brown, R., & Soldano, M. (2005). Bridgeless pfc implementation using one cycle control technique. In *Twentieth annual IEEE applied power electronics conference and exposition*. APEC 2005 (Vol. 2, pp. 812–817).
17. Petrea, C., & Lucanu, M. (July 2007). Bridgeless power factor correction converter working at high load variations, In *2007 international symposium on signals, circuits and systems* (Vol. 2, pp. 1–4).
18. Erickson, R. W., & Maksimovic, D. (2007). *Fundamentals of power electronics*. Springer Science & Business Media.
19. Kim, T.-H., et al. (2012). Analytical study on low-frequency ripple effect of battery charging. In *2012 IEEE vehicle power and propulsion conference*. IEEE.
20. Wang, C.-M. (Feb 2005). Novel zero-voltage-transition pwm dc-dc converters. *IEEE Transactions on Industrial Electronics, 53*(1), 254–262.
21. Jain, N., Jain, P. K., & Joos, G. (Jan 2004). A zero voltage transition boost converter employing a soft switching auxiliary circuit with reduced conduction losses. *IEEE Transactions on Power Electronics, 19*(1), 130–139.
22. Aksoy, H. B., & Bakan, A. F. (Aug 2010). A new zvt-zct-pwm dc x2013;dc converter. *IEEE Transactions on Power Electronics, 25*(8), 2093–2105.
23. Thiyagarajan, A., Praveen Kumar, S. G., & Nandini, A. (2014). Analysis and comparison of conventional and interleaved DC/DC boost converter. In *Second international conference on current trends in engineering and technology-ICCTET 2014*. IEEE.
24. Morrow, K., Karner, D., & Francfort, J. (2008). Plug-in hybrid electric vehicle charging infrastructure review. *US Department of Energy-Vehicle Technologies Program* 34.
25. Patnaik, L., Praneeth, A. V. J. S., & Williamson, S. S. (2018). A closed-loop constant-temperature constant-voltage charging technique to reduce charge time of lithium-ion batteries. *IEEE Transactions on Industrial Electronics, 66*(2), 1059–1067.
26. Marcis, V. A., et al. (2020). Analysis of CT-CV charging technique for lithium-ion and NCM 18650 cells over temperature range. In *2020 IEEE international conference on industrial technology (ICIT)*. IEEE.
27. Surya, S., Marcis, V., & Williamson, S. (2020). Core temperature estimation for a lithium ion 18650 cell. *Energies, 14*(1), 1–1.
28. Surya, S., Bhesaniya, A., Gogate, A., Ankur, R., & Patil, V. (2020). Development of thermal model for estimation of core temperature of batteries. *International Journal of Emerging Electric Power Systems, 1*, ahead-of-print.
29. Surya, S., & Arjun, M. N. (2020). Effect of fast discharge of a battery on its core temperature. In *2020 international conference on futuristic technologies in control systems & renewable energy (ICFCR)* (pp. 1–6). IEEE (10).
30. Surya, S., Vidya, R., & Sheldon S. W. (2021). Comprehensive Review on Smart Techniques for Estimation of State of Health for Battery Management System Application. *Energies* 14.15 (2021): 4617.
31. Surya, S., and Arjun, M. N. (2021). Mathematical Modeling of Power Electronic Converters. *SN Computer Science, 2*(4), 1–9.

32. Surya, S., & Srividya, R. (2021). Isolated converters as LED drivers. *Cognitive Informatics and Soft Computing*. Springer, Singapore, 167–179.
33. Surya, S. (2021). Mathematical Modeling of DC-DC Converters and Li Ion Battery Using MATLAB/Simulink. *Electric Vehicles and the Future of Energy Efficient Transportation*. IGI Global, 104–143.
34. Surya, S., & Sheldon W. (2021). Modeling of Average Current in Ideal and Non-Ideal Boost and Synchronous Boost Converters. *Energies* 14.16: 5158.
35. Surya, S, D. B. Singh. "Comparative study of P, PI, PD and PID controllers for operation of a pressure regulating valve in a blow-down wind tunnel." *2019 IEEE International Conference on Distributed Computing, VLSI, Electrical Circuits and Robotics (DISCOVER)*. IEEE, 2019.

Chapter 8
Overview of Battery Management Systems in Electric Vehicles

K. Vishnu Murthy, K. Sabareeshwaran, S. Abirami, and T. Bharani Prakash

1 Introduction

Nowadays, the usage of batteries has been in a wide range. All the consumer electronics devices, such as laptops and mobile, are powered by the use of batteries. Even batteries are incorporated in the electric vehicle. But the hybrid vehicle consists of both the electric power and IC engine, leading to the state of polluting devices.

To overcome this issue, the globe has been looking for a completely electric vehicle, which doesn't employ any fossil fuels. The full-scale electric vehicle runs on an onboard battery. So, the management and monitoring of batteries have been an essential factor in running the system effectively and in a reliable manner. However, due to the less lifetime and high cost, the growth in the electric vehicle has been dropped. To make the system effective with high life, the battery management part needs to be improved significantly. The battery management system is a system, which has software and hardware part such as various electronic components to perform multiple objective functions. Since the old decade, the batter is a component that has to be monitored in-depth to avoid damage to the human personnel.

The battery management system (BMS) growth is expected to increase by 15% during 2020–2025. Through the Increased sales of electric vehicles, there will be a growth in the automobile sector, which supplies a significant share to the battery management system market globally. Especially for lithium-ion batteries, a BMS system needs to be incorporated that too in electric vehicles. Owing to the safety concerns when using lithium-ion batteries, BMS has a wide application in electric

K. Vishnu Murthy (✉)
PSG College of Technology, Coimbatore, Tamil Nadu, India

K. Sabareeshwaran
Karpagam Institute of Technology, Coimbatore, Tamilnadu, India

S. Abirami · T. Bharani Prakash
Sri Krishna College of Technology, Coimbatore, Tamil Nadu, India

M. Kathiresh et al. (eds.), *E-Mobility*, EAI/Springer Innovations in Communication and Computing, https://doi.org/10.1007/978-3-030-85424-9_8

vehicles. Increased adaptation of electric vehicles and focusing on increased efficient usage of batteries are expected to drive the demand for battery management systems. However, the initial cost of electric vehicles has also restricted the growth of BMS. In China and Japan, the growth of electric vehicles and BMS has exponentially increased.

This increased growth has been due to the continuous efforts taken by the government to mitigate the emission of greenhouse-impacting gases. Furthermore, further discussions about the BMS communication platform are carried out, which plays a vital role in optimizing and controlling battery performance. Finally, conclusions and specific limitations on the battery management system are mentioned.

2 Battery Maintenance Parameters

The type of battery employed for certain applications has a significant impact on battery maintenance as in Fig. 8.1. Since each type of battery has its own operating characteristics, some have a particular reaction behavior evolving with gases concerning their volumetric component. In cases where the battery volume changes, it leads to complexity in sealing the battery; in those cases, various chemicals or regular water are employed to reduce the evolution of gasses. On the other hand, an air-tight sealed battery does not have an option to incur any environmental changes. As a result, these batteries are user-friendly and have less maintenance when compared to the batteries that react with surrounding chemicals. That's why most of the batteries in use are air-tight, as they don't need additional maintenance. Whereas other batteries like lead-acid require significant maintenance at a particular schedule.

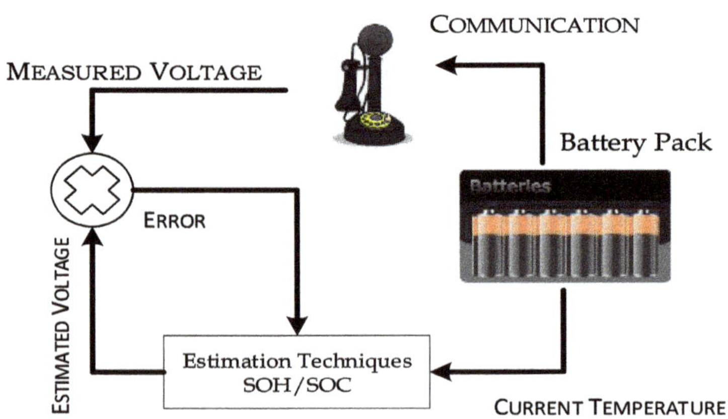

Fig. 8.1 Block diagram for monitoring battery parameters

The battery management system (BMS) is an enlightened package comprising hardware and software systems, which is almost needed for high-module battery packs [1].

The general utility of the BMS consists of:

- Measuring the voltage in cell and control when needed
- Controlling the contactors
- Thermal Monitoring and control actions
- Calculation of State of health
- Monitoring Isolation
- Communicating the operating state or actions needed.

The BMS balances the battery charging accordingly based on the loading and continuous monitoring of the operating state of the battery.

By monitoring the voltage fluctuations, charging capability, and communication with the BMS interfaces, a real-time battery health monitoring technique, the BMS extends the battery life and increases the battery's efficiency.

BMS undergoes different methodologies such as distributed, centralized, and modular.

Distributed BMS

It has a cell board in each cell and has a single point of communication between the controller and battery through a communication cable.

Centralized BMS

It has only one controller, and with the aid of communication wires, it is connected to the battery.

Modular BMS

It has several controllers through which the cells communicate with each other, and it can handle a certain number of cells in a battery pack.

This communication topology is a must needed for thermal management to operate the vehicle safely and adjust the battery life with the vehicle performance. These functions are discussed in the following section from the perspective of lithium-ion battery packs.

2.1 Measurement of Cell Voltage and Controlling

Monitoring voltage is another essential function of BM. The battery management system is designed in a way to monitor the voltage across the group of cells. The increased voltage expedition occurring due to the increased voltage and over-discharging ultimately impacts battery life and safety concerns. The better state of charge (SoC) and state of health needs an effective battery cell balancing algorithm.

For neglecting the impact of overvoltage, the cell voltage should be monitored at most care. Compared to these batteries, lithium-ion type of batteries must not be overcharged. Overcharging of cells will lead to a rise in temperature, and specific chemical reactions will occur. This chemical reaction ruins the battery life.

In accumulation, overcharging increases the cell heat sufficiently, which leads to fire hazards. Overcharges can be reduced by adequately monitoring the voltage and managing the charging state by incorporating the BMS system. The battery pack design must be robust to monitor cell voltage to make certain overcharge and over-discharge conditions in certain rare cases.

2.2 Contactor Control

This is cultivated by utilizing switches at access boards and extra electrical connector sticks that are acclimated electrically to work out if a connector has been eliminated. In direct execution, the coil power shuts the high-voltage contactors travels along a meandering way through every client passageway inside the high-voltage framework. Suppose any access point has been breached or opened. In that case, the conductor is broken, and therefore, the contactor coil power is removed, leading to the opening of the same thereby high-voltage module gets de-energized. This direct execution isn't conventional in current-age vehicles because it generates undesirable failure modes. If such a faulty establishment or switch will de-energize the high voltage module unexpectedly, and this may put occupants in danger while on transit if the vehicle system stops unexpectedly.

2.3 Monitoring the Isolation Parameter

Only when current is given to the human body from an electrical medium, the process works in terms of reflexes, and there's a return path as well to reduce the magnitude of voltage range from the medium. Because of these, the circuit requires two faulty sections even before it reached the human body—one permits electrons to leave away the circuit and flow into the body. A second must permit the electrons to leave the body and rush back to the circuit [2].

These parametric circuits regularly counter files the resistance between the high-voltage system and, therefore, the chassis. Typical guidance denotes the value must be greater than 100 Ω/volt for electricity (DC) systems (ISO 2011) [2]. Thus, system should maintain a resistance above 35 kΩ between the high voltage system and the chassis. This ensures that the current will be as per the IEC standards. There are a variety of methods for measuring isolation on the vehicle. The circuit accustomed to measure isolation shouldn't have any failure modes that allow the isolation monitoring safety circuit to become the source of isolation failure.

2.4 Temperature Control

The BMS is typically accountable for both assembly and thermal behavior of the same, also the temperature is one important consideration in design aspects with regard to safety and life implications [3]. The monitoring strategy for controlling thermal stress on cells is exclusive to every application. But, the BMS should have necessary data about thermal variation across the assemble to permit the control process to regulate heating, cooling, or pack power levels as needed.

2.5 Communication

One important BMS functionality is to have communication with additional controllers in the vehicle system. For data communication BMS often use CAN bus and RS485 because of the interfacing units [4]. The BMS frequently request changes in system operation to bring under the surveillance in battery assembly conditions and might forecast its capability in near future. Transfer of information is also necessary to produce the same to the driving force, like vehicle range, operating mode, and any faults. Finally, with regard to contact establishment with respect to off-vehicle, Off-board DC chargers will be utilized to have an improved charging rate without putting the load on to the vehicle in every possible aspect like weighing capacity and difficulty of inbuilt charges inside the system.

2.6 Battery SoC Measurement Principle

The determination of battery SoC becomes a posh task that depends on the kind of battery and its application. To enhance the reliability, lifetime, and performance of electric batteries, accurate SoC estimation is an important task. SoC is defined because the available capacity is expressed as a percentage of rated capacity or percentage of current capacity (i.e., latest charge-discharge cycle). Because the cell ages, its actual power will start decreasing, and towards the tip of the cell's life, it reaches nearly 80%. Hence for accurate measurement of remaining charge in every battery, the aging and environmental factors should be considered as in Fig. 8.2 [5].

State of Charge Estimation Techniques:

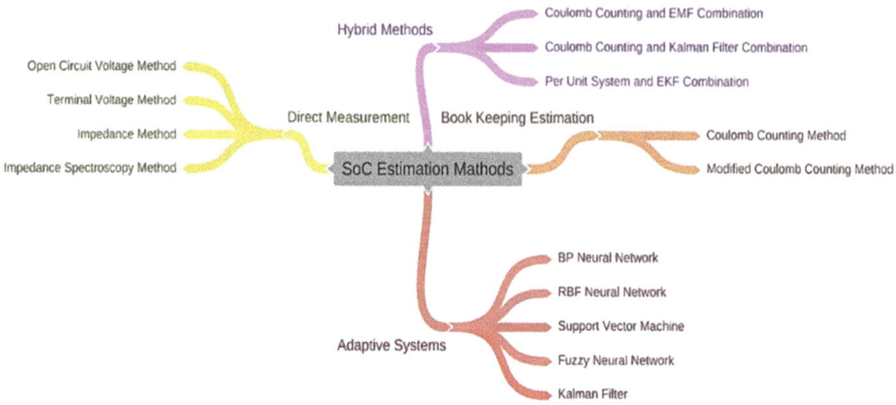

Fig. 8.2 Estimation of SoC methods

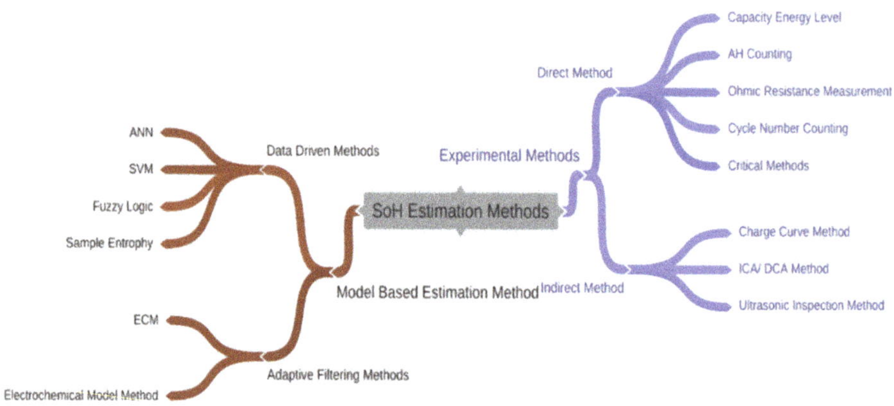

Fig. 8.3 Classification chart on different SoH estimation methods

3 State of Health Estimation Techniques

SOH refers to the performance index calculator of the electric battery, which conveys this state of battery compared with the initial stage of the battery. It notifies the battery aging and ruin state and alerts when the battery needs to get replaced. Different SOH estimation methods are available based upon various features. During this chapter, the SOH estimation methods are classified as experimental methods and model-based estimation methods as in Fig. 8.3 [6].

3.1 Experimental Method

In this method, the battery behavior is estimated within the laboratory founded with a more significant number of continuous experiments to evaluate the SoH, which is also a drawback of this method because it differs from the driven data. But this could be used as a study component to ascertain the model-based estimation methods. Experimental methods are classified as the direct measurement method and indirect measurement method—battery health state pointers like impedance, energy, and ampere counts. Cycle count is the parameter that is held as a base for the direct measurement method. Collection of information, analysis and processing of knowledge are required during this indirect measurement method.

1. Direct Measurement Method:
   (a) Capacity or energy level: The battery capacity data is required to estimate the achievable mileage of an electric vehicle. The energy that may be stored inside A battery is that the critical information of this measurement. So, this measuring technique may be used only in laboratories. But the capacity of the running vehicle can't be determined by this method. If the battery capacity is understood in real-time, then that becomes the particular SOH estimator. Therefore, online prediction methods are available to comprise the battery capacity which can be further discussed.
   (b) AH counting: Using the appropriate technique, the charge transferred during charging and discharging of battery approximation is required for SOH estimation. The computation formula for SOH is SOH $=$ Q_MAX/Q_N $\times$ 100%, where QMAX refers to the maximum available capacity of the current condition. QN refers to the nominal capacity of the battery. Long-run monitoring and continuous updating of battery current are required to obtain accurate results, but it costs a lot of your time and energy. Also, high-care sensor is mandatory to get the precise remaining capacity. Though this method isn't applicable to be utilized in real-time, it may be the ECM method for SOH estimation.
   (c) Resistance/impedance measurement: The internal resistance is a significant parameter that defines the ability of the battery to test and operates within the safety voltage range; the present stage value is applied to its rated voltage thereby, measurement can be done [6]. Different techniques are available to live the resistance. The resistance is additionally influenced by some external factors like temperature and aging. This pulse methodology [7] is employed, which relates the effect of temperature on the electrical phenomenon. Electrochemical impedance Spectroscopy is the other method available to live the electrical phenomenon. This method is employed within the laboratory to test the electrochemical process inside the battery by using the characteristic response obtained when a sinusoidal signal is applied to the cell, which measures the impedance. EIS generally measures the battery impedance in

a vast frequency range, which may measure other parameters like double-layer capacitance, charge transfer resistance, etc.

(d) Cycle number counting: The life model of the battery measure of cycles gone through undercharging and releasing interaction will be estimated by utilizing the counter. Electronic items like telephones, workstations are being used for this SOH sign-in battery. The balance cycle number provides the SOH measurement directly from the compared data of the current cycle number and manufacturer-provided cycle number. Complete depth of discharge is utilized as an essential estimating boundary where conversion co-efficient are wont to convert the various depth of discharge under charging and discharging to the unique, complete depth of charge/ discharge.

(e) Critical methods: The miniature structural changes that occur during the maturing cycle might be controlled by specific techniques like Raman Spectroscopy, X-ray Diffraction, scanning microscope, etc. The structural change happens to the changes within the active material and stress. Dismantling of batteries is needed to breakdown the maturing instrument, which can forever harm the batteries. So these techniques are utilized uniquely inside the research center testing.

2. Indirect Measurement method:

This method involves multistep analysis of parameters related to the battery degradation parameters like capacitance and resistance. From the connection between the health parameters and degradation parameters, SOH estimation will be done.

3.2 *Model-Based Estimation*

Although the scientific method finally at last winds up inexact information, real-time data processing is required, which may be achieved by utilizing this model-based assessment strategy. The model-based estimation method is surveyed upheld the computation technique as adaptive filtering method and information-driven technique. The model-based estimation method is an expansion strategy where continuous preparing happens and assesses different trademark boundaries of the battery using different approaches like filtering and intelligent algorithms.

3.2.1 Adaptive Filtering Methods

The adaptive filtering method utilizes the ensuing models like equivalent circuit model and electrochemical process, which is that the most reformist technique in estimating the boundaries.

Equivalent Circuit Model Method: This technique relies upon the electrical attributes of the battery. The most commonly utilized models are the Rint model, Thevenin model, PNGV model, RC model, and composite model [8].

The ECM-based estimation method involves the subsequent stepwise procedures

- Choosing out the suitable ECM
- ECM parameter identification
- Solving out SOH using some filters.

4 Thermal Management

The occurrence of heat during operation is the most critical parameter to be monitored in all industrial operating devices. The excess production of heat is normal, but it has to be within a specific limit. For example, suppose the rate of rising heat exceeds the nominal operating parameter. In that case, the thermal management system needs to be incorporated to effectively monitor the thermal production and control with the assistance of thermodynamics. The flow of electrons is termed as the flow of current. Each conductor has a resistance in it, which limits the flow of current. If excess current flows, the passage of the conductor has a temperature rise, which in turn develops variation in thermal values and is also not limited to fluctuations. As per the Arrhenius law, the variation in chemical reaction rate leads to an exponential increase in temperature [5, 9].

According to the above-said law, there will be increased power from the battery when the temperature rises. When this criterion is noted from one point, it seems to be an advantage, but indirectly, this scenario causes various chemical reaction changes, which can't be reverted to normal operating conditions, leading to permanent damage to the battery. The perennial destruction occurs to the battery when its operating temperature exceeds the upper limit, and finally, the battery becomes dead. Every battery has its own boiling and freezing point; if these limits are crossed, it leads to the breakdown of insulation and even explosion.

The criterion where the battery yields good reliable performance is when operating above the electrolyte freezing point. Working the battery at freezing point regardless of the mentioned battery specs affects the battery life, which ultimately reduces the battery cycle. These battery characteristics show that it needs to be operated within specific limits for superior battery life. As in Fig. 8.4. for designing a proper thermal management system, the heating and cooling constraints need to be considered.

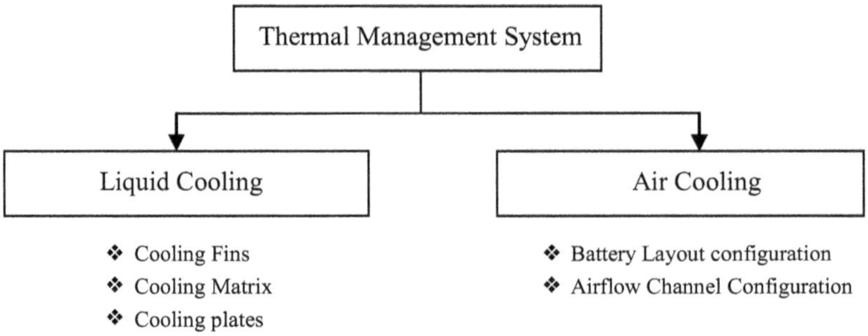

Fig. 8.4 Different thermal management system classification

4.1 Liquid Cooling

The liquid cooling is categorized into cooling fins, cooling matrix, and cooling plates. For pouch and prismatic rectangular type cells, cooling fins are the best thermal management device. For cylindrical cell-type batteries, the cooling matrix can be adopted. Cooling plates are placed below the cells.

4.2 Air Cooling

Air cooling technique has a better safety side with reliable operation. Even though it is not fir to be used for high duty performing batteries, since air cooling provides uneven temperature distribution in the battery pack, batteries can only be adopted for batteries that perform the light-duty operation. The high dense battery modules handle excess power, which requires high specific heat capacity to maintain the thermal limits. As the air has low specific heat capacity, it must precondition the air meant for cooling purposes and other methods of increasing the cooling air velocity for achieving the optimal cooling mechanism [9]. But the air-cooling mechanism can be used for all battery types wisely by undergoing modifications such as altering the airflow valve and battery configuration.

5 Summary

The battery management system (BMS) incorporated for a lithium-ion battery is an intricate system, even though it provides a meaningful contribution to safety and reliable performance. The software and hardware design plays a significant role in overcoming this constraint, while the cost incurred for development is often underrated. The plug-in hybrid vehicles in the present scenario have been designed for recycling energy to some extent. Through a suitable battery management system, the energies from the battery can be fed back to the grid. For battery parameter estimation, various methodologies had been deliberated in this chapter. But based on connected load and accuracy level, adopting the correct technique is an imperative chore. The machine learning techniques can be used for predicting the proper methodology. For the IoT-based devices for processing the big data analysis, BMS is a crucial technique for all-electric vehicles.

References

1. Sun, F., Hu, X., Zou, Y., & Li, S. (2011). Adaptive unscented Kalman filtering for state of charge estimation of a lithium-ion battery for electric vehicles. *Energy, 36*(5), 3531–3540.
2. Hauser, A., & Kuhn, R. (2015). Chapter 12—Cell balancing, battery state estimation, and safety aspects of battery management systems for electric vehicles. In B. Scrosati, J. Garche, & W. Tillmetz (Eds.), *Woodhead publishing series in energy. Advances in battery technologies for electric vehicles* (pp. 283–326). Woodhead Publishing.
3. Berecibar, I., Gandiaga, I., Villarreal, N., & Omar, J. Van Mierlo, P. (2016). Van den Bossche, critical review of state of health estimation methods of Li-ion batteries for real applications. *Renewable and Sustainable Energy Reviews, 56*, 572–587.
4. Wu, F.-B., Yang, B., & Ye, J.-L. (2019). Chapter 3—Technologies for energy storage battery management. In F.-B. Wu, B. Yang, & J.-L. Ye (Eds.), *Grid-scale energy storage systems and applications* (pp. 57–82). Academic Press.
5. Hakeem Akinlabi, A., & Solyali, D. (2020). Configuration, design, and optimization of air-cooled battery thermal management system for electric vehicles: A review. *Renewable and Sustainable Energy Reviews, 125*, 109815.
6. Xiong, R., Li, L., & Tian, J. (2018). Towards a smarter battery management system: A critical review on battery state of health monitoring methods. *Journal of Power Sources, 405*, 18–29.
7. Waag, W., Käbitz, S., & Sauer, D. U. (2013). Experimental investigation of the lithium-ion battery impedance characteristic at various conditions and aging states and its influence on the application. *Applied Energy, 102*, 885–897.
8. Tian, H., Qin, P., Li, K., & Zhao, Z. (2020). A review of the state of health for lithium-ion batteries: Research status and suggestions. *Journal of Cleaner Production, 261*, 120813.
9. Wang, T., Tseng, K. J., & Zhao, J. (2015). Development of efficient air-cooling strategies for lithium-ion battery module based on empirical heat source model. *Applied Thermal Engineering, 90*, 521–529.

Chapter 9
Review on Regenerative Braking System

A. Sathishkumar, R. Soundararajan, T. J. Muthu Vel, M. B. S. Arjith, and G. Sakthivel

1 Introduction

Deceleration is the most important subsystem of an automobile in the safety aspect. The application of the brakes in the moving vehicle is to diminish the speed or to stop the vehicle's movement. This can be achieved by applying the force on the brake pedal. While applying the force on the brake pedals, the vehicle comes to rest at a particular distance which is defined as the braking distance. When the brake pedal is applied, dynamic energy is changed into heat energy in the commercially used braking system. This heat is formed due to the generation of friction between the brake drum and friction pad. There is a wastage of energy because the heat formed is moved away in the air stream. The loss of total energy through this process is depending upon the following factors, (a) how often the brakes are applied, (b) how hard the brakes are applied and (c) how long the brakes are applied [1]. The development of electric vehicles in the automobile sector has gotten consideration from the legislature and the industries. The government and the industries are focused on the development of electric vehicles due to global warming and the fossil fuel shortage. The performance of electric vehicle in various aspects such as driving range, powertrain efficiency, safety and energy-saving is low when compared with conventional vehicles [2]. Regenerative Braking System (RBS) is the developing braking system in hybrid electric vehicle (HEV) as well as in ordinary electric vehicle (EV), where it reduces the heat loss formed while applying the conventional braking system. In a traditional braking system, when the brake pedals are pressed the brake drum gets attached to the friction pads. The dynamic energy is lost as heat to the surroundings because of the friction between the brake drum and friction pad and the vehicle stops. The heat loss due to this system will be minimised by using the regenerative braking system (RBS). In RBS, the dynamic energy is

A. Sathishkumar · R. Soundararajan (✉) · T. J. M. Vel · M. B. S. Arjith · G. Sakthivel
Sri Krishna College of Engineering and Technology, Coimbatore, Tamil Nadu, India

M. Kathiresh et al. (eds.), *E-Mobility*, EAI/Springer Innovations in Communication and Computing, https://doi.org/10.1007/978-3-030-85424-9_9

converted into electrical energy and is stored in the electric vehicle's battery. The driving range and the fuel economy are improved in the electric vehicle which is incorporated with regenerative braking. A mechanical braking system (MBS) is still needed in order to stop automobiles during an emergency. In case of emergency, RBS didn't stop the automobile quickly as done by the commercially used braking mechanism. The MBS and the RBS can function in a same solitary foot pedal. The initial segment controls the RBS to diminish the vehicle's speed and the MBS is controlled by the final segment of the foot pedal to stop the vehicle immediately [3].

The minimal energy loss in the RBS describes the necessity of this braking mechanism in the electric vehicle domain. In addition to that, it also converts the kinetic vitality into electrical vitality and deposited it in the power saving unit. Thus, the RBS increases the functioning characteristics of the EVs. There are various parameters that can be changed to use this system effectively. Various researches have been done in this particular domain for using this braking system in a better way. The electric motor, control system and batteries used are the few parameters in the electric vehicles which influence the performance of this braking system.

2 Principle of Regenerative Braking

The regenerative braking mechanism operates based upon the electric motor's working principle. An electric motor is one of the major components of the electric vehicle. An electric motor rotates when the passage of moving charge is subjected to the coil. Similarly, when the rotor is actuated with the negative torque while applying the brake, the electric motor behaves as a power-producing device and generates electrical vitality. Thus, when the motor rotates in a particular direction it transforms the electrical vitality into mechanical vitality and maximises the acceleration of the automobile. When the electric motor rotates in the converse way, it acts like a power generating machine and generates electricity, which is stored in the batteries. Figure 9.1 represents the functioning arrangement of the regenerative braking mechanism [1].

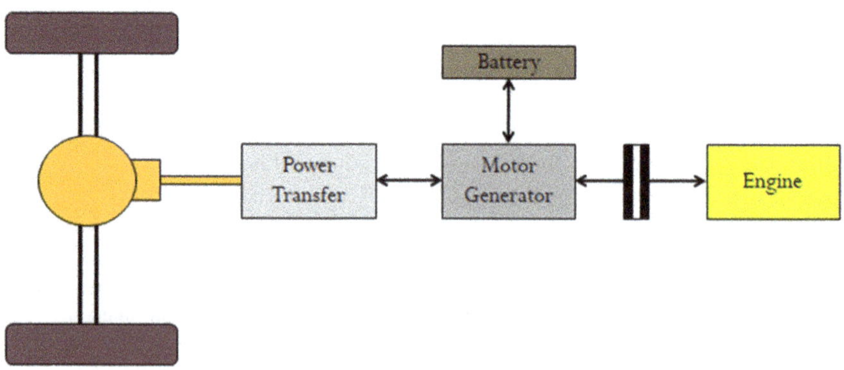

Fig. 9.1 Simple representation of regenerative braking

3 Research Findings on the Regenerative Braking System

A novel approach of a combined braking control strategy (CBCS) was proposed in this research work. This strategy is depending upon another technique of HEV braking torque conveyance, which operates the hydraulic braking mechanism simultaneously with the RBS. In the current study, a non-linear model of a vehicle which has a non-consistent and partly observational prototype of a tire with 5 degrees of freedom is made for the simulation. The simulation is done using MATLAB/ Simulink. Functions of the various systems such as (a) anti-lock braking system, (b) controlling system, (c) system that controls the motor and (d) the units that manage the battery is examined in the simulation system to ensure the long-term performance of braking and regeneration vitality. The CBCS is the combination of two major control strategies such as (a) logic threshold control and (b) fuzzy logic control to secure the long-term braking characteristics of the vehicle. This strategy shields the wheel from being fastened and it regenerates maximum energy adequately. The fuzzy logic computations and the de-fuzzy recognition can be executed by the Mamdani technique and gravity centre technique in this current work by the researcher.

There are two classified parts of braking control design are proposed by the researcher in this present work. The first part is the logic threshold control strategy. This control technique is used for controlling the braking torque of the commercially used system of braking. The hydraulic braking torque of the conventional braking mechanism is adjusted in order to control the braking pressure as increasing, reserving and decreasing. The next technique is fuzzy logic control strategy and the block diagram of the fuzzy logic controller is represented in Fig. 9.2. By utilising the fuzzy logic control technique, the torque generated by the RBS is adjusted corresponding to the target slip ratio. The regenerative braking efficiency is simulated under new European driving cycle. While applying the conventional braking at the time of emergency on the roadways with a low sticking coefficient, its safety is also simulated. The hydraulic system of braking and the regenerative system of braking are worked together with the help of the proposed control strategies. They are worked together to ensure better braking performance and maximum regenerative efficiency, even on roadways having low-bond coefficient when emergency braking is needed [3].

By using the electro-mechanical brake, the researcher investigated the RBS of an electric vehicle in this research. The current innovations in braking technologies are progressed towards the brake-by-wire systems. In order to increase the consumption of fuel by EVs and their safety, the evolution of the electro-mechanical braking (EMB) system will also be a significant one. Figure 9.3 provides the systematic representation of various components and working of the regenerative braking systems in hybrid electric vehicles. The proposed system has an internal combustion engine with 1.4 L of capacity and an electric motor of 24-kW power. These are fastened to one of the axes of the HEV. The torque of the RBS and the EMB system is provided by the vehicle controller based on the following factors: (a) vehicle

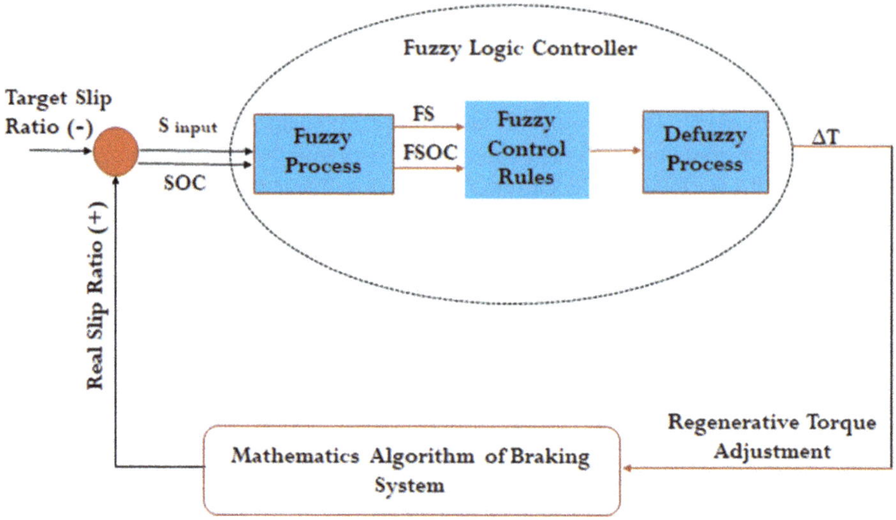

Fig. 9.2 Representation of the fuzzy logic controller using block diagram

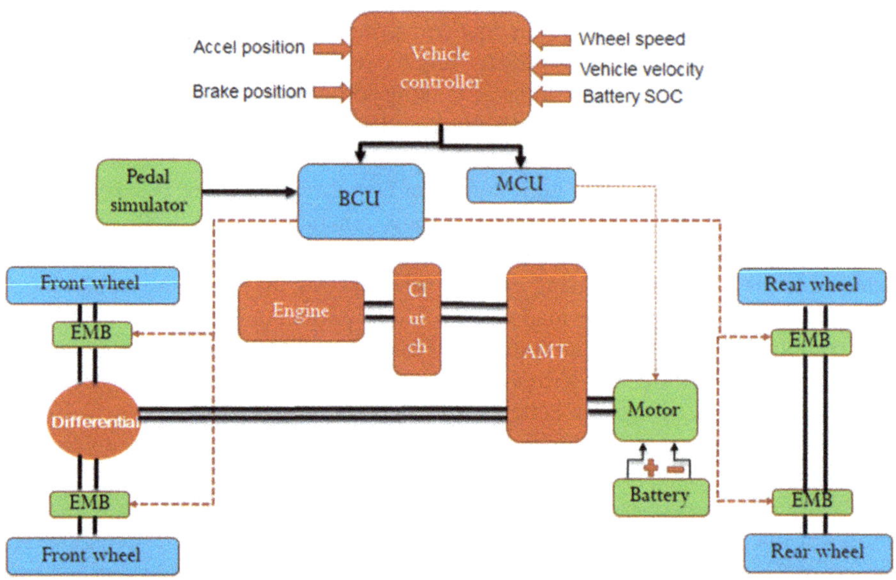

Fig. 9.3 Representation of braking control system in HEV

velocity, (b) battery's state of charge (SOC), (c) driver input and (d) motor characteristics. Based on the demand of the driver, the regenerative braking is working in accordance with the algorithm proposed. The regenerative braking's working is corresponding to the SOC of the battery. This phenomenon helps to increase the life of the battery. When the SOC value is more than 80%, the proposed algorithm does not allow the regenerative braking to work. The performance simulation of RBS in the HEV can be analysed using MATLAB/Simulink in the dynamic model. During different driving conditions, the control performance of the EMB system can be assessed using the simulation of the RBS incorporated in HEV. The outcomes from the simulation show that the proposed HEV which was furnished with an EMB system can regenerate more braking vitality. This can be achieved by utilising the control algorithm which was proposed [4].

A new design of the regenerative braking model is proposed mainly for a parallel HEV in this research work. This new model figures the pressure in the line and pad for the fore and back brakes. It also determines the usage of the generator based on the state of deceleration. This proposed model also has an algorithm, which avoided the wheels from lock-up. The development of a regenerative braking model involves the combination of different types of engines, batteries and electric motors. The product thus formed by the combination of three parameters is then simulated by the Advanced Vehicle Simulator (ADVISOR). ADVISOR is a type of simulator which is based on MATLAB/SIMULINK. This is also a feedback ward simulation especially for the simulations in HEV powertrains. The various parameters such as (a) performance, (b) emissions and (c) the fuel economy can be analysed quickly by using this simulator. The advanced driving schedule of the Federal Urban Driving Schedule (FUDS) is used in this research. The researcher took three types of engines, motors and batteries with various specifications. The displacement, power, maximum torque, mass and efficiency of the three engines differ from each other. Similarly, the power, maximum torque, maximum current, minimum voltage and efficiency of the three electric motors differ from each other. Three different batteries with different capacities, voltages and mass were used in this paper. The various behaviours such as (a) SOC of battery, (b) fuel economy along with (c) emission characteristics are compared with the baseline values and the proposed strategy of RBS [5].

The methodology to find the contribution made by RBS in the improvement of energy efficiency of the electric vehicle was proposed in this work. A pure electric passenger car is used for this research. A synchronous motor with a permanent magnet is acted as a motor and generator. By considering the regenerated braking energy, the energy stream of the electric vehicle was analysed. At the time of propulsion, the battery provides the required energy to all the electronic accessories. The regenerative energy is supplied to the electronic accessories when the regenerative braking is working.

The remaining regenerative energy is stored in the power-saving unit of the electric vehicle. The flow of energy in the electric vehicle in the presence and absence of regenerative braking is schematically represented in Fig. 9.4. The improvement in the efficiency of the electric vehicle by the regenerative braking system can be analysed by conducting three types of experiments with various control strategies. The non-regenerative braking test is the first experiment. The second experiment is carried with parallel regenerative braking strategies and the third experiment is done with serial regenerative braking strategies. Once the experiments are done, the methodologies for analysing the contribution made by

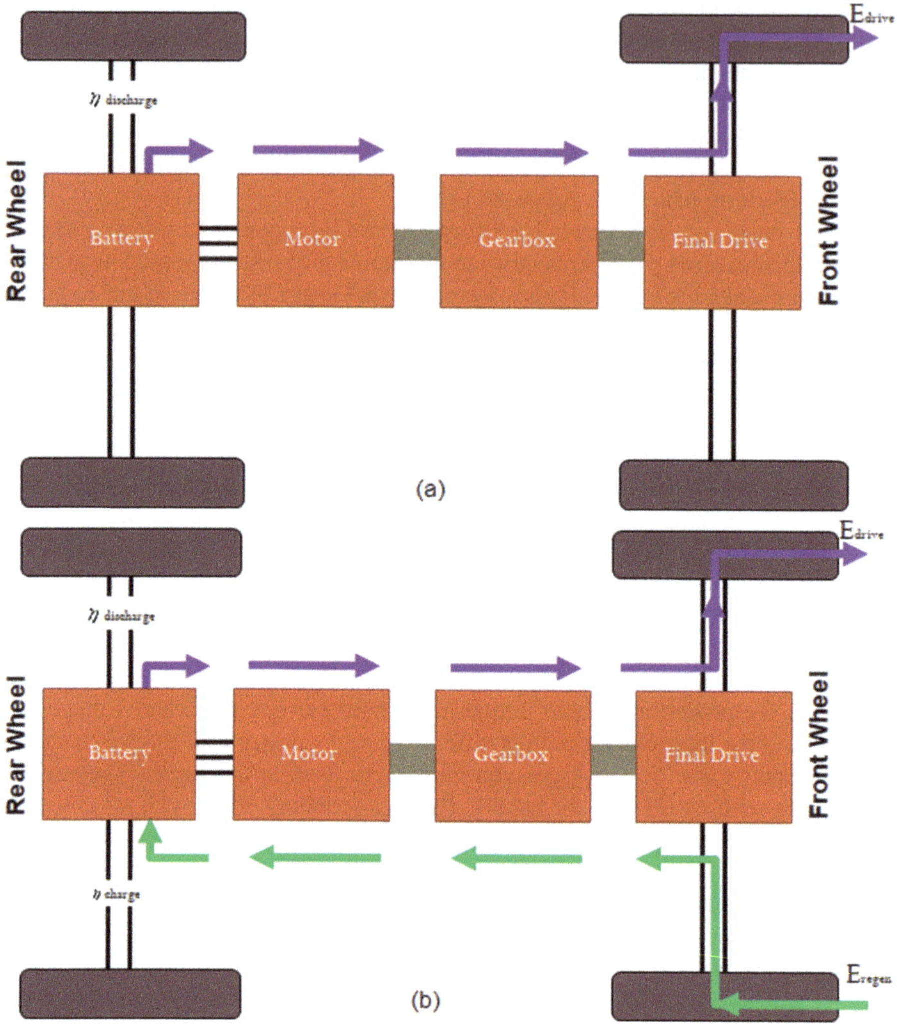

Fig. 9.4 Simple representation of energy flow in the vehicle with the presence and absence of regenerative braking

regenerative braking in the energy proficiency improvement in the vehicle were established. By using three diverse control strategies and standard driving cycles, various tests are done on the proposed vehicle on the chassis dynamometer. The results from the experiments showed that there is an increase of 11.18% and 12.58% of contribution ratios between energy efficiency improvement and extension in driving range made by the regenerative braking. These experimental results were obtained under New European driving cycles [6].

By using the vitality flow, the improvement in the energy efficiency by regenerative braking in the electric vehicle was analysed with the new mechanism in this research work. Novel methodologies are suggested for analysing the vitality contribution made by the RBS. An ordinary passenger electric vehicle is taken for the experimental investigations. In the practical implementation, the input and output power in the wheels are measured using the sensors. The researcher proposed two parameters for the vitality contribution analysis. The first parameter is to analyse the ratio between the vitality contribution and the driving range of the RBS. The next parameter is to analyse the ratio between the vitality contribution and transfer efficiency of the energy produced by RBS while braking is applied. To study the further efficiency enhancement, the author carried out a vehicle test on-road according to China Typical City Regenerative Driving Cycle. The new European drive cycle is adopted for analysing the improvement in the consumption of energy. The researcher proposed three strategies with different modifications on each to analyse the enhancement in the efficiency of the RBS. The author introduced 'serial 2 control strategy' as a new control strategy in regenerative braking. The researcher analysed two parameters based on vitality efficiency. One of the parameters is to analyse the vitality ratio of contribution to the driving range of the regenerative braking. The other one is to analyse the vitality ratio of contribution to improvement in the vitality transfer efficiency. On comparing the regeneration efficiency, the results exhibited that the vehicle with 'serial 2 control strategy' gives maximum efficiency from RBS. The serial 1and parallel strategy offers a minimum efficiency when compared with the serial 2 control strategy. But, the installation of sensors and the modifications made in the brake to adopt this strategy is a very complicated and costlier one [7].

The effects of using the dynamic energy of electric vehicles during braking and regenerating the electric power to charge the vehicles were investigated in this research work. Regenerative braking is an efficient technology that enhances the vehicle's driving range. The conventional braking method leads to a waste of kinetic energy into enormous heat during braking. It gave rise to regenerative braking which is efficient and helps in saving energy and cost. The researcher has mainly focused on the influence of brake energy in the regeneration of electric power. When the brakes are applied to slow down the vehicle, the RBS functioned simultaneously. The main working principle is that, when the rotor rotates in one direction, it acts as a motor. When it rotates in the contrasting direction, it acts as a power-generating device. A brushless DC motor is considered to be a main component of the regenerative electric vehicle. The ultracapacitor is the core technology that produces higher life cycles of a vehicle and also from the Matlab simulation the author has confessed that regenerative braking can save the waste kinetic energy of vehicles up to 8–25%. Figure 9.5 represents the simulation model of regenerative braking developed in Matlab [8].

In this research work, the author proposed a structure of a hybrid vehicle that had a battery or a fuel cell and a permanent magnet DC (PMDC) motor in it and performances were analysed by the researcher. The proposed hybrid vehicle model with various features is diagrammatically represented in Fig. 9.6. A clever controller is used in the hybrid vehicle structure that consists of a PMDC electric controller driven by a fuel cell/battery. This controller performs several tasks of regenerative

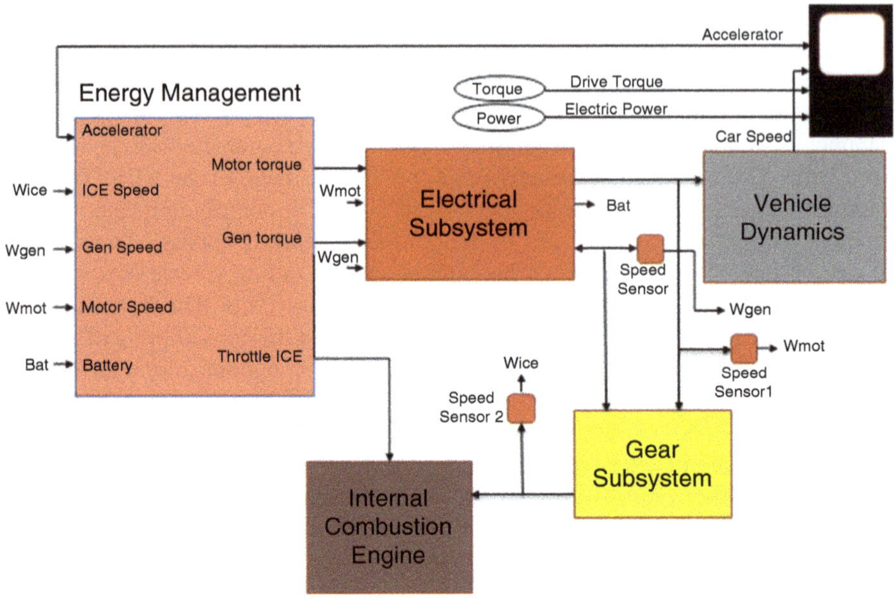

Fig. 9.5 Matlab simulation model of regenerative braking

braking & power management departments. With the help of switching strategy, the power of the fuel cell can be managed by DC to DC converter and the duty cycle is computed with the switching frequencies by a driving controller. Considering power demand, a neuro-fuzzy controller is implemented in order to consider power management issues. A simulation is done with respect to pedal displacement, battery state of charge act, a threshold zone for producing regenerative torque. The setup of the fuel cell and its components are available at the power system of the toolbox in MATLAB. Only longitudinal dynamics are considered in this simulation. By the simulation results, when the acceleration is given around 0–25 km/h, the fuel cell operates in the nominal amount of power and allows the battery to be charged. When the power requested by the driver exceeds the capacity of the fuel cell, the battery is triggered and both the fuel cell and the battery begin delivering the energy. From the regenerative braking test, the results conclude that the mechanical braking torque obtained is better than in the non-regenerative state. The results have clearly shown that when the vehicle is run in the uphill stream, the demand for power is past the ostensible volume of the fuel cell only; the battery ought to go into work with respect to have a commitment on power supplying [9].

An effective control system for regenerative braking is proposed by the researcher in this current investigation. HEVs are designed dependent upon suitable modification on a traditional hydraulic braking system furnished with an antilock braking system (ABS). To resolve limitation and pressure complications, a pressure coordinated control system (PCCS) is applied. The system model has been implemented in the simulation software. Assuming the opening perspectives of the high-speed

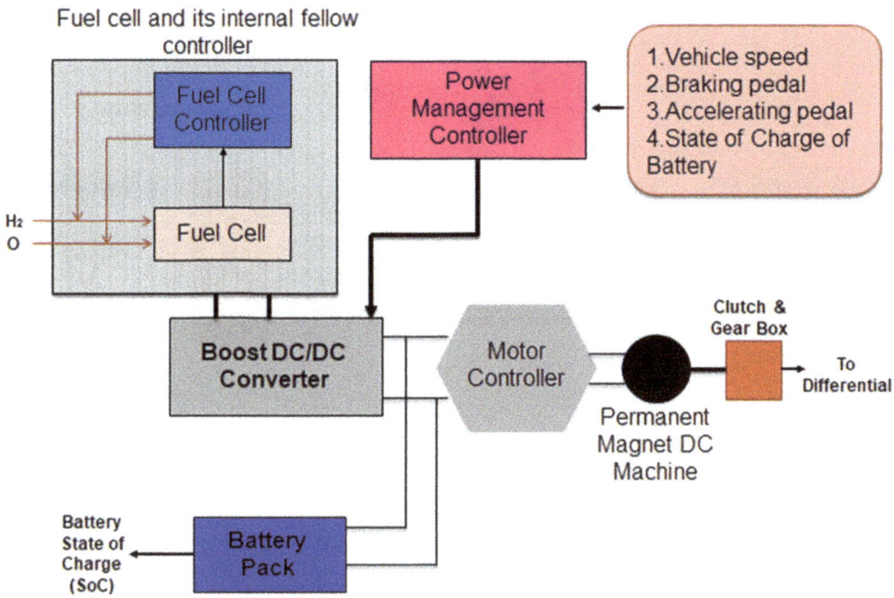

Fig. 9.6 Diagrammatic representation of the proposed hybrid vehicle

switching valve, to look after the booster valve and reducing the valve's duty cycle, a PID-bang bang controller is utilised in this study. At typical driving cycles, the forces obtained through braking not only satisfy the required braking of a vehicle but also increase regenerative braking efficiency by 12.5%. For the parallel braking approach, a minimal amount of energy has been recovered since its design has a simple structure. A control method called Pulse Width Modulation (PWM), utilised by PCCS, insisting a quicker reaction of braking force and it holds appreciable real-time tracking performance. In respect to PCCS, the wheel velocity precisely tracks the slip ratio, braking distance by the wheel & force by braking. If the distance of braking is at 50 m, the slip ratio differs lightly almost at the anticipated rate of 0.2, at the same time braking force changes in the ABS process. Thus, it indicates that the PCCS accomplishes a substantial performance of ABS [10].

For an automatic transmission (AT) based HEV, a brake system has been created and accompanied by a regenerative braking cooperative control algorithm. An electronic hydraulic brake (EHB) has setup on the rear wheels that are also used in friction brake system (FBS), so that the system need not expect a pedal simulator, and an electronic wedge brake is used on the vehicle's front wheels as represented in Fig. 9.7. A regenerative braking cooperative control algorithm is determined based on the model of FBS. Using the dynamic model of the braking system, this algorithm is made to distribute friction force and regenerative braking force. A simulator is developed based on this algorithm and the results are taken based on the simulation. From the result, it has attained the demanded braking force and based on its gradient with respect to pedal stroke, the energy is recovered. Assuming that gradient

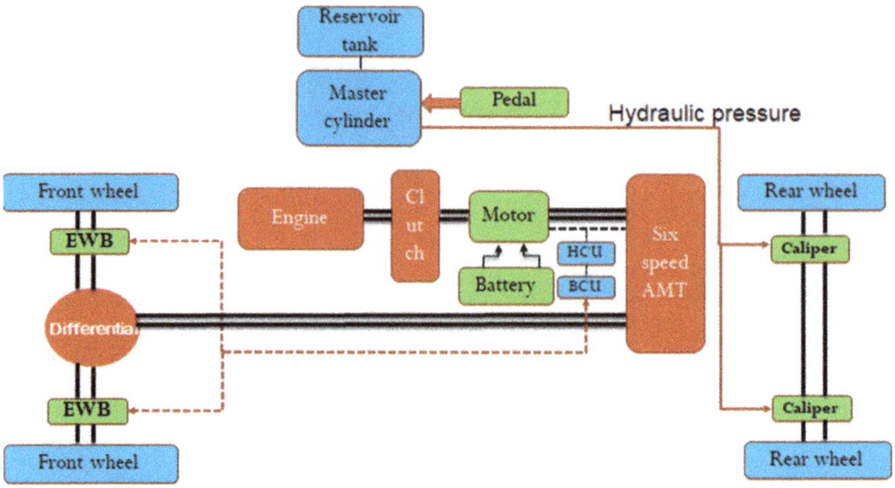

Fig. 9.7 Structure of the brake system

obtained with respect to total demanded braking force increases with respect to the pedal stroke, the braking force obtained due to friction at the rear wheel becomes lesser, additional force is provided due to extra braking that is transferred to the front wheels and thus there is an increase in the recovery of energy at regenerative braking [11].

This proposed research study develops a unique integrated brake pressure control algorithm based on two high-speed on–off valves to ensure precise and timely pressure monitoring for the regenerative braking system adopted in electric commercial vehicles. Two distinct models are being used in this study, first type is a valve control system made based on the mathematical model that is comprised of various sub-models. In another type, A PID Controller combining a fuzzy controller with pulse width modulation (PWM) is acquired separately with the incorporated algorithm to fill in with the traditional PWM approach. The tracking of the pressure of air cylinder is done and also the behaviour of 2/2 high-speed valve is read and with the help of thermodynamic and mechanical property of filling and depletion process of airflow is done periodically. With the results obtained from the simulation in PID Controller, a pressure response is obtained and they are capable to meet the target with respect to the response obtained in experimental results. From the results, it is concluded that the brake safety and brake feelings are improved significantly [12].

A method named dynamic low-speed cut-off point detection that is to be combined with EV is proposed in this research. In this type, one doesn't need to alter the hardware and the architecture of vehicle braking and can be applied exclusively by altering the brake controller. A strategy has been developed to control the diffusion of brake force in EV between RBS and FBS which depends on fuzzy logic control. Some factors like deceleration rate, mass at the vehicle, speed of the wind and slope angle with respect to the road were analysed individually in a simulation model based on the displacement of breaking cut-off point.

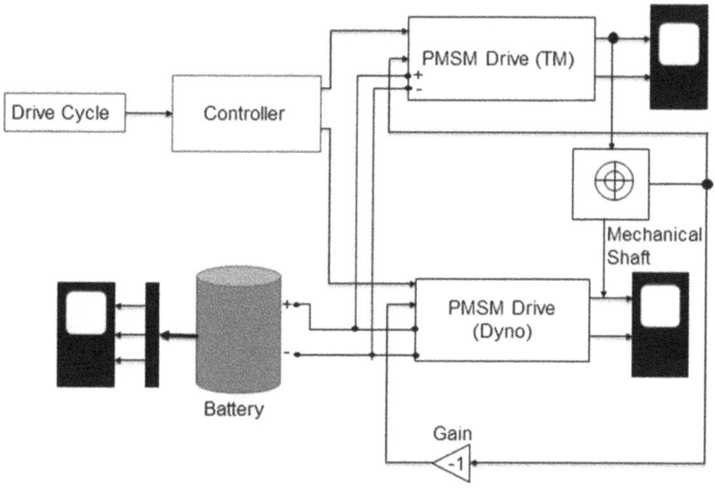

Fig. 9.8 Simulation model diagram

Figure 9.8 represents the proposed EV simulation model. From the simulation results, it is understood that if a deceleration rate is faster or when the vehicle mass is high, it increases the low-speed cut-off point (LSCP) while an increase in speed of the wind and in the uphill driving it ends up in LSCP. With the brake controller considering dynamic LSCP shows that, designing a brake controller with the change of state of LSCP, the loss of energy during RBS can be reduced or avoided and can enhance the driving range of the vehicle. The experimental results tell us that by grabbing the advantage of the proposed method, the energy recovery in RBS can be enhanced greatly assuming that having a constant low-speed threshold during the process of braking [13].

A brake controller which is almost similar to the actual EV brake system is added to its design with the consideration of limitations in both RBS and FBS. The designed brake controller is coordinated into the controller of the EV hardware-in-loop (HIL) test bench. With the help of the advanced vehicle simulator, several results of experimented HIL energy consumption are obtained and investigation of consequences on adding a brake model is also done. The two main limitations are considered while integrating EV braking and determining the share of regenerative braking. The limitation that is considered first is the capability of RBS to maximise the recovery of energy. This is considered as a major factor as it can affect energy recovery in RBS a lot during harsh deceleration.

The second limitation is efficiency contemplation and also the consumption of energy is the reason why the electric motor doesn't function as a generator that regenerates the battery of a vehicle at low energy consumption. One of the main elements is the controller that controls the drive motor and dynamometer under various circumstances. It is capable of calculating the speed reference with respect to drive motor and resistive torque reference with respect to the dynamometer. The drive motor used is a 15kw Six Pole Permanent Magnet Synchronous motor

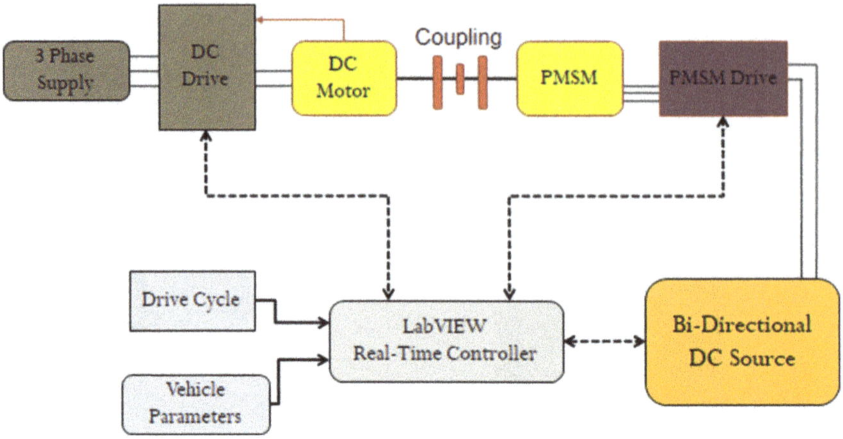

Fig. 9.9 Experimental setup

(PMSM) and the experimental setup for the proposed model of brake controller is presented in Fig. 9.9. The result shows a realistic approach towards braking limitations with respect to EV emulation on motor/dynamometer model & almost gives an exact assessment of the performance of an EV [14].

An electromagnetic brake designed as a miniature that is similar to conventional brake structure, designed with a 3D map, called capacitance and rotating speed module map, comprising of capacitance, the rotating speed and stator resistance, can predict whether the self-excitation occurs which results in the feasibility of the designed process. Electromagnetic brakes are also used in heavy vehicles along with the usual friction brakes. Effect of four influenced parameters namely magnetising inductance, resistance in the rotor, leakage inductance in the stator and also leakage inductance in rotor have been studied with an electromagnetic regenerative braking system. Using the finite element analysis, one can determine the magnetising inductance and leakage inductance. Avoiding hysteresis loss, the actual shaft torque will be smaller with respect to actual air gap torque. Then the voltage and current are being obtained with the help of the LeCroy Oscilloscope. The experiment results provide the value, that is almost equal to the 3D analysis and hence the minimal loss is caused provided that it is due to hysteresis loss. The Capacitance Map greatly helps to determine whether self-excitation happens or not in an electromagnetic regenerative braking system for several speed and capacitance [15].

The updated model of a novel electro-hydraulic brake system named Regenerative electro-hydraulic brake (REHB) is proposed in this research work. Simulators dealing with stroke are recommended to increase the pedal feel quality. The researcher coordinates the regenerative braking with hydraulic braking. In order to combine these two, the pressure containing in the master cylinder should not be equal to pressure containing in the wheel cylinder. This method can be attained only by brake-by-wire feature. To achieve this method, a solution based on high-pressure accumulator and a solution based on an electric booster is utilised. To ensure the highest amount of force due to regenerative braking, brake force due to friction should be linked which can be done by designing a blending control strategy for the braking force. The performance of the proposed braking strategy was analysed by Hardware In Loop bench test (HIL). Keeping fundamental control in mind, to obtain better precision of hydraulic pressure modulation and to eradicate noise caused due to vibration, a method called current amplitude modulation is adopted. With the help of the MATLAB-AMESism co-simulation platform, a model is constructed for high fealty models of vehicle and brake systems. From the analysis it is concluded that REHB with a very straightforward structure, improves the regenerative brake by wire system execution when compared to EABS and iBooster generation [16].

To carry on solidity and to refine the regenerative braking during unpredictable weak situations, a knowledge-based structure is designed which is called as a hierarchical control structure. The energy recovery analysis was also made in this research work. The researcher revealed that battery capacity is one of the major aspects that influence braking efficiency. The wheel-braking force which is made by braking actuators and the braking force generated by the tire-road interaction were analysed. The author also made a comparison on the simulation of the braking on high adhesion roads and on the slippery road. While reaching the optimum anti-skid braking performance, this feature enormously assists to maintain a strategic distance to determine the optimum slip ratio. The outcome of this method provides that this feature stops the locking of the wheel and makes regenerative braking effective as it can retrieve kinetic energy as best as it can. To observe the efficacy, the experiment and simulation are executed. The setup and working flow of the hybrid braking system used in this research work are represented in Fig. 9.10 [17].

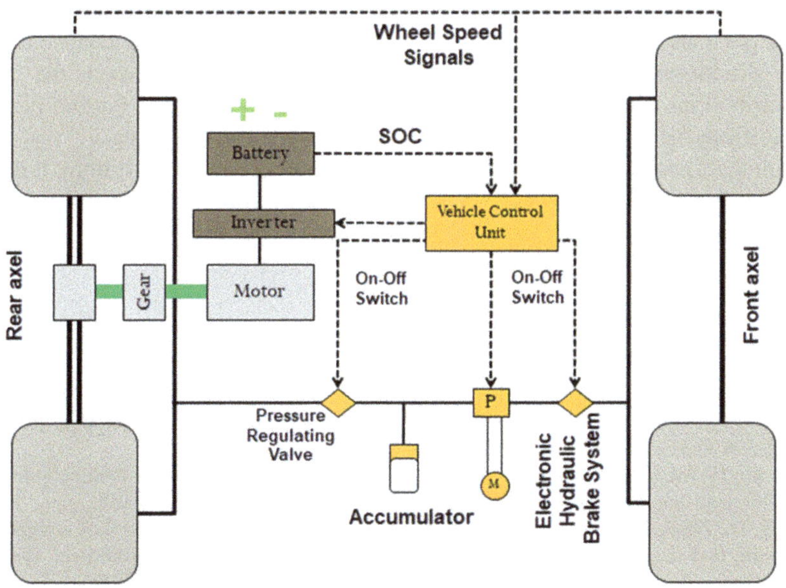

Fig. 9.10 Hybrid braking system

Creative RBS is adapted to brushless DC (BLDC) motor in the present study, which stresses on the conveyance of the force due to braking along with BLDC motor control. BLDC employs proportional-integral-derivative (PID) control during the distribution of braking force by utilising fuzzy logic control. At various brake pedal inputs, simulation analysis is done on dynamic performance and also with brake pedal distribution. Also, the researcher made the simulation on EV speed curve, energy recovery efficiency along with SoC (State of Charge). The researcher also adopts PID control to regulate the BLDC motor pulse width modulation duty in order to receive the same amount of brake torque. The combination of PID control and fuzzy control can make a smooth transient. Comparing with other solutions, the new one has shown finer conduct with respect to efficacy, robustness and realization [18].

4 Summary

- Combined braking control strategy (CBCS) operates the hydraulic braking system simultaneously together with the RBS, preventing the wheels from being fastened thereby effectively regenerates more energy.
- ADVISOR is a MATLAB/SIMULINK-based analyser is used for analysing (a) performance, (b) emissions and (c) fuel economy of the electric vehicle quickly.
- Regenerative braking to energy efficiency improvement and to driving range extension was up to 11.18% and 12.58%, respectively. Also, save the waste kinetic energy of a vehicle up to 8–25% and effectively regenerate the energy of braking by 12.5%.
- The parameters such as analysing the ratio between the vitality contributions to the driving range of the RBS and analysing the ratio between the vitality contributions to transfer efficiency of the regenerative braking is used primarily to analyse the vitality contribution made by regenerative braking.
- With the dynamic approach of the RBS cooperative control algorithm, it is made to distribute the friction and regenerative braking force which helps to recover energy based on the gradient flow with respect to the pedal stroke.

References

1. Patil, S. K. (2012). Regenerative braking system in automobiles. *International Journal of Research in Mechanical Engineering & Technology, 2*(2), 45–46.
2. Xu, G., Li, W., Xu, K., & Song, Z. (2011). An intelligent regenerative braking strategy for electric vehicles. *Energies, 4*, 1461–1477. https://doi.org/10.3390/en4091461
3. Peng, D., Zhang, Y., Yin, C. L., & Zhang, J. W. (2008). Combined control of a regenerative braking and antilock braking system for hybrid electric vehicles. *International Journal of Automotive Technology, 9*(6), 749–757. https://doi.org/10.1007/s12239-008-0089-3

4. Ahn, J. K., Jung, K. H., Kim, D. H., Jin, H. B., Kim, H. S., & Hwang, S. H. (2009). Analysis of a regenerative braking system for hybrid electric vehicles using an electro-mechanical brake. *International Journal of Automotive Technology, 10*(2), 229–234. https://doi.org/10.1007/s12239-009-0027-z
5. Panagiotidis, M., Delagrammatikas, G., & Assanis, D. (2000). Development and use of a regenerative braking model for a parallel hybrid electric vehicle. *SAE Technical Paper Series, 01*, 0995. https://doi.org/10.4271/2000-01-0995
6. Lv, C., Zhang, J., Li, Y., & Yuan, Y. (2015). Mechanism analysis and evaluation methodology of regenerative braking contribution to energy efficiency improvement of electrified vehicles. *Elsevier Energy Conversion and Management, 92*, 469–482. https://doi.org/10.1016/j.enconman.2014.12.092
7. Qiu, C., & Wanga, G. (2016). New evaluation methodology of regenerative braking contribution to energy efficiency improvement of electric vehicles. *Energy Conversion and Management, 119*, 389–398. https://doi.org/10.1016/j.enconman.2016.04.044
8. Malode, S. K., & Adware, R. H. (2016). Regenerative braking system in electric vehicles. *International Research Journal of Engineering and Technology (IRJET), 03*(03), 394–400.
9. Naderi, P., & Azizianfard, M. (2013). A fuel cell/battery hybrid vehicle modelling and power management/regenerative braking controller design. *Journal of Fuel Cell Science and Technology, 10*(011008), 1–12. https://doi.org/10.1115/1.4023384
10. Yang, Y., Zou, J., Yang, Y., & Qin, D. (2014). Design and simulation of pressure coordinated control system for hybrid vehicle regenerative braking system. *Journal of Dynamic Systems, Measurement, and Control, 136*(051019), 1–8. https://doi.org/10.1115/1.4027283
11. Ko, J., Ko, S., Son, H., Yoo, B., Cheon, J., & Kim, H. (2015). Development of brake system and regenerative braking cooperative control algorithm for automatic-transmission-based hybrid electric vehicles. *IEEE Transaction on Vehicular Technology, 64*(2), 431–440. https://doi.org/10.1109/TVT.2014.2325056
12. Pi, D., Cheng, Q., Xie, B., Wang, H., & Wang, X. (2019). A novel pneumatic brake pressure control algorithm for regenerative braking system of electric commercial trucks. *IEEE Access, 7*, 83,372–83,383. https://doi.org/10.1109/ACCESS.2019.2924739
13. Heydari, S., Fajri, P., Rasheduzzaman, M., & Sabzehgar, R. (2019). Maximizing regenerative braking energy recovery of electric vehicles through dynamic low-speed cut-off point detection. *IEEE Transactions on Transportation Electrification, 5*(1), 262–270. https://doi.org/10.1109/TTE.2019.2894942
14. Fajri, P., Lee, S., Prabhala, V. A. K., & Ferdowsi, M. (2016). Modelling and integration of electric vehicle regenerative and friction braking for motor/dynamometer test bench emulation. *IEEE Transaction on Vehicular Technology, 65*(6), 4264–4273. https://doi.org/10.1109/TVT.2015.2504363
15. Bae, J.-N., Kim, Y.-E., Son, Y.-W., Moon, H.-S., Yoo, C.-H., Jung, T.-C., Lee, J., Huh, K., & Jin, C.-S. (2014). Design and analysis of a regenerative electromagnetic brake. *IEEE Transactions on Magnetics, 50*(11), 8103204. https://doi.org/10.1109/TMAG.2014.2327621
16. Yuan, Y., Zhang, J., Li, Y., & Li, C. (2018). A novel regenerative electrohydraulic brake system: development and hardware-in-loop tests. *IEEE Transaction on Vehicular Technology, 67*(12), 11,440–11,452. https://doi.org/10.1109/TVT.2018.2872030
17. Xu, G., Xu, K., Zheng, C., Zhang, X., & Zahid, T. (2016). Fully electrified regenerative braking control for deep energy recovery and maintaining safety of electric vehicles. *IEEE Transactions on Vehicular Technology, 65*(3), 1186–1198. https://doi.org/10.1109/TVT.2015.2410694
18. Nian, X., Peng, F., & Zhang, H. (2014). Regenerative braking system of electric vehicle driven by brushless DC motor. *IEEE Transactions on Industrial Electronics, 61*(10), 5798–5808. https://doi.org/10.1109/TIE.2014.2300059

Chapter 10
Vehicle to Grid Technology

R. N. Ravikumar and S. Madhu

1 Introduction

In the present world, the major requirements for higher productivity both in the area of industries and agriculture are needed for uninterrupted electricity. The electricity generally is generated by using fossil fuels and other alternatives are by solar, wind and nuclear energy. In India, the majority of electricity is produced by using hydropower, coal and nuclear power. There is a need for utilization of power as and when required. Whenever excessive power is generated, it is stored in the batteries but the biggest challenge is that it requires a huge number of batteries. To avoid this and to maintain the power supply whenever higher power demand is present, we can focus on vehicle to grid technology (V2G). The V2G technology utilises the energy stored in the battery which can be supplied to the grid and whenever required, it is possible to take the energy from the grid to charge the battery of the electric vehicle (EV). This concept will be realised by using plug-in electric vehicles. In the current scenario, every country's requirement is focused on reducing CO_2 emissions and the best alternative for reducing CO_2 emission is by promoting the usage of EV. Here, an attempt is made to highlight mainly on how effectively we can establish the V2G communication in EV. One of the major requirements for electrical engineering is the creation of grid technology which is smart. By creating an eco-friendly advanced grid technology, the economic growth of individuals and the nation as a whole can be enhanced. To make electrical energy mobility, we require a strong, safe, smart and intelligent communication between

R. N. Ravikumar
Department of Mechanical Engineering, BMS College of Engineering, Bengaluru, India
e-mail: rnravi.mech@bmsce.ac.in

S. Madhu (✉)
Department of EEE, BNM Institute of Technology, Bengaluru, India
e-mail: madhus@bnmit.in

M. Kathiresh et al. (eds.), *E-Mobility*, EAI/Springer Innovations in Communication and Computing, https://doi.org/10.1007/978-3-030-85424-9_10

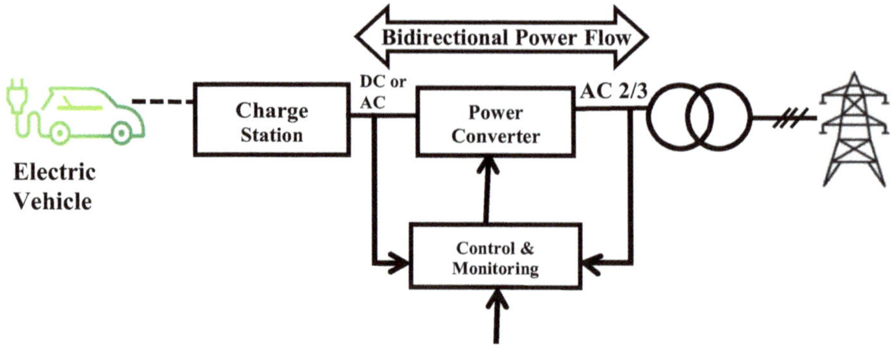

Fig. 10.1 Smart charging communication protocol

plug-in EV, charging station and the grid. It also needs a two-directional energy transfer between all of the three. All the stakeholders such as EV owners, EV manufacturers, power grid, charging stations, Government & private authorities, etc. should be defined with their roles to play with the understanding of the importance of communication and coordinated system operation. To realize all these and to have commonly accepted regulations, there is a need for standard communication models and protocols which will strengthen the application of V2G technology.

The manufacturers of the EV provide first information on availability of charging stations and charging process. But there is no communication available for the consumer and smart grid for charging in both directions. Presently, EV uses the smart connect technology which enables them to connect to charging stations and other utility services. There are no proper standards for V2G technology. Some of the standards are developed by automotive societies, manufacturers and service providers.

Fig. 10.1 outlines the general structure of V2G technology. The connection between charging stations, electric vehicles and grid is standardized by the International Electrotechnical Commission (IEC) 61851 and ISO 15118 [1].

The Society of Automobile Engineering (SAE), American National Standard Institute (ANSI), the International Electrotechnical Commission/International Standards Organization (IEC/ISO), National Institute of Standards and Technology (NIST) and several other industrial organizations are aiming towards the development of interoperability standards. On the basis of V2G communication interface standardization and study of the integration of the EV to grid, the V2G interface needs to be analysed. Along with all these, there is also a need for standardisation of the V2G interface and the structure of message patterns.

2 Vehicle to Grid Communication Flow Structure

V2G basically needs an effective two-way communication protocol, less time delay and higher reliability. The SAE J2931 under V2G facilitates the various stakeholders to communicate. The process is grouped into serially connected blocks as shown in the Fig. 10.2. The groups are Utility-Independent system operator (U-ISO) to Aggregator EV Service Provider (A-EVSP), A-EVSP to EV Supply Equipment (EVSE), EVSE to EV-Battery Management System (BMS).

SAE J2931 under V2G technology allows different stakeholders to operate in different modes. The software needs to support all the customers simultaneously without any confusion but with synchronisation. They are allowed to communicate with the next level in either receiving or sending message mode. This means that the stakeholder can receive or send a message to the next stakeholder as in Fig. 10.2. The example messages are also mentioned under the stakeholder in the Fig. 10.2. These modes are considered with respect to some factors such as the price of charging, availability of energy, availability of charging point, etc. This V2G technology allows the operator of the vehicle to program when to charge the battery, time of recharge and duration of the charge. They can also get information about the cost of energy, various tariffs, etc. It helps the utility also with various options such as price flexibility, user preferences and managing the bulk number of vehicles simultaneously.

In addition, it has to consider the stability of the grid and efficient use of available power in the grid and there should be provision for taking energy through renewable energy which is intermittent in nature. The users of energy for EV can have

	Utility-ISO	**Aggregator-EVSP**	**EV Supply Equipment**	**EV-BMS**
SEND messages	· Registration In/Out signals · Energy Price · Capacity parameter	· EVSE control parameters · Performance statistics · Participant parameter	· Change signal · verification & confirmation signal · Recorded data	·Battery Capacity ·Charge/ discharge rate ·Battery SOC
RECEIVE messages	· Verified Aggregator Control response · Power Grid balance signals	·ISO/ RTP signal · Valid EVSE control response · EV's energy need command · Market tariffs	· Aggregator control signal · EV requirement priority input	·EVSE battery charge controlling signal

Fig. 10.2 Representation of communication path under V2G technology

Table 10.1 SAE standards for communication between various stakeholders

SAE Protocol	Details about the communication
J2836 (Part 1 to 5)	Power transfer between PEV and the GRID & also reverse flow of power
J2847 (Part 1 to 5)	Apply of use cases as defined under J2836
J1850	Vehicle and EVSE network
J2293	V2G bidirectional power flow, updation of communication medium to PLCC or wireless
J2931	Communication needs for EVSE with home, EMS and GRID

information on when the grid is receiving energy from the renewable resources and they can select the charging time during such periods, thus reducing the emissions and pollution. With such communication, the renewable energy resources can be effectively utilised, which otherwise may be left wasted. With advancements of technology, it is possible to develop standards and protocols for effective utilisation of the system for all the stakeholders. The standards and the protocols that regulate the use of the V2G technology are also being formulated by SAE. Some of the SAE standards related to the communication are defined under the codes as given in the Table 10.1.

The standards provided in Table 10.1 facilitate standard requirements to ensure that the different manufacturers of EVSE develop products that are in agreement with the standards to their network interfaces (DOE, 2010). The NIST, National Institute of Standards and Technology standard and the IEEE 1547 standard give use cases for the communication, monitoring and management of distributed sources related to the power grid.

The standards and protocols developed are discussed in the preceding section.

3 Codes and Standards

Although a lot of technical research and policy debates have been conducted to verify the V2G concept, many practical standards for flat grids, vehicle-related standards and V2G equipment need to be revised or a practical framework to be created for the support of the V2G business model. On the contrary, they are specifically developed for the flow of power in one direction.

In the fields of equipment installation, communication, security, intercommunication, billing, approval, etc., also formulate standards of behaviour and practice. For many reasons, it is necessary to adopt a common set of norms and standards. Each participating public charging station requires an electronic identity certificate, so that the participating electric vehicle owners can establish a connection and participate in V2G operations that require electronic quotation and customer agreement to power contracts. For the purpose of billing, each vehicle needs an identifier which can report the global position of the EV, the meters implemented and the agreement between vehicle operator and the utility company. It is designed to enable vehicle and network operators to control the energy transmission rate and limit the energy that can be extracted from the vehicle. This enables network operators to do predictions and adjust traffic load while receiving data in the real time. Systems which are recording the information (such as customer requirements and cancellations) and support customer interactions with utilities require powerful and predictable management capabilities. To encourage the owners of EV who are participating in the energy exchange market, some standards should be established for prices that change over time. Network security standards should be established which are universally accepted, especially when protecting many charging stations located in public places such as malls and public parking places. Since many cars are connected to the network and left unattended, personal information needs to be protected from intruders. Cover the vehicle to maintain the ability to interrupt the charging process when the battery condition is unsafe, even if there is a charging command from the main power source. When the transition from internal combustion engine vehicles to automobiles begins, electrical, automotive, and utility companies need to work together to help provide compatible and safe systems for their mutual customers.

The codes, protocols and standards are usually related, although they have different meanings. The standard specifies that the necessities of all the stakeholders must be satisfied and supported by regulatory requirements and compliance. The competent authority is responsible for reviewing permits and other documents to ensure compliance and compliance with relevant regulations. Regulations usually aim to protect safety, health, collaborations and are approved by local governments or regulatory agencies. The prime goal of the standard development organizations is to have secure, robust, safe and common charging methods which can be integrated and can work in synchronization with the smart grid. The various standard developing organizations are IEEE, SAE, NEC, IEC, ISO and so on.

There should be interactive communication methods between the electric vehicle and smart grid for charging and to have collaborative operation. No interruption to be tolerated during communication. This has to be considered while designing the charging stations and network sharing. With an improved communication, this can be assured in V2G communication. Mismatched communication and miscommunication are to be avoided between different service-providing companies. For this to happen, there is a need for communication standards which are to be followed by all the vehicle manufacturers.

There are standards that are to be maintained and considered for effective communication which was developed by IEC (International Electrotechnical Commission). The parameters considered for the standardizations are plug, communication, vehicle couplers, vehicle inlet, communication network, switchgear, charging method, EV conductive charging system for both AC & DC, UPS, safety, etc. The standards for these under IEC are IEC 62196-1, IEC 62169-2, IEC 62196-3, IEC 61850-x, ISO/IEC 15118, IEC 61851-21, IEC 61851-22, IEC 61851-23, IEC 61851-24, IEC 61851-21, IEC 61140, IEC62040, IEC 60529, IEC 60364-7-722, ISO 6469-3 [1, 2]. These standards are described in Table 10.2.

Table 10.2 IEC/ ISO standards for the PEV charging components

IEC 62196-1	Charging of EV up to 250 A AC and 400 A AC with conductive charging, vehicle couplers, plugs, vehicle inlets and socket-outlets
IEC 62196-2	Socket-outlets, plugs, vehicle inlets and connectors, charging of EVs, compatibility with dimension, and exchangeable needs for contact tube and AC pin accessories
IEC 62196-3	With rated operating voltage 1000 V DC and current up to 400A with exclusive DC charging. vehicle couplers, socket-outlets and plugs Conductive charging of EV, compatibility with dimension exchangeable need for contact-tube coupler and pin.
IEC 61850-x	Systems and network communications in distribution substations.
ISO/IEC 15118	Communication interface for V2G.
IEC 61439-5	Control gear assemblies and low-voltage switchgear and assemblies for distribution of power in public networks
IEC 61851-1	General requirement for EV charging.
IEC 61851-21	AC/DC supply for EV requirements and conductive charging system for EV.
IEC 61851-22	Charging station for AC EV.
IEC 61851-23	Charging station for DC EV.
IEC 61851-24	Controlling communication protocol between EV and on board DC charger- EV conductive charging system.
IEC 61140	General aspect of protection from shock during installation of equipment.
IEC 62040	Uninterrupted power supply (UPS)
IEC 60529	Level of protection supported by the safety enclosures (IP code)
IEC 60364-7-722	Electrical installations of low voltage, need for special installations or locations for supply of EV
ISO 6469-3	Protection from electric shock for persons and EV related with V2G

The other important technology, policy and advocacy organizations which are contributing to the developments of the standards for V2G are National Resources Defence Council, Electric Power Research Institute, National Renewable Energy Laboratory, California Energy Commission, Edison Electric Institute, Electric Drive Transportation Association, U.S. Department of Energy, Pacific Northwest National Laboratory, California Public Utilities Commission, Idaho National Laboratory, Rocky Mountain Institute, National Association of Regulated Utility Commissions, Argonne National Laboratory, Oak Ridge National Laboratory.

The vehicles and electric vehicle supply equipment (EVSE) are going to be principally standardised through the IEEE, SAE and/or NIST proceedings which all the automotive makers should follow. As per the native jurisdiction, the car makers can follow any of the standards and rules that may drive V2G technological developments.

4 Protocols

The field of communication protocols for network operators to control electric vehicles is developing rapidly, and each protocol is becoming more and more extensive. The Information Technology Certified Associate (ITCA), a standard development organization, continues to formulate standards and certification procedures for protocol selection.

Although messaging protocols must technically support messages for specific use cases, this is not enough. First, the availability of the product IEEE 2030 can be used for disaster recovery, but requires a lot of industry investment to achieve a strong interoperable ecosystem, or use cases that are likely to be inconsistent (or interoperable) between utility and utility and between provider and provider such as authentication. The new DNP3 AN-2019 provides a technical specification for the inverter curve and AC Regulation 21 configuration required for transmission via IEEE 1815/DNP3. Manufacturers now support it in their DER, but there is currently no certification program to ensure interoperability. These systems can interact with each other without certification; however, certification alleviates interoperability issues caused by misunderstandings in standard terminology, which manifest as delays or change requests.

There are various protocols for V2G Technology framed by different standard development organizations. Some of them are as follows.

1. OCPP—Open Charge Point Protocol
2. OCPI—Open Charge Point Interface
3. OICP—Open Intelligent Charging Protocol
4. OSCP—Open Smart Charging Protocol
5. OCHP—Open Clearing House Protocol
6. Open Interchange Protocol—OICP
7. E-MIP—e-Mobility

4.1 OCPP—Open Charge Point Protocol

The OCPP is a communication application protocol, which is used for communication between central management systems and vehicle charging stations. It is developed internationally and is freely available. It is open source and seller-independent.

Open Charge Alliance has developed this protocol for the market of EV infrastructure. These protocols are the real application of standards for interoperability, infrastructure market for all the manufacturers of charging equipment. It is also taken as standard by charging network operators, software and system providers and research institutions. The protocol has proved that it minimises the cost and risk for the investment in the development of infrastructure and is easily accessible by EV owners.

The present version of OCPP is OCPP 2.0. It has more improved and newly added features for transaction handling, device management, display, security, smart charging functionalities, messaging and many additional developments as per the need of the EV charging community. Additionally, OCPP 2.0 provides the facility to plug-in and charges the EVs, thus supporting the ISO 15118 protocol.

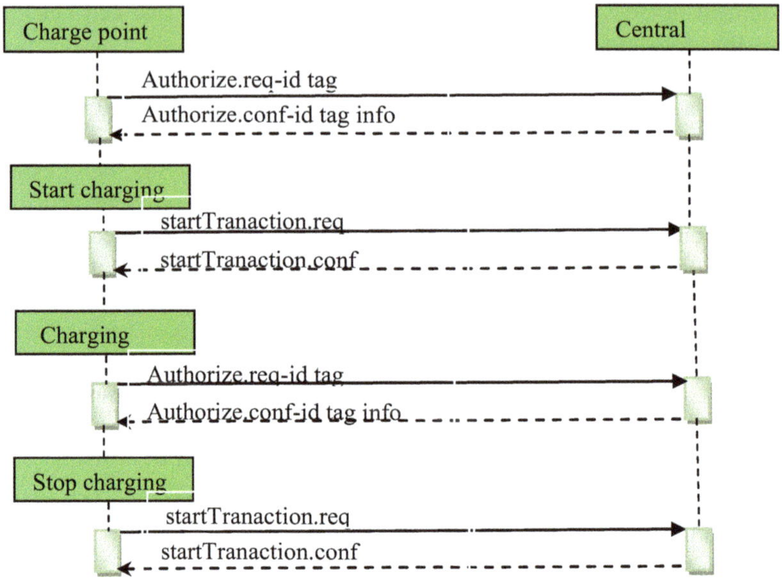

Fig. 10.3 Sequence diagram: Starting and stopping transaction example

This protocol supports communication between the charging point and the central system, when an agreement requires one or the other to take a specific action or response [3]. Figure 10.3 describes the general process between the charging point and the central system in two cases. One is the charge point requisition for card authentication and sending the charging status and the second one is the request by the central system for the charge points for updation of the firmware. It also indicates the sequence diagram showing the start and stop of a transaction. The sequence diagrams are the ones which are presented in the standard documents of IEC and ISO for various protocols. Here, one of them is presented to give an idea about the process flow due to any event.

When the charging station is ready to charge the EV, the user must be authenticated first and then the charging process can start. If the user is authorised, then after completion of charging the battery, it checks for the completion of task and checks whether the user belongs to the authorised group, then allows them for completion of battery charging. The central system is kept informed by the charging point that it has stopped charging the EV.

Figure 10.4 presents the sequence diagram for the updation of the firmware. When the charge point has to match with the updated new firmware, the central system sends the information to the terminal device about the update of the new firmware. During each step of updating the software, the charge point should send information to the central system, receive authentication and complete the installation process.

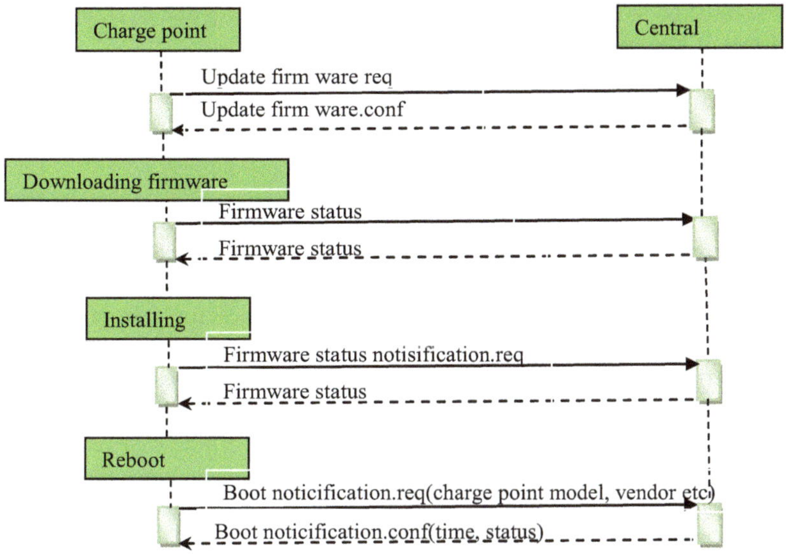

Fig. 10.4 Sequence diagram: a firmware update example

4.1.1 Local Authorisation and Offline Behaviour

If there is a communication problem or it may be a central system, when the charge point is autonomous. This situation means that the charge point is not connected. To improve user experience, the purpose of authorisation may be maintained with a local authorisation ID, an associated authorisation cache of the degree of victimisation, and/or a list of allowed areas.

This helps for two purposes. They are the authorisation of offline users and a quick authentication time when the link between the central unit and charge point is slow.

The configuration key *'LocalAuthorizeOffline'* determines the authorisation of the charge point even when offline. This can be done by use of local authorisation cache. Whether the charge point can use this option to initiate a new transaction without any delay can be controlled by the configuration key '*Local Preauthorize*'.

4.1.2 Authorisation Cache

The local authorisation list can be an identifiers' list to be checked with the central processing system. This list consists of an authorisation status and authorisation status/expiration date of all (or more) identifiers. Identifiers in the own authorisation list are considered valid, expired, (temporarily) blocked or blacklisted, similar to the Id Tag Info status value. These values usually also provide additional details. Users (such as viewing messages) are authorised locally. The authorisation list of this machine must be saved in nonvolatile memory for loading and must be maintained during restart and power failure.

4.1.3 Local authorisation list

This list has identified users who are all synchronised with the central processor. The list of users will have different tags based on whether they are valid, expired, blocked or unauthorised users. This process helps in authorising the charge point at the stage for confirmation of identity so that they receive permission for charging. The central system can take action based on the identity of the user being connected as indicated in the Fig. 10.5. Similarly, a detailed process, actions implemented and authorisation schemes are explained in detail in the standard document on OCPP. The relation between authorising agent and the local list, unknown offline cases are all dealt in detail. These sequence diagrams are from the standard document of OCPP. More detailed structures and sequence diagrams can be referred from the document mentioned in the references list [3].

The transaction in relation to energy transfer period, ID tokens, transaction-related messages, connector numbering, parent id tag, reservations, vendor-specific data transfer, smart charging, time zones all have rules and conditions to be followed under this protocol. The document on this protocol provides a detailed description of these cases with illustrative diagrams and examples.

The transaction for energy transfer period has protocols to be followed between the EV and EVSE during the charging period and the not connected mode.

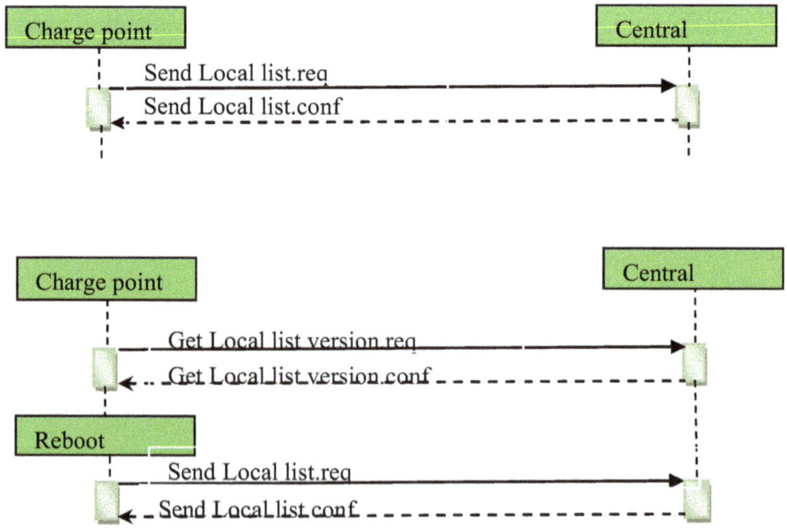

Fig. 10.5 Sequence diagram: a differential local authorization list update example

The transaction messages deal with giving and receiving messages between the charge point and the central unit. All types of messages had a tag which indicates the information. The messages are defined and also considered in some predefined order for every status that is to be considered during all the processes.

The error responses are also dealt with by certain defined rules and order. Upon identification of any failure in delivering messages, the stats should be handled with predefined preferences which are all defined in this protocol.

The numbering of the connectors is also considered to be one of the important considerations. The connectors should have identifiable numbering. A connector Id 0 has been reserved for the main controller of the charge point.

The ID tokens deal with the UID value of the physical card that the customer will have. An ID is assigned for every charge which is considered as the virtual transaction code. Also a central account code known as parent ID will be provided. The data type of the ID also has predefined constraints.

A parent ID tag refers to the ability of the central system to have token sets as a 'Group' to allow any one member in the group having that token to make communication for starting or stopping the transaction. This protocol helps in using the token ID by a family with multiple drivers using the same vehicle with a single contract number.

Reservations help in reserving the charge point up to the required but a valid time. This allows reserving a particular connector in the charging station.

The vendor-specific data transfer enables for the interchange of information which may not be defined under OCPP, thus enabling a possible implementation of additional functions if the central system accepts the function.

One of the important functionalities under this protocol is smart charging. The central system has capability for controlling the charging parameters like power and current to a particular EV. It can monitor the energy consumption by each EV, at each charging point, grid connectivity, energy available from the grid, the type of connector, specific limits, charging profiles, charging purposes, etc. Under this protocol, three types of smart charging options are facilitated. They are load balancing, smart central charging and smart local charging.

The above mentioned processes can be well understood with the sequence diagram for the process given in reference [3].

All the above mentioned categories have detailed processes and tagged transactions. All of these are having different tags for each of the processes which the charge points should follow and satisfy. The operations that can be started by the charge point are Authorize, Meter values, Boot notification, Firmware status notification, Data transfer, Heartbeat, Start transaction, Status notification, Diagnostics status notification, Stop transaction. Each of these operations is defined with the protocols that need to be followed. The charge points should follow the protocols as mentioned under each of these processes so as to have collaborative transactions. By doing this, a standardised messaging structure is assured.

Similarly, the operations that are supported by the central system are Cancel reservation, Change configuration, Change availability, Clear Cache, Clear charging profile, Data transfer, Get composite schedule, Get configuration, Get diagnostics,

Get local list version, Remote stop transaction, Remote start transaction, Reserve now, Reset

Send local list, Set charging profile, Trigger message, Unlock connector, Update firmware. These steps involved with respect to the central systems enable for the monitored transactions under V2G technology.

4.2 OCPI—Open Charge Point Interface

The Open Charging Point Interface (OCPI) aims to exchange charging point information between charging point operators and electric vehicle service providers to realise the scalability and automatic roaming of electric vehicles. Specifically, they provide session information, duration including location information. It supports send commands remotely such as backup commands. Provide detailed billing records (CDR) for billing. Use token exchange to authorise charging sessions. E-mobility operators and service providers support authorisation, exchange of charging point information (including real-time status updates and transaction events), exchange of transaction details, billing records, remote charging point command, exchange of smart payment-related information between parties. An inter-network (international) roaming solution that avoids costs and frustrations through innovations related to modern handheld solutions or central roaming centres. In this way, it can help electric car drivers charge for rapid development in a fully informed market and help seafarers. Guide participants so that they can implement their business model in the best way.

Main features of the protocol:

- Real-time information on location, prices and availability
- Better roaming system (two-way use and/or via a hub)
- Unified data exchange (notification data collection and accounting data collection) after, during and before the transaction.
- No prior registration is required to access mobile remote support for each charging point.

OpenADR-an open and secure information interoperability exchange framework to promote automatic query operators (DSO), utilities energy management and control the system to balance the peak energy demand, which led to the development of OCPI in 2014, which has supported the OCPI international group of companies initiated by EV Box. There are several organisations and platforms such as The New Motion, GreenFlux, ElaadNL, Freshmile, BeCharged, Plugsurfing, and Last Mile Solutions, ihomer and Siemens who are involved in participation. The Dutch Freight Infrastructure Knowledge Platform (NKL) supports and coordinates the project to ensure progress and feasible results [4].

4.2.1 EV Charging Market Roles

In the EV charging infrastructure under V2G, various market roles are identified as mentioned below. The following Table 10.3 presents various roles taken up by the typical modules. The Table 10.4 shows the typical communication role such as sender, receiver and/or both.

4.3 OpenADR 2.0

The protocol OpenADR 2.0 provides standardisation of distributed energy communication (DER), demand response (DR) and automated DR/DER processes. It provides protocols to simplify the customer's energy management and removes unused assets.

Table 10.3 Market roles of service providers

Role	Description
Charging Point Operator (CPO)	Operates a network of charge points
E-mobility service provider (eMSP)	Provides EV drivers to access to the charging services
Hub	Multiple CPOs are linked to multiple eMSPs
National access point (NAP)	Provides a national-level database with all charging
Navigation service provider (NSP)	Provides EV drivers with location information of charge points
Smart charging service provider (SCSP)	Provides smart charging service to other stakeholders and can use a lot of different sources for energy to calculate smart charging profiles

Table 10.4 Typical communication role: Receiver, Sender or Both. S-Sender, R-Receiver, B-both

Modules	CPO	eMSP	Hub	NSP	NAP	SCSP
CDRs	S	R	B			
Charging profiles	R		B			S
Commands	R	S	B			
Credentials	B	B	B	B	B	B
Hub clients info	R	R	S	R	R	R
Locations	S	R	B	R	B	
Sessions	S	R	B			R
Tariffs	S	R	B	R	B	
Tokens	S	R	B			
Versions	B	B	B	B	B	B

OpenADR: A protocol that enables automatic demand response (ADR) procedures, along with the potential for collector intervention. It is an open protocol standard which is based on Energy Interoperation version 1.0 (EI) of Organisation for the Advancement of Structured Information Standards (OASIS). The OpenADR 2.0b [5] has some of the services mentioned in the EI. The prime features implemented by OpenADR 2.0 b are EiEvent, EiReport, EiRegisterParty and EiOpt. The services permit the stakeholders to register to an event, change resources, reports and options to participate in the events [6]. According to the document, the OpenADR protocol is mainly aimed at the electricity charging market (but it can also be used to exchange real power, reactive power, etc.). With this in mind, the OpenADR protocol attempts to implement EI. The focus of implementation is demand response (DR) management. There are several implemented interfaces that provide the functions provided by the OpenADR standard.

4.4 OSCP—Open Intelligent Charging Protocol

It is one of the important open protocols used for communication between the charging point control system and the site owner's power management system or DSO system. The real-time estimation of the grid energy capacity can be transmitted to the charging point operator through the agreement. OSCP provides capacity-based intelligent charging of electric vehicles.

4.5 OSCP—Open Smart Charging Protocol

This is an open communication protocol that can send a 24-h local available capacity forecast to charging point operators. Service providers will adjust the charging configuration of electric vehicles within the framework of available capacity. It is an agreement between the charging point management system and the power management system or DSO system of the site owner. So this applies to website owners and DSOs.

4.6 OCHP—Open Clearing House Protocol (e-clearing.net)

The OCHP is an open source protocol that allows for simple and consistent communication between freight management systems and information exchange systems. OCHP allows unlimited charging of electric vehicles through a network of charging stations (electronic roaming). With OCHP, electric vehicle service providers can connect with electric vehicle charging providers and operators to access their networks. The location of the electric vehicle clearing house, especially the

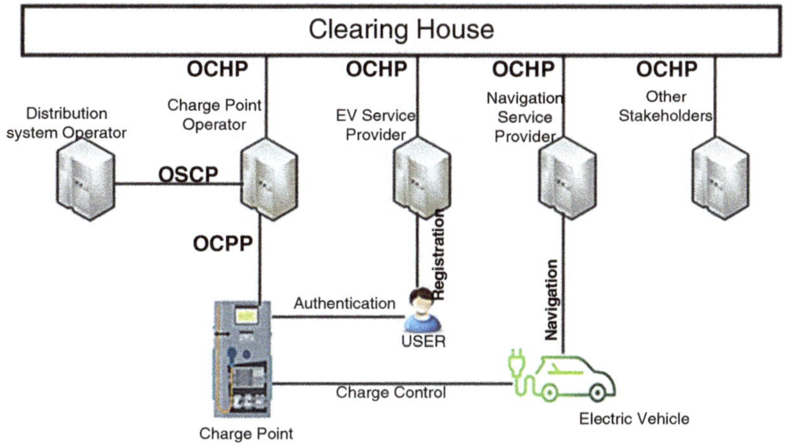

Fig. 10.6 Relationship between the clearing house and the electric vehicle service provider

location of the European electric vehicle clearing house, is shown in Fig. 10.6 and it shows three types of stakeholders, but these procedures can be conveniently extended to other cases of similar types.

The following listed stakeholders are directly related to clearing houses: Other (local) clearing houses, which are of two types: Clearing houses that do not conduct financial transactions, so related (secondary) parties must take over financial transactions. Currently, this type includes ECH, the clearing house that handles financial transactions, and acts as counterparty for transactions in its infrastructure without delay, such as Ladenez.de.

For an electric vehicle mobility service provider with multiple customers, the preliminary idea is that the clearing house provides 'roaming support' for every electric vehicle mobile service provider connected directly or through other clearinghouses. The ultimate goal of EV consumers is to be able to easily charge their EV at any charging station of any EV service provider. The clearing house provides the support of roaming and with this the complexity of the relationship can be reduced from the many-to-many two-way relationship with the electric vehicle service provider to the single-to-multiple relationship between the clearing house and the electric vehicle service provider.

4.7 OICP—Open Interchange Protocol

OICP was developed on a platform called Hubject and is a communication standard implemented between the systems of electric vehicle service providers (EMSP) and charging point operators (CPO) through this Hubject platform. The agreement is based on the contractual relationship between EMSP and Hubject's CPO to achieve

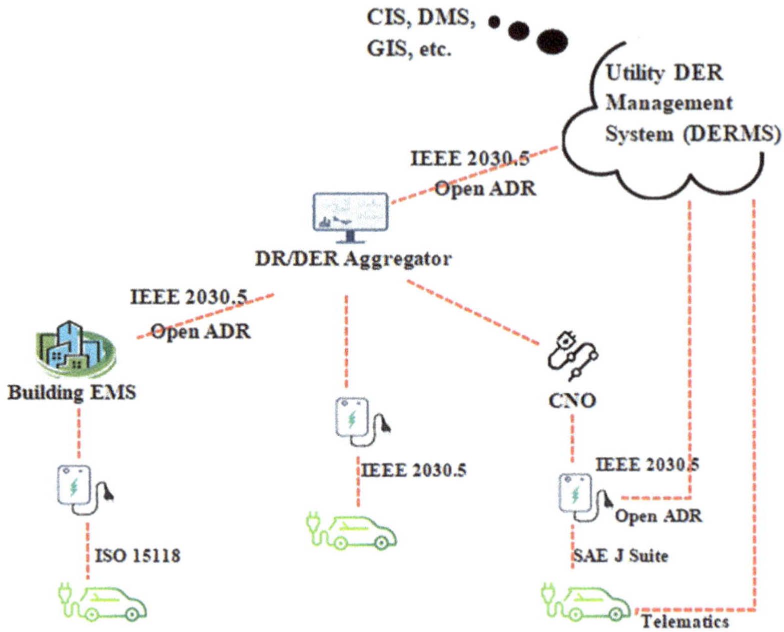

Fig. 10.7 Communication standard implemented between the systems of electric vehicle service providers (EMSP) and charging point operators (CPO)

information exchange, which enables them to provide reliable roaming services for EV drivers (Fig. 10.7).

Interoperability requires flexibility and collaboration between market participants. The open standard aims to open up the electric vehicle charging market. Two of them are initiatives of their own associations for the development of OCPI and OCH protocols. Also by using proprietary standards for PEV, it can be used by everyone at free of cost. Open standards for roaming electric vehicles are proprietary, but they can also be used publicly and free of charge. GIREVE and Hubject establish user friendly and new service network for charging. Hubject and GIREVE on-site facility providers, automobile manufacturers, power suppliers and authorities provide a wide range of cost-effective solutions and simplified communication advantages [7].

Hubject designed OICP in 2012 by an established German automobile professional. In addition to the roaming platform, Hubject gives roaming platform, technical facility and prescribed provisions for electronic roaming. The general construction of OICP is partitioned into two sections, in particular e-Mobility specialist co-ops and charging administrators; however, the item range is continually growing. Hubject said that OCIP is 'the most generally utilized standard for communication among e-MSP and CPO frameworks in Europe'. The most recent adaptation 2.3 was delivered in October 2020. The most recent updated protocol addresses market issues: it enhances the nature of POIs in the extended market, which grows legally, compliance, data processing performance, and ensures greater transparency for different load applications. In addition, Hubject provides many useful CPO-related remote tests to fully explore various use cases. The agreement is free and open and does not require registration. As of May 2019, OICP is a free and open source agreement that is provided on the website of Hubject as open source supply systems. The main objective is to have more customers to involve in the development of communication systems. This protocol works in real time. It is a real-time protocol and also provides asynchronous operation. Also there is a backup database; Hubject does not support uploading to charging stations. The platform records data of transactions.

The established market role of documentation in OICP protocol as mentioned below.

Service provider e-Mobility needs to make charging points available to owners of electric vehicles. In most cases, using a valid agreement with e-MSP and transmits the information to the charging point operator through the platform. Hubject provides connectivity through the e-MSP aggregator to communicate with the platform. This means that multiple eMSPs can be integrated through one e-MSP aggregator, so that sub-partners do not have to work with Hubject.

- Charging point operators (CPO) manage charging points. Through the Hubject platform, the operator can sign a contract and exchange any information. Like e-MSP, multiple CPOs can be integrated through the CPO aggregator. Therefore, the sub-partner does not have to register with Hubject as the communication with Hubject is through the CPO aggregator.
- Hubject, which is a roaming centre, establishes connection between e-MSP and CPOs through the EV platform.

Functionalities Supported by OICP

- Roaming via hub—Using web-based service e-MSP and CPO are connected through the platform.
- Ad hoc payment—This allows vehicle owner to make contactless payment for changing battery.
- Authorization—Hubject binds e-MSP or CPO ID and SSL certificate information. Then each author is allowed to download the session. The Hubject database is used as a backup, but it does not allow downloading from its database to the charging site.
- Reservation—Owners of electric vehicles can reserve charging points through the e-MSP application. Hubject tests the compatibility of charging points with electric vehicles. If it matches, the CPO request is sent, and if the reservation is successful, the CPO request will respond. You can also withdraw your reservation.
- Billing—Total duration of charging has been recorded and stores in the system for the billing process
- Charge point information—The information stored in the database is downloaded and transferred between e-MSP and CPO which include all the data, like contact

point ID, name, location, time, availability, price, method of payment, performance and all the real-time status information.
- Real-time charge point information—This provides availability of charge, status and price information at present time.
- Session information—This will provide information on charging duration, ID-like session, service, start–stop authorisation, meter, used energy.
- Remote start/stop—Using this application, it is possible to control the complete process without human intervention, that is, without fully automatic.

4.8 e-MIP—e-Mobility Interoperation Protocol

The e-Mobility Interoperation Protocol has been developed by French start-up company French players in Electric mobility in 2013. The main objective of this protocol is to facilitate an 'open access protocol to vehicle charging stations'.

The e-Mobility interworking protocol facilitates you to move through clearing-house data. The protocol provides access to the database of the charging point and helps for smart charging functions.

In 2018, the GIREVE platform complies with the OCPI standard. The roaming platform is currently active in 28 countries/regions. The latest version of e-MIP 1.7 was released in late July 2019. GIREVE also provides certification services. Platform, e-MIP platform is not suitable for point-to-point connections in practice. To provide a latest quotation and high degree of architectural openness, GIREVE negotiated with stakeholders on possible future functions. GIREVE assumes full responsibility. In contrast to the above methods, the e-MIP protocol does not have a formal association of e-MIP members. E-MIP is based on Simple Object Access Protocol. Basically, the e-MIP is designed as a real-time protocol but it allows asynchronous operation. The characteristics of the e-MIP architecture make it highly adaptable and flexible: the definition table can be used to quickly add each new type of data message or identification method.

Market roles of e-Mobility include charging vehicles, vehicle sharing, charging infrastructure and data on the flat form.

The e-Mobility Inter-operational Protocol supports the following functionalities:

- Roaming through hub
- Authorisation
- Synchronous Authorisation
- Asynchronous Authorisation.
- Asynchronous Authentication Data Exchange and Synchronous Authorisation
- Reservation
- Billing
- Real-time charge point information
- Static charge point information
- Charge point search

- Session information
- Real-time session information
- Remote start/stop
- Platform monitoring

A single network will be effective because EV owners and service companies can improve their services, offering better services for customers. The company develops the protocol for electric vehicles that use their charge infrastructure with specific simple instructions. The cross-border networks are available, and the support of road experts and commercial leaders promises the unit and resilience of the network. Please note that both grids have developed their own custom protocols (OCP and EMIP), they also have contributed to the development of OCHP and OCPI.

5 Present V2G Communications and Protocols

EV charging using V2G technology is an important setup which helps to obtain a net zero case, but this technology is difficult to be utilised by EV owners before the end of this decade.

- According to the National Grid ESO's prediction on future scenarios on energy, up to 45% of the domestic consumers may actively participate in V2G services by 2050. This survey has been provided in July 2020.
- Many EVs are currently unable to participate in V2G technology due to the interfacing issues, as predicted by the experts.

The Figure 10.8 shows how the developments in the field of smart charging will lead to the increase of V2G utilization by 2050.

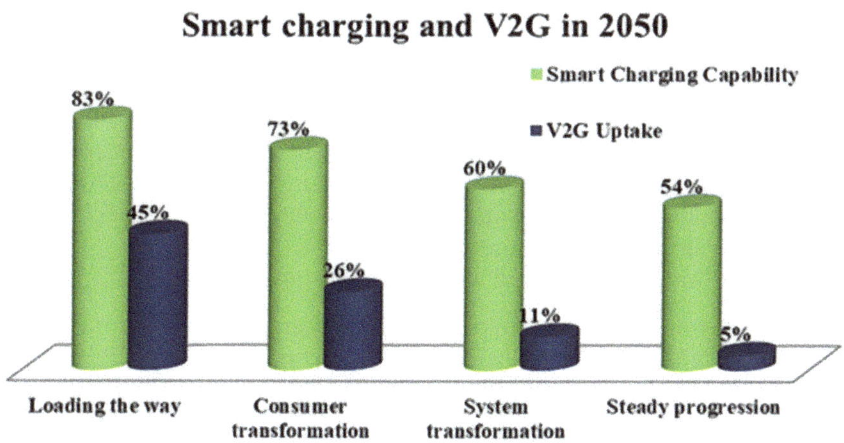

Fig. 10.8 Smart charging and V2G in 2050

When all electric vehicles are connected at the same time, such as during peak load at 5:00 pm, the rapid increase in the number of electric vehicles on our roads leads to higher demand on the local grid. 7:00 in the evening. Controlling the charging time of electric vehicles, as tested by Electric Nation's original project, helps avoid this.

However, V2G charging will be more efficient than smart charging because electric vehicles can be connected and a large amount of energy can be fed back to the grid during peak hours, just like in a huge decentralised power plant. In addition to reducing the demand on the power grid and providing cleaner and cheaper power to electric vehicle drivers, the power grid is also required to provide additional power that is usually produced by fossil fuels during peak hours.

As per the National Grid Electricity System Operator (ESO) future national grid scenario, as many as 45% of the companies will be actively involved in V2G communication services by 2050, and V2G services can provide 5.5 million vehicles with up to 38 GW of flexibility. However, the range of results is wide. For the V2G technology in the ESO national grid scenario, the lowest estimate of the number of households that will actively provide vehicle-to-grid services in 2050 is 4.6%. V2G is also considered to be slow to market, and there will be a 5- to 15-year delay between buying an electric car and participating in V2G.

The take-up ranges from 5% to 45% throughout the transition to V2G, with 45% being the 'Leading the Way' early delivery of Net Zero in 2050. The two other Net Zero compliant studies show 11% and 26% in 2050, exhibiting the uncertainty in the implementation of V2G technology.

5.1 Guidelines for Setting Up of Control Architecture for Messaging

The control architecture of the electric vehicle communication infrastructure describes the communication architecture used to receive messages from/to network operators and controlled terminal devices. For example: one model assumes that the actual endpoint is controlled by the building energy management system (EMS) that actually communicates with the Utility-DERMS. In this case, the operator will contact the EMS, which is responsible for addressing the terminal device [8]. A similar and supporting discussion on control architectures and quality logic has been done earlier by the authors of references [9, 10].

The Control architecture related to EV communication issues are:

- **Direct control**: The utility program sends configurations or requests directly to the EV or EVSE. This currently only applies to V1G use cases.
- **Pass-thru Aggregation**: The smart inverter settings and emergency call transmission commands are designated by the energy supply company for each inverter and 'passed' to the aggregator, CNO or other inverter gateways.

- **Smart DR/DER aggression:** (Demand Response and Distributed Energy Resources): This process is usually used by the utilities to manage DR and DER, in which smart gateways are implemented to convert the utility guidelines to check what settings or instructions are to be sent to the stakeholders receiving the energy. The communications information may be advisory types which are specific mandatory needs of DER or DR or for both.
- **Third-party charging network operator:** This is the setup made between the utility and the third party who implements the charging stations with some intelligent system for managing the EVSE, V1G and V2G vehicle charging setups. Here, only the EVs are managed and the utility grid is not controlled.
- **Vehicle measurements:** It is a substitute path for communication between the EVs and utility DERMS without the involvement of the EVSE. This network is implemented for managing EVs in either V1G or V2G technology.

From the point of view of the protocol, the architecture implemented decides the requirements of messaging, its functional features, and messaging protocols that are implemented in the architecture. It varies from architecture to architecture with varying protocols and facilities included.

5.2 *Requirements of Communication for EV*

The main intention of messaging requires that it gives the information about the type and the content of messages which are used to exchange information between the two parties for some application. These are obtained from the different cases. For example: in many control architectures, the EVSE and the EV will communicate among themselves to make an agreement about the charging session and other issues related to it.

The requirements for information exchange are ownership information, billing method, battery-related parameters, energy price, charging time and schedules, authentication-related certificates, agreements between the two parties, start and end of charging time, etc. After the required tasks are identified, which the architecture should satisfy, the various applicable protocols can be determined. Various types of messaging preferences, architectures and messaging protocols can be defined based on the various aspects of requirements. And there is no one single fixed technique or type for messaging requirements.

One of the examples of messaging types and groups could be as shown in Table 10.5. This example considers DER administration, grid requirements, prices, DER operations, groups, PEV messages, reporting and monitoring, various other transactions, etc.

Various technologies have different strengths and options for messaging and communication with the other stakeholders. The table gives summary of how the messaging is applied under different groups (also called as use cases) with respect to various factors as mentioned. Considering the cases as V1G-residential, V1G-workspace and V2G (AC or DC) as shown in the Table 10.6, it can be observed

Table 10.5 DER and DR message groups for V1G and V2G applications

DER Administration	Grid Requirements	Prices/Events
• Enrolment/registration • Asset owners/utility programs • Individual DER device knowledge	• Volt/freq support • Emergency dispatch • Notifications/alarms • 61850-7-420	• Price signals: Time of use, CCP • RT pricing CPR • Events/schedules
DER operations (real-time operations) • DER settings/schedules for automatic responses • Emergency dispatch • Notifications/alarms • 61850-7-420 information mode	**Reporting/Monitoring** • DER information/status • Configuration • Metering/performance • Telemetry	**PEV-Specific Messages** • PEV SOC/status • Start/end times • Energy required • Ramping/charge rate • Restart
Targeting/Groupings • Group Assignments • Aggregators	**Built-in Cyber Security**	**Transactions** • Bids/negations • Forecasting • Settlements

that there are certain similarity and also some differences for the messaging requirements by these cases. We can observe that the V2G technology does not involve price signals but V1G has to utilize that with incentives upon participation.

There are some conceptual groups which can be used relating to prices and certain events under DER managements, but this selection depends on the rule under the selected model of DER management. In Table 10.5, the transaction group is included. This group is considered whenever pricing and related negotiations are involved between the EV owners and the grid.

5.3 Selection Guidelines for Messaging Protocol

The suitable protocol chosen for EV communications can be based on the capacities of the messaging protocols and other important factors and implications of a selected protocol. The existing ecosystem should be first understood with respect to the possibilities for improvement, initiatives that can be taken and the possible time for its implementation for usage.

5.4 Other Factors in Protocol Selection

Although messaging protocols must technically support messages for specific use cases, this is not enough. The availability of the product and the authentication should be checked [8].

Table 10.6 Messaging requirements for some use cases

Messaging types	V1G Residential	V1G workspace	V2G AC or DC
DER Administration	✓	✓	✓
DER operations	-	-	✓
Target groups	✓	✓	✓
Reporting & monitoring	✓	✓	✓
PEV specific messaging	✓	✓	✓
Prices	✓	✓	-
Transactions	✓	-	-
Cyber security	-	-	✓

The factors to be considered in protocol selection include:

1. **Maturity of protocol ecosystem:** Ecosystem is a collection of equipment, systems, procedures and stakeholders related to the agreement. This protocol may be very suitable for applications; however, mature ecosystems generally have lower costs and faster recovery than less mature ecosystems. Immature ecosystems can guarantee cost, time, and opportunity; however, they may not have been fully field tested.

    (a) **Product availability**: Implementing a protocol means the protocol must be supported by EVs, EVSEs, and associated management systems. If little to no technologies and vendors support the protocol (including devices or control systems), then additional R&D or costs may be required for products to be available in time for a utility program.

    (b) **Conformity assessment tools**: Even though protocols may be robust, there is always room for misinterpretation or vague language. Tools and processes to validate that EVs, EVSEs, and associated management systems can facilitate this.

    (c) **Industry experience**: The more stakeholders have experience with the protocol, the more likely interoperability will occur at interconnection. Maturity in this factor includes individuals with experience across utilities, aggregators, manufacturers, consultants, and other stakeholders.

    (d) **Use-cases understood**: The protocol requirements to implement a specific application (variants of V2G and V1G) vary. Application specific profiles narrow down the complexity of implementation by providing guidance on how to implement programs consistently so that products can be made to support them.

(e) **Certification program(s):** These are critical for ensuring that vendors implement the protocol consistently for the applications certified.

2. **Mandates/adoption**: Protocols are adopted based on natural industry adoption but outside motivators like mandates can nudge the industry in a specific direction.
3. **Cyber security**: Any time connectivity is added to a system or device, the security of the asset and associated data needs to be considered. Different protocols support different cyber-security capabilities. For instance, IEEE 2030.5 and OpenADR both include cyber security in the standard while IEC 61850 and DNP3 rely on separate cyber-security standards.
4. **Internal factors**: Adopting new protocols can be an expensive and time-consuming endeavour. It is natural that EV/EVSE manufacturers, utilities, and management system providers will gravitate to protocols they have the most experience in or already have on their roadmaps.
5. **Applications addressed**: While a protocol may be a good technical and business match for a specific use case application, consideration should be given to other DR/DER use cases that are likely to be implemented. The fewer protocols required addressing the anticipated DER use cases, the less time and expense will be involved in implementing them.

6 Summary

To make electric vehicles as a feasible solution for transportation, smart communication between electric vehicles, charging stations and infrastructure, as well as two-way energy transmission (V2G and Grid for Vehicle, G4V) and end-to-end communication are needed to eliminate the manual billing and make it more effective. Considering all these requirements, safety is of the utmost importance, especially during the charging process, which will go out of control over time. In the field of signalling and voice communication, knowledge, experience and standardisation are required to help in the development. Various concepts and technological understandings such as the electrical characteristics of the battery's state of charge, the expected charging time (idle time) of the battery, user authorisation and prevention of abuse, safe supply of charging stations by different operators in different countries, standardization of connections and interfaces and implementation of new value-added services are to be known.

Electric vehicles have the significant capability to reduce the consumption of imported oil and create many high-paying jobs by creating various businesses. To realize this potential and fully penetrate the consumer market, electric vehicles must undoubtedly be safer, cheaper, and meet the expectations and needs of users. This chapter focuses on light plug-in electric vehicles (PEV) that are charged via electrical connections and supports PEV charging, including battery all-electric vehicles (AEV) and sometimes vehicles (BEV) and hybrid electric vehicles (PHEV) are called battery electric vehicles. An electric vehicle (EREV) that works as an AEV also has the function of extending the vehicle's range beyond the battery (e.g. by using gasoline generators and other functions). A traditional hybrid electric vehicle (HEV) charged by an internal combustion engine is a different type of electric vehicle. Although it is not the subject of this roadmap, it will be highlighted when there are related safety issues and other considerations. In view of the current range limitations of plug-in electric vehicles that only use battery power, it is important to support the charging infrastructure so that the vehicles can be charged at office, at home and in public places. Regardless of the PEV or charging system used, the infrastructure must be reliable and largely compatible. It is also important to establish a strong and comprehensive support service department, including training for emergency personnel, technicians, electricians and inspectors, as well as training for competent authorities, owners and consumers. However, although the times seem particularly promising, electric vehicles face major challenges as they become mainstream. Electric vehicles are not as safe as traditional internal combustion engine vehicles and have unique safety challenges and risks, which need to be understood and considered as part of the vehicle life cycle. Widely and extensively penetrate into the consumer market. Safe charging anytime, anywhere will greatly improve the driver's flexibility and convenience and the function of technological development.

Impact on environment: The requirements from customers and manufacturers for environmentally friendly vehicles which have the highest efficiency and reduce the usage of fossil fuels. The PEV should meet standards, codes and regulations along with conformance and educational programs to take the advantages of PEV for development of home PEV and enterprise in infrastructure for a cleaner environment.

7 Future Challenges and Implementations

The protocols and the standards are framed so as to have a common platform for successful communication between the grid and the vehicle. Hence, its implementation for synchronised operation is very crucial. Constant innovations are being identified in this field and are successfully being standardised and implemented. It is important to identify the challenges in this regard so that future planning can be taken up by the researchers to propose innovative ideas in manufacturing of vehicles and also framing protocols to have reliable V2G technology and communication.

Technology challenges: The disadvantages of V2G communication are having major drawbacks such as technological problems related to communication and monitoring and also improper communication, reduced battery life [11].

The increases in the cost of V2G services of PEVs are due to a number of parameters such as power electronics converters, communication and control. And others pointed out that the commercialisation of V2G may depend on technological improvements in dispatching, modelling, and freight communications.

Many other engineering studies have confirmed that there are considerable technical obstacles due to design considerations and patents required for enhanced communication, control and coordination systems. Some of the research works at some Universities show that they can be solved.

The effect of PEV demands on the mid-voltage distribution network is not clear, and there are many bottlenecks and real risks of lowering charges, especially low-voltage transformers and line interruptions.

Adding to this, the spread of PHEV will have a 'strong impact' on many distribution networks. Second, the provision of V2G services will inevitably shorten battery life; the question is how much is the relationship between battery consumption when driving alone. The only quantitative answer published concluded that the provision of V2G will require continuous battery replacements throughout vehicle service.

Using accelerated aging tests to model 100 BEVs with two different configurations of lithium-ion batteries, it was found that the battery performance would change significantly due to battery chemistry, weather and weather conditions, temperature and processing methods. In some cases, it is expected to exceed. But they will not meet each other. Also larger the battery, smaller the marginal advantage of V2G vehicles. Diesel vehicles are more suitable for large vehicles such as vans (based on cost) evaluated based on types of PEV and ranges. There is a need for higher range and maximum battery power level.

As per the available data, the battery can be used upto 70% to 80% of the capacity after that it cannot be used in PEV and that leads to shorter life of the batteries which in turn increase the capital cost [11].

Finance: The first cost barrier: The financial outlook of the V2G system is not absolute, and is still limited by the first cost barrier: V2G-enabled PEV may be more expensive than traditional PEV, and traditional PEV is already more expensive than traditional alternatives.

Revenue/share: Compared with the predictions of the rational participant model, the savings in fuel or electricity costs are greatly underestimated. In fact, a survey found that no one estimated the current value of fuel economy when making a new car purchase decision. Another study of vehicle owners found that none of the respondents consistently analysed vehicle fuel costs. Few people track gasoline expenditures over time, and few see transportation fuel expenditures. . The International Energy Agency (IEA) found that consumers who are concerned about fuel economy when buying a vehicle expect to spend the first 3 years or less improving vehicle efficiency within a period of time. Consistent with consumers, they will consider only the present cost or investments on vehicle and fuel. Also revenue can be generated by establishing V2G infrastructure.

Social Ecology: Negative externalities is another category of issues belonging to the category of social environment, including negative externalities related to the V2G system, especially issues related to the wider use of PEV. For example, switching from internal combustion engines to electricity may increase power consumption, which may have a negative impact on water distribution, required for power generation using fossil fuels and nuclear power. Thermal power plants utilise huge amounts of water for producing steam and cooling purposes. This leads to more water storage which increases the complications in transportation of electricity.

Also increase in the BEV production and usage leads to problems of battery disposal as it may cause harm to the environment and disposal of other components of BEV.

Behaviour: Inconvenience, suspicion, confusion and worry about the ranks.

Important obstacles for establishing a V2G program include error in the program which leads to problems in connecting PEV at any point of time to estimate the range of PEV. That is, the V2G program that sells electricity and the effect on the charging process. The availability of power at the grid during an emergency will affect the consumer mind or lead to usage of ICE then electricity in PHEVs. Most consumers worry about the range of BEV that can travel per charge and high capital cost of batteries. Also there is a need for proper understanding between different service providers and vehicle owners. V2G communication programs should assure the customers about their secure customers' privacy data. Consumers think about immediate requirements of instant battery charging during long-distance travel and proper communication without any failure in the programme.

The fourth potential obstacle to the use of V2G is consumers' fear of battery degradation. Studies have shown that this problem currently only occurs in PEV, because customers have very minimal technical knowledge about the battery life.

8 Research Opportunities

The views and analyses on technical procedures in V2G will offer complete scope of its utilisation. Additionally, the research gaps should also be identified to enhance the features of the technology and improve upon its regular usage. There are four important fields of research with respect to V2G technology. These topics are discussed in the preceding section. The research opportunities are need for widening of the VGI cases, overcome failure of transformation, practical models of V2G and exploring interdisciplinary methods.

8.1 Need to Widen the VGI

Future research on V2G can be extended to a wider range of case studies, that is, the vehicle layout, users, and system functions that can be transferred to the vehicle to grid integration (VGI) system. As mentioned, the existing literature on V2G tends to focus on automobiles like BEVs but is not focusing on PHEVs, focus on V2G but not on V1G (V1G is energy transfer from grid to EV only). To study and model the various advantages and disadvantages of different vehicles, more extensive comparison work is needed. The type of vehicle (light/medium and/or heavy duty electric vehicles), types of owners (owners of light-duty vehicle and cargos), agreement of ownership (private and car sharing), type of technology (PHEV and BEV and degree of automation), degree of VGI (types of V1G and V2G), and method of V2G participation (time-sharing price, revenue sharing, controlled charging plan or voluntary participation.) Researchers, policy makers, and other stakeholders should prioritise V2G development work to seize opportunities that are more feasible and more likely to bring social or economic benefits in different time frames.

8.2 Overcoming Transformative Failures

The next scope of research is to understand how to accept the high level of transition towards V2G from the existing practices.

Further research can be explored based on the Weber and Roracher framework [12], which is composed of 12 kinds of errors that hinder change and categorized into three groups.

First, the market failures which consider knowledge diffusion, effects, and short-sighted investors should be considered which leads to misutilization of the various resources leading to wastage in the technology and innovations. In short, this leads to under investment in V2G innovation, which naturally takes a longer time to mature and generate revenue.

Second, the system has structural flaws, including the lack of infrastructure; hence, the companies, research institutions, and service providers need to support the large-scale transition to V2G. The innovative research may aim for improvising the existing systems, understand and evaluate the effectiveness of new methods to overcome the shortcomings. For example, a study has been conducted on the efforts of the California independent system operator in the interest of a wide range of public and private stakeholders to establish a V2G roadmap, including V2G case definitions. Elaborate case study can be done to understand the efforts in incorporating guidelines and codes into all institutions [13].

Third, the deficiencies of the transformation system due to the lack of a common vision among key stakeholders or 'the focus is wrong'. Different stakeholders like utilities and vehicle manufacturers may have diversified views on VGI and different views on the possibility of V2G success in the future. For example: the different design standards of EVSE [charger] may result in restricted access to V2G services [13]. Such gaps in views and other obstacles are to be minimised by following a common policy, including Weber and Roracher's concept of policy coordination [12].

To reduce greenhouse gas emissions, a high-carbon tax can stimulate the ethical transportation practices and also power sectors can be encouraged to innovate in a low-carbon direction, which may include the development of V2G. Studies suggest that carbon tax alone may not be enough but an encouragement can be made to incline towards the renewable resources of power. Although these measures can be complementary, they rarely target different sectors. Hence, a detailed analysis on implementation of V2G technology has to be studied which can be successfully implemented and operated. Many discussions are still in the documentation level. A thorough understanding is needed for its practical implementation. So there is an encouragement required to provide incentives to do innovations helping for easy and smooth transition from current practice to V2G technology.

User complexity: There is little research information on the consumer aspect of V2G Most V2G modelling studies generally assume the estimated number of PEVs participating in the V2G program, and PEV is calculated based on a hypothesis. The problem considers only one side constraint such as to optimise network operation or minimise the download cost of individual PEV owners. The study that has been conducted has made assumptions regarding the number of PEVs, charging time, grid conditions, etc. But considering the constraints from the point of view of consumers in the model is important. The consumer perceptions and participations are more critical and challenging to make the system more flexible and collaborative in nature.

Although there are many conceptual studies to learn from references [14, 15], some provide illustrative examples of the concepts of Axsen and Kurani, which were first developed to classify consumers' perceptions of PEV from two dimensions [16].

The first one is functional and symbolic. The PEV technology and V2G can provide both cost benefits and functional profits like cost savings and symbolic benefits, such as informing consumers that they are environmentally friendly or care for the environment. The second consideration is private participation and the societal dimension. We can realise that social benefits are used by the whole society, for example, by reducing greenhouse gas emissions and air pollution. Axsen and Kurani [16], in their work have distinguished two types of societal frames and have presented the functional and symbolic aspects for both. The functional frame gives the impact of the vehicle on the environment, resources usage pattern and energy-related aspects and the symbolic frame relates the ability of the vehicle to encourage others to participate in such activities to impact the society positively. The structure combines the capabilities of tools to incentivise other users, various stakeholders, companies, and governments to participate in activities, which in turn has an impact on society in a broader sense, and supports or amplifies existing negative impacts (such as low-carbon fuel) [5, 17].

Due to the complex dynamics, passenger cars can be seen as an integrated commodity with public and private dimensions, specifically secondary/alternative fuel vehicles and transportation modes where pollution causing impact to nature is usually the main driving factor for development [18]. This concept is a convenient method for collecting a wide range of consumer opinions on V2G, but this will not assume that all PEV owners optimise their behaviour based only on functional personal motivations (such as cost savings) [19].

8.3 Exploring Interdisciplinary Methods:

Cross-disciplinary modelling methods are to be implemented and a model to be developed considering technical, social, economic and financial dimensions. The final loophole and potential priority of future V2G research are to develop in the direction of interdisciplinary and multidisciplinary efforts. Technology, finance, social ecology and personal/behaviour related are the four important categories between which relations can be drawn. The widely considered connection is between technical- and finance-related aspects or between V2G technical and economic evaluations. Few studies include both complex behavioural models and technical, economic or ecological models. As mentioned above, V2G modelling research usually relies on some type of modelling and discipline and makes assumptions on consumers such as vehicle usage rates, participation pattern and PEV adoption rates of PEV owners. Often, there is little or no recorded evidence to validate the assumptions. In some cases, greater integration (and better understanding) can be achieved by using more than one method. For example, although optimization models dominate V2G modelling, energy economic 'simulation' models can be implemented to represent the task of stakeholders and the customers in a given political context, while taking into account their preferences and views [20]. One innovation is to directly combine the empirical results of surveys and interviews with the V2G participation model to model the technical, economic and ecological impact of these systems. The study depends on endogenous and consumer-informed views on PEV purchasing behaviour and V2G share and optimisation models representing the power grid. More comprehensive research study is needed that also adds an overall institutional component to describe the transformative system-level interruption that is an obstacle for the transition to V2G.

8.4 Conclusions

In short, moving to V2G can bring many benefits to the community. Convincingly transform the vehicle from the centre of the traffic problem to part of the solution. This transformation has enabled the EVs to increase the efficiency of grid along with increasing the profits to the power company. With PHEVs, there is a reduction in greenhouse gas emissions, adaptation to low emissions of CO_2 energy and it is cost-effective for owners, drivers and other users. However this transition is quite difficult as it must face many obstacles related to technical aspects such as vehicle sub-systems, batteries, configurations, and communication systems, financial aspects such as purchase prices and the negative environmental impact of initial costs. Behavioural problems, including discomfort, self-confidence, confusion and fear of hierarchy. In addition, the net impact of a V2G system may depend on which goals are prioritised; for example, a V2G system that cannot guarantee minimisation of costs will reduce environmental impact, especially when politics has not yet considered negative external environmental impacts. Therefore, when we consider the future prospects of converting to V2G, we must not only consider vehicle structure/configuration, batteries, vehicles, and power plants but also the entire social technology system. We need to expand the research agenda of V2G to examine more cases, overcome transformative errors, understand user complexity and apply interdisciplinary and hybrid methods. If we accept the choice of cars for reasons other than 'rational' or 'technical', then the direction of transportation research and development aimed at promoting new modes of transportation must undergo tremendous changes. Although billions of dollars have been invested in R&D, procurement, tax relief, grants, regulations and grants, there are many obstacles and more sustainable modes of transportation. As long as the barriers are addressed in such a way that researchers, scientists and engineers overcome the challenges, the prospects for new transportation or energy systems like the extensive and socially beneficial V2G program will continue to exist. The stakeholders' involvements, their values, expectations and roles are all very essential parameters to obtain an improved technology for determining the V2G services.

References

1. S. Vadi, R. Bayindir, A. M. Colak, and E. Hossain, "A Review on Communication Standards and Charging Topologies of V2G and V2H Operation Strategies," 2019.
2. 650 series IEC 61850 Communication Protocol Manual. (2011).
3. "Open charge point protocol 1. 6 Table of Contents." 2015-10-08.
4. *Open charge point interface*. Retrieved from https://github.com/ocpi & https://ocpi-protocol.org.
5. Axsen, J., Orlebar, C., & Skippon, S. (2013). Social influence and consumer preference formation for proenvironmental technology: the case of a U.K. workplace electric-vehicle study. *Ecological Economics, 95*, 96–107.
6. OpenADR Alliance OpenADR 2.0 Profile Specification B Profile. (2013).
7. *Open ADR*. Retrieved from https://solidstudio.io/blog/ev-roaming-with-open-standards-oicp-emip.
8. Mater, J., Myers, E., Ealey, B., & Weyl, V. (2020). Communication protocols for managed EV charging and V2G applications. *EVS33 international electric vehicle symposium* (pp. 1–12).
9. *Distributed energy resources communication workshop pre-reading: Use cases and requirements*, Draft Report, April 14, 2017, Joint OpenADR/EPRI Work Group.
10. *Working DRAFT: Distributed energy resources communication workshop prereading: Requirements and protocols,* Draft Report, April 14, 2017, OpenADR and EPRI.
11. Sovacool, B. K., Axsen, J., & Kempton, W. (2017). The future promise of vehicle-to-grid (V2G) integration: A sociotechnical review and research agenda. *Annual Review of Environment and Resources, 42*, 377–406.
12. Weber, K. M., & Rohracher, H. (2012). Legitimizing research, technology and innovation policies for transformative change: combining insights from innovation systems and multi-level perspective in a comprehensive "failures" framework. *Research Policy, 41*(6), 1037–1047.
13. California Independent System Operator. (2014). *California vehicle-grid integration (VGI) roadmap: Enabling*.
14. *Vehicle-based grid services*. Sacramento, CA: CAISO.

15. Jackson, T. 2005. Motivating sustainable consumption: a review of evidence on consumer behavior and behavioural change. *A report to the Sustainable Development Research Network*, University of Surrey, Guildford, UK.
16. Axsen, J., & Kurani, K. S. (2012). Interpersonal influence within car buyers' social networks: applying five perspectives to plug-in hybrid vehicle drivers. *Environmental Planning A, 44*(5), 1057–1065.
17. Noppers, E. H., Keizer, K., Bolderdijk, J. W., & Steg, L. (2014). The adoption of sustainable innovations: driven by symbolic and environmental motives. *Global Environmental Change, 25*, 52–62. https://solidstudio.io/blog/ev-roaming-with-open-standards- oicp-emip.
18. Brown, M. B. (2001). The civic shaping of technology: California's electric vehicle program. *Science Technolody Human Values, 26*(1), 56–81.
19. Delucchi, M. A., Yang, C., Burke, A. F., Ogden, J. M., Kurani, K., Kessler, J., & Sperling, D. (2014). An assessment of electric vehicles: technology, infrastructure requirements, greenhouse-gas emissions, petroleum use, material use, lifetime cost, consumer acceptance and policy initiatives. *Philosophical Transactions of the Royal Society A, 372*, 37220120325201 20325.
20. Jaccard, M., Loulou, R., Kanudia, A., Nyboer, J., Bailie, A., & Labriet, M. (2003). Methodological contrasts in costing greenhouse gas abatement policies: optimization and simulation modeling of micro-economic effects in Canada. *European Journal of Operational Research, 145*(1), 148–164.

Chapter 11
Smart Charging in Electric Vehicles and Its Impact on the Evolution of Travelling

D. Ruth Anita Shirley, B. Siva Sankari, Rajakumar S. Rai, D. A. Janeera, and P. Anantha Christu Raj

1 Introduction

The automotive industry is undergoing a major transformation as the growing need for a greener environment and better security has become the norm with which the automobiles are built. Technological advancements are progressing at a rapid speed, evolving the way transportation takes place. There is much need for the Indian auto industry to emerge as one among the Top 3 automobile manufacturer in the upcoming decade, based on the Automotive Missile Plan 2016–2026. To meet the ever-increasing consumer expectation and need for digitalization, automotive business has a lot of criteria to fulfill and the introduction of smart charging paves the way to quickening electric mobility. The Indian market takes the globalization of automotive industry to be one of the best opportunities to participate and excel. In the current scenario, decarbonization has emerged as one of the prime aspects that has triggered the revolution of automobile industry. All the countries have changed a number of long-term policies to meet the ecological goals of decarbonization [1]. The introduction of e-mobility has gone a long way in ensuring decarbonization.

D. R. A. Shirley (✉) · D. A. Janeera
Department of Electronics and Communication Engineering, Sri Krishna College of Engineering and Technology, Coimbatore, Tamil Nadu, India

B. S. Sankari
Department of Electronics and Communication Engineering, SNS College of Technology, Coimbatore, Tamil Nadu, India

R. S. Rai
Department of Mechanical Engineering, Karunya Institute of Technology and Sciences, Coimbatore, Tamil Nadu, India

P. A. C. Raj
Robotics Engineering, Karunya Institute of Technology and Sciences, Coimbatore, Tamil Nadu, India

M. Kathiresh et al. (eds.), *E-Mobility*, EAI/Springer Innovations in Communication and Computing, https://doi.org/10.1007/978-3-030-85424-9_11

Moreover, the electrification policies in the automobile industry will bridge the gap between the use of conventional vehicles and technologically enhanced vehicles, paving way to electric charging of vehicles, resulting in creating awareness among the public for the evolution of electric vehicle technology. The policy changes implemented are framed taking into consideration the evolutionary nature of e-vehicle technology along with the investment required. Norway, Netherlands, China, and United States have grown leaps and bounds in their technological advancement in electrification of automobiles, resulting in a strong momentum of manufacturing ecosystem, charging infrastructure [2]. Though India has stepped into the path of electrification at a later period of time, a concerted policy will go a long way in aiding directing India on the electrification path.

The National Electric Mobility Mission Plan 2020 was launched by the Indian Government in the year 2013. Based on this plan, FAME India Scheme was launched and on completion, it was further extended on 1st of April, 2019 for a period of 3 years. A total of 529 crores was utilized during Phase-I and an outlay of Rs. 10,000 crores has been outlaid. Based on the progress of electric vehicle technology and its advancement in the market, there is a high demand to reduce the amount of energy consumed while simultaneously curbing decarbonization caused from vehicles. In 2017, NITI Aayog's transformative mobility report laid out a clear-cut solution to implement pure electric vehicles using shared connected electric mobility. The report showed that using this as the key, it is possible to incorporate it in 40% private vehicles and 100% public vehicles which will guarantee that all vehicles run on electricity by 2030 [3]. The introduction of electric vehicles creates a huge shift in paradigm in both power and transport sector with the ability to decrease carbonization of the environment by coupling the two sectors. Though in the current scenario, renewable energy is being used at a lower share, the transportation sector is transforming and the number of EVs on the road is increasing.

- A survey report shows that there are about 5.8 million EVs in Germany at the start of 2019.
- There are 1.1 million electric vehicles in the United States and
- China accounts for 2.5 million electric vehicles.

By 2040, if more number of vehicles sold were electric, it is possible that over a billion electric vehicles will be crowding the streets by 2050 [4] as shown in Fig. 11.1. The 9 TWh of stationary batteries will eventually be replaced by 14 TWh EV batteries by 2050.

The use of electricity to power vehicle serves as an attractive means of travel considering the cost reduction in renewable power generation in the recent years. This has led to an increase in people buying electric vehicles, paving way to scaling of power sector. The electric vehicles can be used as a means to store vast amount of electricity. They can be used as a decentralized storage resource and flexible load which can provide flexibility to aid power system operations. Using smart charging, it is possible to enable the vehicles to make their pattern of charging flexible in order to make sure that there is no peak demand, ensure grid balancing by changing

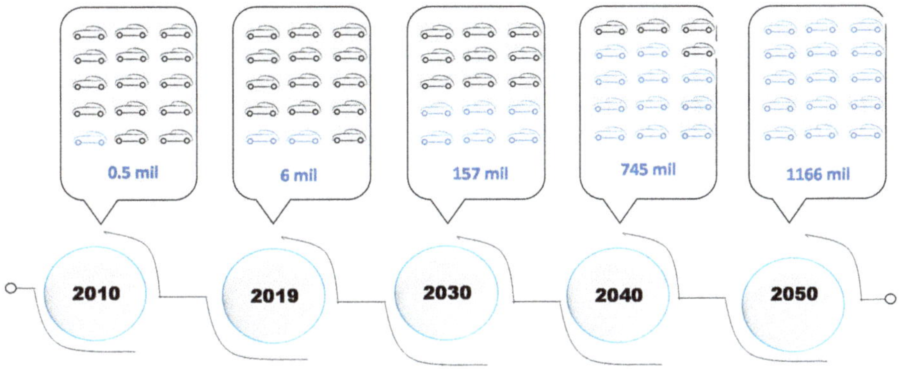

Fig. 11.1 Electric vehicles prediction chart 2050

charging level as well as to fill load valleys [5]. This methodology of using smart charging for electric vehicles will decrease the requirement of investment in fossil fuel power plants which offered flexibility, but have a higher intensity of carbon level.

When smart charging is used, it will serve as a means to adapt the charging cycle of the vehicles such as with respect to the user requirement and power system consideration, ensuring collaboration of the electric vehicles while taking care of mobility needs at the same time.

Thus based on customer requirements, availability of local renewable resources, and distribution grid constraints, it is possible to optimize the charging process, using smart charging. It gives the administration a level of control on setting the default constraints in charging such as technical charging options and pricing.

There are ways of motivating the consumers to change their timing of charging such that they will be unable to charge simultaneously during peak periods. In general, a more advanced approach of smart charging is required to provide a solid solution that will have better levels of penetration level to provide a good balance in real time as shown in Fig. 11.2.

There are many such types of charging available which include V2H, V2B, V1G, and V2G (as shown in Fig. 11.3).

- V1G mechanism—This uses unidirectional control of vehicles that gives possibility to decrease and increase the charging rate.
- V2G mechanism—This indicates bidirectional vehicle to grid, which when in discharge mode gives the electric vehicle the ability to service the grid.
- V2B and V2H mechanism indicates vehicle to building and vehicle to home, respectively, which could be used as backup power that can be used.

This book chapter is organized as shown below:

- Literature survey: This section describes the previously existing literature on smart charging and smart grid to charge electric vehicles in an optimum manner without overloading the grid.

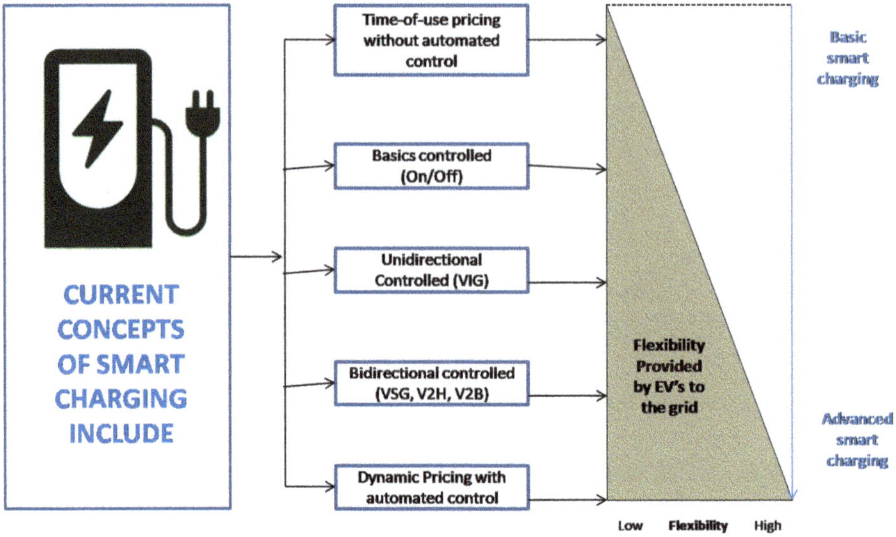

Fig. 11.2 Flexibility through smart charging

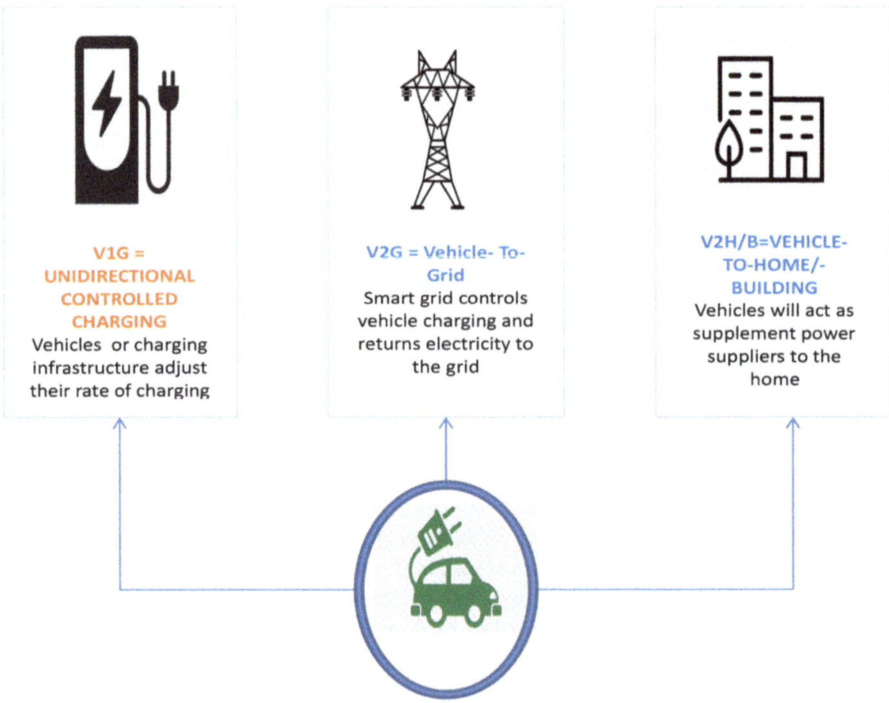

Fig. 11.3 Types of smart charging

- Smart grid: A brief outline on the working of smart grid and its introduction in electric vehicle is described.
- Proposed methodology: A hybrid methodology combining the genetic algorithm and intelligent scatter search algorithm is proposed to make it flexible for smart charging in smart grid system and its constraints are laid out.
- Results and discussion: The analysis of results obtained is recorded and compared with other approaches, indicating the optimal use of smart grid for smart charging purpose.
- Conclusion and future scope: A conclusion on the optimal methodology to be followed for smart grid technology with respect to peak overload is determined and future enhancements and suggestions are made.

2 Literature Survey

Ever since the concept of electricity was introduced into automobiles, electric vehicles with a battery and an electric motor have been around. These vehicles are relatively quieter and cleaner, have lower cost of maintenance, lesser number of moving parts, and are also higher energy efficient, when compared with the other vehicles that are powered by gasoline. Since their introduction, some of the disadvantages of the initial phases of electric automobiles like high cost consideration, slow rate of charging, and charge range limitation. But due to advancement in technology and availability of electric charging station, the performance of the electric vehicles as well as their accessibility have improved dramatically over the course of the time. According to Williams et al. (2015) [6], Ryan and Lavin (2015) [7], Donohoo-Vallett et al. (2016) [8], and Baumhefner et al. (2016) [9], the development of electric vehicles like buses, bikes, and scooters will result in reducing the emission of harmful global warming. These vehicles will use batteries that can be charged in a clean manner and direct emission will be absent. According to Reichmuth, Anair, and Nealer (2015) [10], the use of electric vehicle is much cleaner when compared to even the best conventional vehicle as well as other internal combustion engines in most of the nation.

If all the vehicles used in India were electric, it might increase the demand for electricity by many Kilowatt-hours, every year. Constant travel demands are assumed to perform calculation; however, when dealing with autonomous vehicles, there will be a rise in the vehicle miles. This will lead to decreasing the travel cost as the transit time can be spent on other essentials like entertainment or work. According to Stephens et al. [11], there will be more than 6 trillion vehicles racing the streets by 2020. Even though not all of them are electric vehicles, there is a significant improvement in the number during the past decade. This in turn is affecting the demand and supply of electricity thereby affecting the nation's power grid. In general, power grids operate such that when the use of electronic supply is required, electricity demand is created and it needs to be met by supply

[12]. However, the development of electricity generators involved higher fuel cost and lower capital cost or lower fuel cost and higher capital cost.

Hence, a system that is comprised of many operation layers is required in order to supply electricity:

- Peaking units can be used for a couple of hours in a year.
- Load following is used to generate ramp down and up resulting in a flexible system to be equivalent to the hourly demand of electricity.
- Base load plants with low-operating cost and high capital.

The future of grid lies in the development of solar photovoltaic system. In fact, according to Liebreich in reference [13], within the period of 1975–2015, the cost of PV panels has fallen dramatically (up to 99%). The reason for this is its implantation in advanced niches such as nanoscale electronics and satellites along with it being deployed in adopters of grid-tied systems and off-grid power as stated in reference [14] by Geels. In fact, adopters have contributed greatly to reducing harmful gas missions in the previous years. National and state also supported the deployment of adopters during the 1990s and 2000s, paving way to solar panels installing even at times when they were not the optimal choice of electricity with respect to cost (Leon 2013) [15] (Kimura and Suzuki 2006) [16]. Due to early adopters and introduction of policies, the market of solar panel grew at a very fast pace such that their cost price will be in par with that of fossil regions (Lazard 2016) [17].

There is a solid overlap between photovoltaic systems and electric vehicles. In fact, it is possible for owners of electric vehicles to install that of photovoltaic system and vice versa in order to lay a constraint on the emissions further. Using grid-scaling, it is possible to eliminate services like demand-response aggregators. In general, all EVs should take into consideration both discharging as well as charging power rate limit, battery capacity, final energy demand, customer traveling habits, and initial state of charge (SOC) [18]. However, there are certain limitations on the power system which include frequency, voltage, transformer capacities, network structure, generation units, etc. Power grids are built with complex volatilities and complexities making them more practical and applicable to real-time charging schemes. To address EV scheduling issues, a number of algorithms that use various solving techniques have been developed. In general, similar to the way power supply will set the upper and lower limit on charging and discharging of EVs, all other functions will also be linear in nature. On the other hand, load smoothing and cost minimization functions can be solved as convex function.

Thus, the EV scheduling issue can be addressed using CPLEX [19], GAMS [20], and CVX [21]. Similarly, other alternative methodologies have also been introduced such as tree-based dynamic programming and multipliers direction method in order to address other issues of EVs. This approach will be useful in solving problems with multi-objectives and multi-constraints. However, curse of dimensionality might make these methodologies to have some hiccups during operation [22], resulting in it affecting the EV charging at a large scale. A number of heuristic algorithms like genetic algorithm (GA) [23, 24], chemical reaction optimization (CRO) algorithm [22], and particle swarm optimization (PSO) algorithm [25] are also used in this work to compare and contrast the results of the hybrid model. The outputs of these algorithms indicate that they can provide better results in terms of computational overhead. Although a number of algorithms have been incorporated to schedule charging of EV, there is a limitation that shows that these approaches that only focus on the charging direction can monitor the charging status [25, 26] to manage the charging pattern of the EV or will work based on the arbitrary value set for the EV charging system.

3 Smart Grid for Smart Charging

3.1 Grid Constraint

Fluctuation in power occurs when there are multiple electric vehicles being charged through the day at a fixed location. This will cause the batteries to be filled up till charging stops gradually. When the demand for charging of electric vehicles surpasses the maximum capacity of charge in a location, it might result in power failure. Hence, there is need for adjusting the charge speed as well as moment taking into consideration electricity demand of the different charge points as well as that of the building [27]. A careful observation of the habitat of people as well as the environment leads us to the conclusion that using natural division over time, it is possible to meet this demand. People residing at home, returning from work, and leaving to work, at different time is taken into account and the results show that the demand for charging affects the total power only at a few instances of time. The rest of the time, it is possible to charge the electric vehicles at full capacity. Thus, smart charging paves way to a quicker and more efficient manner of charging when compared with the conventional alternatives [28] that are not smart and will always require us to restrain all charging. The charger capacity is depicted in the figure below, Fig. 11.4.

In the current scenario, taking grid constraints into consideration, smart charging is the safest bet to meet the business demand for charging of electric vehicles. Moreover, care should be taken to determine that there is no overload during the charging of electric vehicles which might also affect the overall output of the circuitry.

3.2 Charging Network Scaling

In today's scenario, a simple data cable serves as a means of connection for multiple charging technologies in order to build a local network. The data cable will belong to a particular manufacturing company and communication is established based on the protocol of the company. However, this will result in a number of drawbacks [29]. Since it is not possible for the system to communicate beyond its predetermined local network, it will result in increased energy prices as well as user requirements.

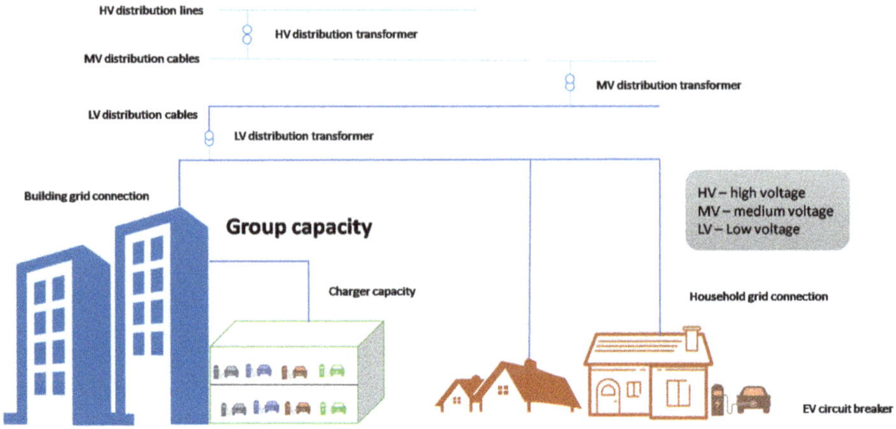

Fig. 11.4 Charger capacity and voltage considerations

Moreover, there is a limitation on the radius within which the chargers should be located along with the total number of stations to which it can be confined to. Moreover, since communication is not dependent on the international and open channel, vehicle is locked-in to a particular charging station's manufacturer and will not be compatible with the stations of other manufacturers. Due to this factor, there is also possibility for the charging aspect to become slightly more expensive or there are no charging station manufacturers, as time goes by. As long as the server capacity is available, smart charging can be scaled infinitely with help from a cloud-based system that controls unlimited chargers. Taking the user inputs into consideration, this will be feasible with any charging station. Moreover, since it is a part of ecosystem, it is also possible to modify the algorithm and interface it with the other operating systems.

3.3 Local Renewable Energy

The choice of using the local renewable energy in the place of the energy from electricity grid may or may not have an impact on the financial gain based on the specific country's legislation. In countries like the Netherlands, because of net metering, local renewable energy is not optimally used by EV holders. Most people might not be equipped with net metering. However, the cost of getting energy from the grid as well as feeding energy back to it will be the same and hence will not aid the utilization of local energy. Similarly, in Germany, there is a constraint placed which ensures that not more than 80% of peak power is fed back into the grid, taking into account the energy from the solar panels on the roof. As a result, the remaining 20% of energy is optimally used during sunny days. Apart from these examples, there are many different ways of charging the vehicles in various countries throughout the globe. However, when local renewable energy is used, people get to produce as well as use the energy that they produce, giving them positive "feel good" attitude toward it [30].

3.4 Smart Charging

An electric vehicle is "smart charging" [31] when the charging system interacts with your vehicle, the company that provides the charge as well as the charging operator by means of data connection. This means that when the electric vehicle is connected or plugged to the charger, it will transfer the data gathered from the vehicle in order to optimize the charging aspect. This will give the charging operator the ability to monitor as well as control the amount of energy that the EV connected requires. The charging operator may include an operator with multiple charging stations or an individual using the charging system at home. However, the total amount of electricity used varies based on the total number of people who are using it at a particular time period, resulting in exhausting the grid [32]. Moreover, based on the limit on tariff of energy and the local grid capacity, smart charging will also stop charging of the EV when it steps over the maximum energy capacity of the building. It is also possible to lay a constraint on the energy consumed by the utility companies so that the grid does not suffer any kind of overload by using more energy than the amount which is produced. This will result in saving money, time, and energy, thereby protecting the resources of the earth.

3.5 Charging of the Smart Electric Vehicle

The charging station stores and retrieves data from the cloud using an intelligent back-end solution that monitors and brings the data on charging events and devices to the attention of the charging station's administrator. When the charging stations use the cloud, they can be used depending on different signals like local electricity consumption, fickle energy production, total utilization of energy by electrical devices as well as other electric vehicles [33]. A smart EV charging system will be able to build a reliable energy system that works on renewable source of energy. Figure 11.5 gives the layout of a typical intermittent distributed smart charging.

3.6 Process of Smart Charging

To initiate the charging process, the electric car has to be identified at the charging station. On identification, the details about the charging event, charging point as well as the electric vehicle will be backed up to the cloud. Based on this information, the

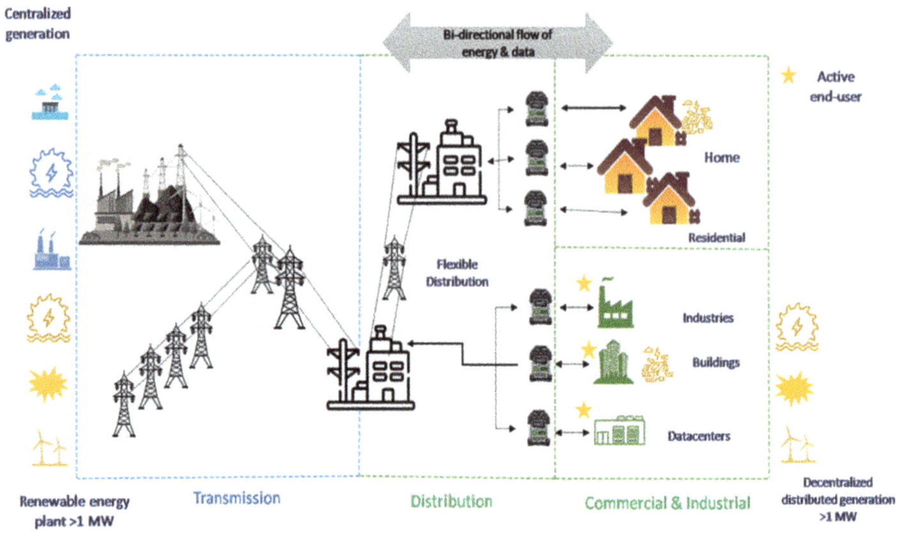

Fig. 11.5 Distributed network for smart charging

customer will be charged the apt price and payment will be transacted from the customer to the charging station owner [34].

The biggest advantage of smart charging is that it takes place in an automatic manner without the need for human interference. To identify the electric vehicle that is to be charged, a mobile application will be used to keep track of the charging process, previous charging records, and balance charging. A simple push of the button in the mobile application or scanning of the RFID will initiate the charging process at any charging station. The roaming networks make smart charging easier, helping coordination between the charging service providers as well as the electric vehicle drivers.

4 Proposed Methodology

One of the prime issues in electric charging is scheduling of the EV load demand. Over the years, the charging schemes developed fall in either one of the following four categories: V2G-G2V charging schemes, smart charging schemes, conventional controlled charging schemes, and dumb charging schemes. The purpose of this chapter is to flatten the load of current in the distribution network by optimizing EV charging. To reach that objective, let us consider that the EVs have a pattern of identifying the vehicles to be charged based on the availability pattern which will aid the smart grid to find the EV charging schedule. Further, technical network constraints should be taken into account such as the voltage violations in the buses and cables' thermal limit [35]. To find an optimal solution, a low-voltage network topology along with a residential network which has round-the-clock demand is used as the system's input.

4.1 *Function of Objective*

The everyday load demand in the residential area is diagnosed at a regular pace to obtain optimal scheduling of the charging which will aid in flattening the loading profile. As the need for electricity from every household increases or when the electric car from a household [36] is connected to the charging point, it will lead to loading at the transformer. To level the load profile of the system and optimize the EV charging schedule, it is necessary to develop a novel methodology which will be able to bridge the gaps of load profiles of the residential area during lower load demands and further prevent peak load EV charging hours in Eq. (11.1). To attain this, we need to provide an objective function which will provide a value of deviation observed in the load profile using the following equation:

$$\min F(x) = \sum_{h=1}^{24} (T_h - T_{h-1})^2 \tag{11.1}$$

where T_h denotes the main transformer loading at "h" hour.

4.2 *Limitations*

There are some limitations set in view of optimizing the solution, in various categories.

4.2.1 Network Limitations

- Ratio of power through the lines of the transformer with respect to that of the substation transformer will account for the thermal limit of the network. This can be represented as follows:

$$T^{\mathrm{h}}_{\mathrm{transformer\ substation}} \leq T_{\mathrm{nominal}} \tag{11.2}$$

$$I^{\mathrm{h}}_{\mathrm{lines}} \leq I_{\max} \tag{11.3}$$

- Buses' voltage level must not exceed the set limits as per grid voltage regulation.

$$V_{i,\min} \leq V_i^h \leq V_{i,\max} \tag{11.4}$$

4.2.2 Charging Limitations

To comply with our power system operation, EV discharging and charging is added to meet the demand in the residential area. This must not increase the load demand of peak transformer substation. Smart charging approaches and conventional controllable charging schemes are added as constant power load while V2G and G2V options will represent the injected energy.

$$\begin{aligned} &\text{Peak transformer substation load demand (h)} \\ &\quad \geq \text{Residential Electricity demand (h)} + \text{power losses} \\ &\quad\; - \text{EVgeneration}_{\text{V2G}}(\text{h}) + \text{EVdemand}_{\text{G2v}}(\text{h}) \end{aligned} \tag{11.5}$$

4.2.3 Power Grid Limitations

Power grid limitations are applied for all the buses as seen in the equations below. Here $BA_{gi}$and BR_{gi} represent the reactive and active power of generations while ΔBR_i and ΔBA_i represent bus reactive and active power mismatch equations

$$BA_{gi} - BA_{li} - BA_i = \Delta BA_i \tag{11.6}$$

$$BR_{gi} - BR_{li} - BR_i = \Delta BR_i \tag{11.7}$$

$$BA_i = V_i \sum_{i=1}^{N} V_k (G_{ik} \cos \theta_{ik} + B_{ik} \sin \theta_{ik}) \tag{11.8}$$

$$BR_i = V_i \sum_{i=1}^{N} V_k (G_{ik} \sin \theta_{ik} - B_{ik} \cos \theta_{ik}) \tag{11.9}$$

The total power demand is determined as the total sum of variable demand of EV charging along with residential demand of power.

4.2.4 Electric Vehicle Demand

Based on the initial SOC of battery, maximum SOC of battery, battery size, and power system to which the EV is connected, the energy demand of the vehicles should be configured. Under the assumption that all energy demands of the vehicles are met within a period of 24 h, the charging process initiates such that:

$$\text{EV demand (topology)} = N * \sum_{i=1}^{\text{number of EV}} \left(SOC_{i,\max} - SOC_{i,\text{initial}}\right) \tag{11.10}$$

Here, the total number of electric vehicles charging is represented by "*N*" which is a constant. Hence, when there are no vehicles that are connected to the charging point, $N = 0$, while when there are multiple vehicles that are charging at the same time, N will be assigned a value 1.

Similarly, there is also a limitation on the SOC of the battery such that it satisfies the following condition:

$$SOC_{i,\min} \leq SOC_i(h) \leq SOC_{i,\max} \tag{11.11}$$

Depth of discharge (DOD) plays a crucial role in maintaining the maximum limit of the battery such that it will not permit a discharge greater than 80% of the rated battery size.

4.2.5 Pattern of Parking

At the customer point of connection (CPOC), their electric vehicles should be charged based on the availability the charging point. A parking pattern is formulated such that an estimate of number of EVs connected is determined showing the hours of connection essential. Based on the EV demand, the charging availability is to be checked such that the following condition is satisfied.

$$\text{EV Charging}_{\text{h}} \leq \text{EV parking availability}_{\text{h}} \tag{11.12}$$

4.3 Hybrid Algorithm Implementation

In recent years, a number of intelligent algorithms have been formulated, focusing on power system optimization issues. Use of genetic algorithm seems to fit the description for an optimized charging scheduling of the electric vehicles. This algorithm was initially developed by Holland and later on used to address optimization issues that were constrained and unconstrained. The natural process of evolution forms the basis for GA. It has its origin in the main driver and operates based on the natural sequence of the issue at the. GA is preferred because of its ability to structure and modify the operation of the algorithm using a set of possible solutions (known as population).

Based on the output of every stage, GA evolves into a new generation of individuals using operations like crossover, mutation, and reproduction within the solutions. GA is preferred over the other algorithms due to the following reasons:

- Based on the offspring and parents' generations, the optimal individual is chosen, thereby paving way to enable it into becoming a global optimum.
- With the use of iterative characteristics and multiple points of population, this algorithm will be able to experiment with many regions, resulting in the conclusive study and identification of important distinctions that set it apart from the other algorithm, wherein only one particular direction is to be pursued.
- As this algorithm runs on coding (also known as chromosomes), it will not require prior information about the system and will be able to adapt based on the right constraints and appropriate objective functions.

The major objective of this work is to decrease the deviation in load profile by optimizing the EV charging time. Here, the demand of electric charge which varies on an hourly basis is the variable factor considered. Hence, this approach will be making use of GA chromosome made up of 24 genes. Real numbers are used to encode the genes according to the hourly total demand represented in Eqs. (11.1)–(11.12). Hourly load demand will involve EV generation using V2G scheme, EV demand due to G2V, and home electricity demand.

$$\begin{aligned} \mathrm{load_{demand}(h)} &= \mathrm{EVdemand_{G2V}(h) - EVgeneration_{V2G}(h)} \\ &\quad + \mathrm{home\ electricity\ demand\ (h)} \end{aligned} \tag{11.13}$$

The GA algorithm is as follows:

- Encoding is the initial step of GA implementation. Boundaries are fixed between the demands of EV charging ahead as well as the actual load demand during the 24hours, in order to determine the apt solution of the population. To improve the convergence rate, a large size of population is chosen. This proves to be crucial determining criteria in the implementation of genetic algorithm, thereby further increasing the computation time. Thus, this will help in formation of the population.
- This is followed by the evaluation process using low grid topology. The limitations of the system defined by Eqs. (11.1)–(11.12) are used to find the feasibility.
- The population evolves based on the scaling and evaluation of the elements using genetic operators in order to transform the population.
- The cross operators are assigned a rate of 50% and a 50% mutation rate is also chosen in order to improve the evolution process. At every generation, two children are sustained to ensure that the optimum solution is kept in track of.
- Thus, the process will continue until the limitation set is attained.

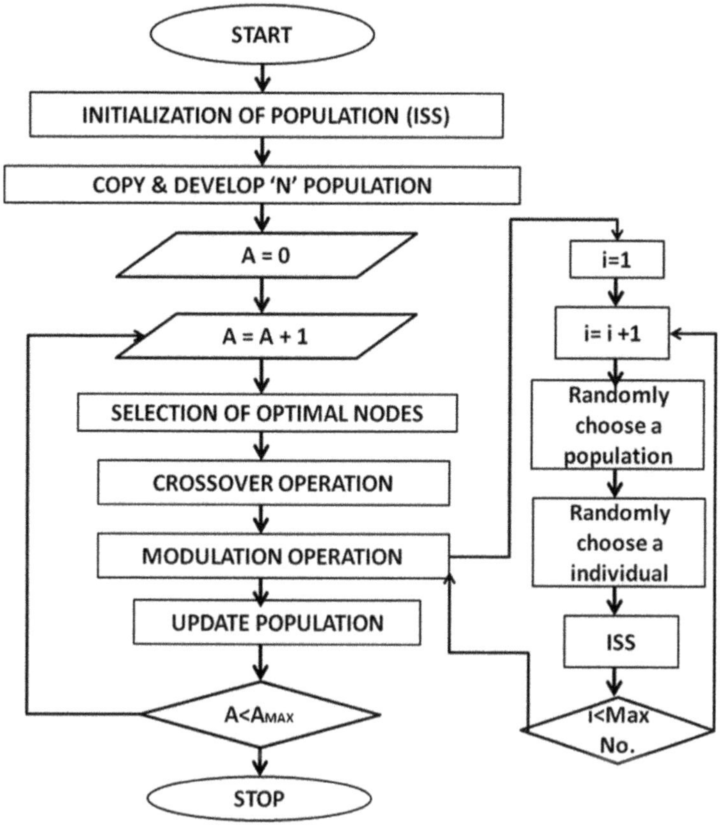

Fig. 11.6 Proposed methodology

- The final output of the GA process will result in the optimal solution to EV charging management.

When there is massive need for charging EV, vehicles will be charged simultaneously and will impose the need for quicker computation. However, there are some limitations when using deterministic techniques that already exist, such as:

- Due to complexity of the issue, the computation time of the algorithm is long.
- The amount of data to be maintained and saved is very vast resulting in the CPU running out of space.

Hence, a new methodology of hybrid model based on intelligent scatter search (ISS) and genetic algorithm (GA) is proposed in this paper (Fig. 11.6.)

.

The ISS methodology is used for its quick convergence and searching capability while the GA algorithm is implemented for its robustness and strong global search ability. Using this hybrid algorithm, it is possible to handle a large fleet of EVs using the crossover and mutation process of GA while a solo EV can be found with the help of ISS algorithm. To quicken the process, ISS will start the population and two children will be optimized inside the loop. Since ISS has the advantage of optimization and speed, new population is refined based on the previous population. Since ISS uses the information of the previous population, it will quicken the process since the previous population would have been optimized already.

5 Results and Discussion

5.1 Single EV Charging

The ISS algorithm is applied for charging single EV and can be used to find the schedule of optimal charging rate and also to ensure that the load of the system does not peak over and operates smoothly. To determine the performance of ISS, the following considerations are made:

- The algorithms PSO and GA are used to check the charging rate modes based on the output. GS method is also used to find the best result as reference as it takes into consideration many solutions.
- However, it is not possible to solve the MINLP problems with GS, hence these two algorithm can be used to determine the charging rate.
- 15 is fixed as the local solvers' iteration number and 25 is set as the population number of the ISS algorithm. Hence, 100 is fixed as the population number for PSO and GA algorithm. Moreover, the iteration is also fixed at 100.

Based on the observation in Table 11.1, it is seen that

- Rather than constant charging rate, a flexible system of charging rate provides better benefits.
- Rather than charging-only cases, the discharging ability provides a better way to flatten the load and decrease the charging costs.

Table 11.1 Mean values for single EV charging in various methods

Objectives	Type	Proposed ISS	PSO	GA	GS
OF1 (linear price)	CD-F	369.0	370	370.5	369.0
	C-F	369.3	370.2	369.7	369.71
OF2 (time of use price.)	CD-F	45,388.33	45,389.15	45,389.66	45,388.30
	C-F	44,280.3	44,281.5	44,282.1	44,280.1
OF3 (Standard Deviation)	CD-F	41,568.50	41,570.26	41,570.77	41,568.46
	C-F	41,574.35	47,574.83	41,575.40	41,574.32

5.2 Group EV Charging

The proposed hybrid algorithm of GA-ISS is applicable when there are many electric vehicles connected to the charging station and will help to decrease the cost of charging and will also smoothen the load profile. Taking into consideration the three objectives, CD-F (charging/discharging using a flexible charging rate) is implemented to observe the output. To determine the advantages of scheduling, the following observations are made:

- A dumb charging methodology is used to create a scenario where all EVs will be charged when they are connected to the charging station.
- A GA-PSO hybrid methodology is selected to compare the output as this can provide a good solution coupled with limited processing time and variation when compared with other similar methods.
- The iteration number and population size for GA-ISS and GA-PSO is fixed at 55 and 25, respectively.
- A CVX toolbox is used to do the EV charging as it will be able to determine the constraints as well as variable and further determine the optimized result.

The Table 11.2 given below represents the results obtained for the 20 simulations. The observed results show that the proposed hybrid GA-ISS methodology outperforms all the other methods, thereby proving its effectiveness.

On successful implementation of the proposed hybrid algorithm, the time taken to charge an electric vehicle based on the available power at the terminal output is shown in Fig. 11.7.

Similarly, based on the power at the terminal output and the charging of the electric vehicle for a period of one hour, it is observed that it can travel maximum kilometers on fast charging as shown in Fig. 11.8.

6 Conclusion and Future Scope

As the need for optimal charging of electric vehicles increases, an appropriate way to charge the vehicles is proposed in this book chapter using a hybrid methodology. The smart grid operates in such a way that a schedule for the electric vehicles to be

Table 11.2 Results of different solving methods

Objectives		GA-ISS	Global control	GA-PSO
OF1 (linear price)	Result	291	290	298.02
	Time	113	93	338
OF2 (time of use price.)	Result	44560	44540	44,615
	Time	106	88	292
OF3 (standard deviation)	Result	41230	41223	41274.5
	Time	117	94.2	360

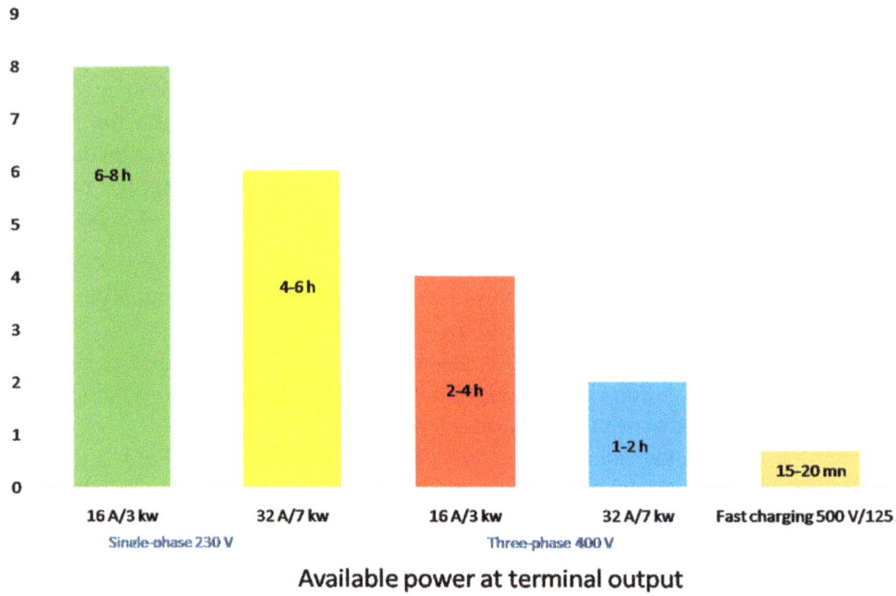

Fig. 11.7 Time taken to charge the electric vehicle

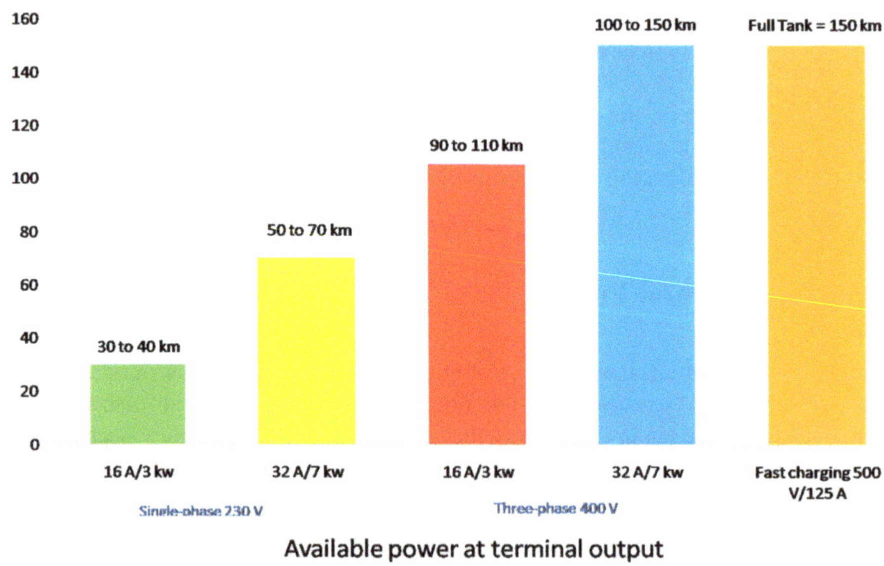

Fig. 11.8 Kilometers that an EV can cover with an hour of charge

charged is framed in order to avoid charge peak overload. This will also decrease the cost price at which electric charge is provided to the electric vehicles. The proposed hybrid methodology observes promising results in terms of accuracy and efficiency, thereby proving to be an optimal technique to implement smart charging by combating peak overload.

The rate at which electric vehicles are sold has increased exponentially during the past decade and it will continue to grow in the future. This will lead to an increasing demand for electricity to charge the vehicles. In view of this, we should ensure that this need for electricity is met using natural sources of energy and efforts should be made to make the use of solar panels more prominent in all nooks and corners of the country. Future work in smart vehicle charging can be incorporated on taking into account a number of aspects for electric vehicles like EV battery degradations, SOC impact on charging rate, charging rates for single EV battery, and so on.

References

1. Rehman, U. U. (2020). A novel decentralized load scheduling algorithm for electrical vehicles in smart grid. *Pakistan Journal of Engineering and Applied Sciences, 27*(2).
2. Li, W., Sun, D., Ou, Q., Yao, X., Gao, S., Wang, Z., & Ma, W. (2020). Integrated human-machine intelligence for EV charging prediction in 5G smart grid. *EURASIP Journal on Wireless Communications and Networking, 139*, 1–15.
3. Sherkat Masoum, A. (2020). *Online coordinated charging of plug-in electric vehicles in smart grid to minimize cost of generating energy and improve voltage profile*. Doctoral dissertation, Curtin University.
4. Einfalt, A., Brunner, H., Prüggler, W., Schultis, D. L., Herbst, D., Beidinger, T., Hauer, D., & Lugmaier, A. (2020, February). Efficient utilisation of existing grid infrastructure empowering smart communities. In *2020 IEEE power & energy society innovative smart grid technologies conference* (ISGT) (pp. 1–5).
5. Williams, J. H., Haley, B., Kahrl, F., Moore, J., Jones, A. D., Torn, M. S., & McJeon, H. (2015). Pathways to deep decarbonization in the United States. The US report of the deep decarbonization pathways project of the sustainable development solutions network and the Institute for Sustainable Development and International Relations. Revision with technical supplement, Nov 16.
6. Ryan, N., & Lavin, L. (2015). Engaging utilities and regulators on transportation electrification. *The Electricity Journal, 28*(4), 78–91.
7. Donohoo-Vallett, P., Gilman, P., Feldman, D., Brodrick, J., Gohlke, D., Gravel, R., Jiron, A., Schutte, C., Satyapal, S., Nguyen, T., Scheihing, P., Marshall, B., & Harman, S. (2016). *Revolution ... now: The future arrives for five clean energy technologies—2016 update. DOE/EE-1478*. United States Department of Energy, Office of Energy Efficiency and Renewable Energy.
8. Baumhefner, M., Hwang, R., & Bull, P. (2016). *Driving out pollution: How utilities can accelerate the market for electric vehicles*. Natural Resources Defense Council. Online at www.nrdc.org/resources/driving-out-pollution-how-utilities-can-accelerate-market-electric-vehicles.
9. Nealer, R., Anair, D., & Reichmuth, D. (2015). *Cleaner cars from cradle to grave: How electric cars beat gasoline cars on lifetime global warming emissions*. Union of Concerned Scientists.
10. Stephens, T. S., Gonder, J., Chen, Y., Lin, Z., Liu, C., & Gohlke, D. (2016). *Estimated bounds and important factors for fuel use and consumer costs of connected and automated vehicles. NREL/TP-5400-67216*. National Renewable Energy Laboratory. https://doi.org/10.2172/1334242
11. Das, S., Acharjee, P., & Bhattacharya, A. 2020. Charging scheduling of electric vehicle incorporating grid-to-vehicle (G2V) and vehicle-to-grid (V2G) technology in smart-grid. In

2020 IEEE International Conference on Power Electronics, Smart Grid and Renewable Energy (PESGRE2020) (pp. 1–6).
12. Liebreich, M., & McCrone, A. (2016). *Electric vehicles—it's not just about the car*. Bloomberg New Energy Finance. Blog, August 22.
13. Geels, F. (2002). Technological transitions as evolutionary reconfiguration processes: a multi-level perspective and a case-study. *Research Policy, 31*(8-9), 1257–1274. https://doi.org/10.1016/S0048-7333(02)00062-8
14. Leon, W. (2013). *The state of state renewable portfolio standards*. Clean Energy States Alliance. Online at http://cesa.org/assets/2013-Files/RPS/State-of-State-RPSs-Report-Final-June-2013.pdf.
15. Kimura, O., & Suziki, T. (2006). 30 years of solar energy development in Japan: Co-evolution process of technology, policies, and the market. *Presented at the Berlin conference on the human dimensions of global environmental change*, Berlin, November 17−18. Retrieved from http://citeseerx.ist.psu.edu/viewdoc/download?doi=10.1.1.454.8221&rep=rep1&type=pdf.
16. Lazard. (2016). *Lazard's levelized cost of energy analysis—Version 10.0*. New York, NY. Retrieved from www.lazard.com/media/438038/levelized-cost-of-energy-v100.pdf.
17. Heredia, W. B., Chaudhari, K., Meintz, A., Jun, M., & Pless, S. (2020). Evaluation of smart charging for electric vehicle-to-building integration: A case study. *Applied Energy, 266*, 114,803.
18. Bourass, A., Cherkaoui, S., & Khoukhi, L. (2017). Secure optimal itinerary planning for electric vehicles in the smart grid. *IEEE Transactions on Industrial Informatics, 13*, 3236–3245.
19. Zhang, X., & Zhong, Q. (2016). Improved adaptive-series-virtual-impedance control incorporating minimum ripple point tracking for load converters in DC systems. *IEEE Transactions on Power Electronics, 31*, 8088–8095.
20. Mao, T., Lau, W. H., Shum, C., Chung, H. S. H., Tsang, K. F., & Tse, N. C. F. (2018). A regulation policy of EV discharging price for demand scheduling. *IEEE Transactions on Power Systems, 33*, 1275–1288.
21. Malikopoulos, A. A. (2014). Supervisory power management control algorithms for hybrid electric vehicles: A survey. *IEEE Transactions on Intelligent Transportation System, 15*, 1869–1885.
22. Kang, Q., Wang, J., Zhou, M., & Ammari, A. C. (2016). Centralized charging strategy and scheduling algorithm for electric vehicles under a battery swapping scenario. *IEEE Transactions on Intelligent Transportation System, 17*, 659–669.
23. Alonso, M., Amaris, H., Germain, J. G., & Galan, J. M. (2014). Optimal charging scheduling of electric vehicles in smart grids by heuristic algorithms. *Energies, 7*, 2449–2475.
24. Soares, J., Morais, H., Sousa, T., Vale, Z., & Faria, P. (2013). Day-ahead resource scheduling including demand response for electric vehicles. *IEEE Transactions on Smart Grid, 4*, 596–605.
25. Sun, B., Huang, Z., Tan, X., & Tsang, D. H. K. (2018). Optimal scheduling for electric vehicle charging with discrete charging levels in distribution grid. *IEEE Transactions on Smart Grid, 9*, 624–634.
26. Antony, M. M., & Whenish, R. (2021). Advanced driver assistance systems (ADAS). In M. Kathiresh & R. Neelaveni (Eds.), *Automotive embedded systems. EAI/Springer innovations in communication and computing*. Springer. https://doi.org/10.1007/978-3-030-59897-6_
27. Marino, C. A., & Marufuzzaman, M. (2020). Unsupervised learning for deploying smart charging public infrastructure for electric vehicles in sprawling cities. *Journal of Cleaner Production*, 121,926.
28. Bilal, M., & Rizwan, M. (2020). Electric vehicles in a smart grid: a comprehensive survey on optimal location of charging station. *IET Smart Grid*.
29. Fachrizal, R., Shepero, M., van der Meer, D., Munkhammar, J., & Widén, J. (2020). Smart charging of electric vehicles considering photovoltaic power production and electricity consumption: a review. *eTransportation, 100,056*.
30. Frendo, O., Graf, J., Gaertner, N., & Stuckenschmidt, H. (2020). Data-driven smart charging for heterogeneous electric vehicle fleets. *Energy and AI*, 100,007.

31. Antony, M. M., & Whenish, R. (2021). Advanced driver assistance systems (ADAS). In M. Kathiresh & R. Neelaveni (Eds.), *Automotive embedded systems. EAI/Springer innovations in communication and computing*. Springer. https://doi.org/10.1007/978-3-030-59897-6_10
32. Dhingra, K., & Singh, M. (2020). Smart charging station to cater the sudden ingress and egress of EVs while supporting the frequency of microgrid through VSG. *Arabian Journal for Science and Engineering*, 1–13.
33. Morstyn, T., Crozier, C., Deakin, M., & McCulloch, M. D. (2020). Conic optimisation for electric vehicle station smart charging with battery voltage constraints. *IEEE Transactions on Transportation Electrification, 6*(2).
34. Rehman, U. U. (2020). A decentralized dynamic marketing-based demand response using electric vehicles in smart grid. *Arabian Journal for Science and Engineering, 45*, 6475–6488.
35. Korotunov, S., Tabunshchyk, G., & Okhmak, V. (2020). Genetic algorithms as an optimization approach for managing electric vehicles charging in the smart grid. *CMIS*, 184–198.
36. Szinai, J. K., Sheppard, C. J., Abhyankar, N., & Gopal, A. R. (2020). Reduced grid operating costs and renewable energy curtailment with electric vehicle charge management. *Energy Policy, 136*, 111,051.

Chapter 12
Artificial Intelligence-Based Energy Management and Real-Time Optimization in Electric and Hybrid Electric Vehicles

D. Pritima, S. Sheeba Rani, P. Rajalakshmy, K. Vinoth Kumar, and Sujatha Krishnamoorthy

1 Introduction

Every household is dependent on automobiles for transportation. With the rapid development in technology, the efficiency and comfort of automobiles have been increasing considerably. The petrol engines used in most vehicles operate on internal combustion engines (ICEs) and offer the required mobility as well as high reliability to the drivers [1]. The evolution of technology has led to the development of electric vehicles (EVs), which are gaining popularity due to the reduced energy consumption and environmental pollution. Extensive attention has been drawn by fuel cell technology in automobiles [2]. The dynamic performance and energy utilization

D. Pritima
Department of MCT, Sri Krishna College of Engineering and Technology, Coimbatore, Tamil Nadu, India
e-mail: pritimad@skcet.ac.in

S. S. Rani (✉)
Department of EEE, Sri Krishna College of Engineering and Technology, Coimbatore, Tamil Nadu, India
e-mail: sheebaranis@skcet.ac.in

P. Rajalakshmy
Robotics Engineering, Karunya Institute of Technology and Sciences, Coimbatore, Tamil Nadu, India
e-mail: rajalakshmy@karunya.edu

K. V. Kumar
Department of EEE, New Horizon College of Engineering, Bengaluru, India
e-mail: dr.vinothkumar@newhorizonindia.edu

S. Krishnamoorthy
Department of Computer Science, Wenzhou Kean University, Wenzhou, Zhejiang, China
e-mail: sujatha@wku.edu.cn

M. Kathiresh et al. (eds.), *E-Mobility*, EAI/Springer Innovations in Communication and Computing, https://doi.org/10.1007/978-3-030-85424-9_12

efficiency can be improved and cost can be reduced in hybrid electric vehicles (HEVs) by using fuel cells for the conversion of fuel into electrical energy.

The Fuel cell hybrid electric vehicle (FCHEV) reduces the waste rate and improves the energy consumption of the system [3]. Future transportation can be potentially based on FCHEV. Cost reduction and performance of power system plays a major role in energy management and devising an optimal control strategy in EVs. Plug-in Hybrid Electric Vehicles (PHEVs) take advantage of the power grid for charging along with the inherent features of the conventional HEVs, thereby providing an all-electric range [4]. The various energy sources like battery or ICE has to be balanced for appropriate allocation of energy in PHEVs using efficient strategies for energy management.

For improving the lifespan of batteries, reducing emission, optimizing the fuel economy, and improving the overall performance of the vehicle, it is essential to design an efficient energy management system (EMS). For developing this EMS, we exploit the features of artificial intelligence. The input and output variables significantly influence the generalization, utilization, network performance, and energy efficiency of the model [5]. Artificial intelligence schemes, optimization methods, analytical algorithms, and rule-based algorithms are proposed for energy management control and optimization in electric vehicles [6].

1.1 Structuring Your Paper

There is rapid development in PHEVs, FCHEVs, proton exchange membrane fuel cell (PEMFC) electric vehicles, HEVs, and pure EVs. There has been a 9% increase in the delivery of plug-in vehicles globally during 2019, compared to the previous year [7]. China, Europe, and the United States have been the largest markets in this development. There has been an accelerating growth of 44% in the European EV market in 2019. With the COVID-19 situation, EVs and smart automobiles are expected to be much researched and sought after in the near future to enable the contactless and safe solutions to the community [8]. However, the closure of industries and quarantine has affected the market due to insufficient supply of parts and lesser transportation. It is expected that by 2030, the EV sale will exceed 20% which is around 3.5 million vehicles annually. EVs are being used for both private and public transportation globally since 2015 [9]. Figure 12.1 provides the statistics of deliveries, sales, and growth of Battery Electric Vehicles (BEVs) and PHEVs globally during 2018 and 2019.

1.2 Features of Electric Vehicles

Methanol, biodiesel and ethanol-based biofuels, biogas, natural gas, hydrogen, and various such energy sources are studied for the development of electric vehicles and

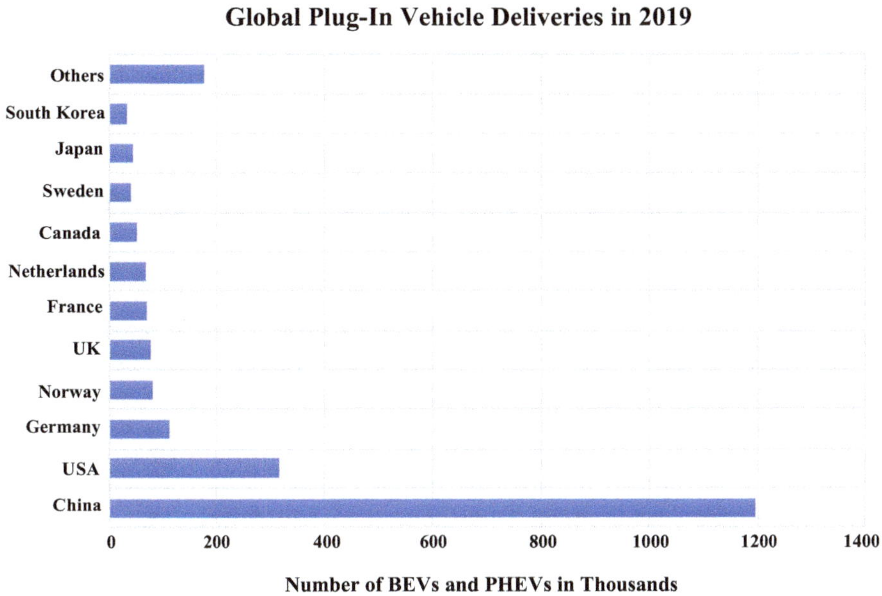

(a)

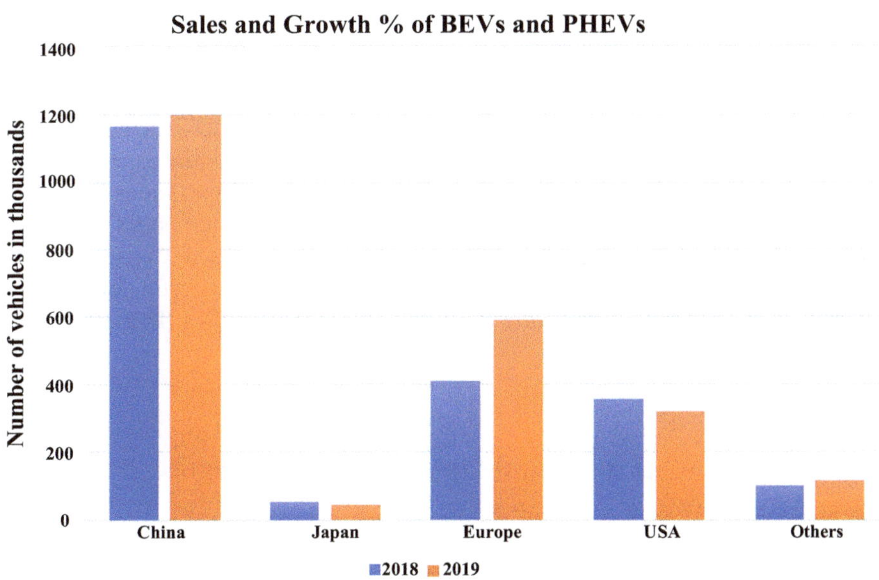

(b)

Fig. 12.1 (**a**) Global BEV and PHEV deliveries in 2019. (**b**) Sales and Growth comparison of BEV and PHEV during the past 2 years

reducing the emission of carbon dioxide as well as for optimizing the energy consumption [10]. Based on the energy source, the EVs are classified as BEV, HEV, PHEV, FCHEV, and so on. Motor-driven BEVs operate without ICE. Around 100kWh of energy can be stored and it can be charged with an external power supply [11]. Safety issues and cost are the major drawback of BEVs despite its huge storage capacity. Motors as well as the traditional ICEs are used in HEVs as power sources. They are also referred to as oil-electric hybrid vehicles. Ethanol and such alternative fuels are used by the modified engines in this kind of vehicle. HEVs are run by ICE in pure electric mode especially during acceleration, deceleration, starting or stopping the vehicle due to the small size of the battery. It is not possible to charge these batteries externally [12]. EVs are less likely to roll over as the center of gravity is lower for these vehicles. The durability and body construction provides safety during the collision and the risk of explosion or fire is also low in EVs.

The BEV and traditional ICE technology are combined together in PHEV. These vehicles offer large energy storage solutions when compared to HEVs [13]. Grid-based stored energy can be used for charging these vehicles thereby enabling operation in pure electric mode. The battery size of PHEVs is smaller than BEV and ICE vehicles. The energy has to be distributed between the battery pack and engine in order to achieve extended battery life and fuel economy. PEMFC is used for powering the fuel cell in EVs [14]. The electrochemical reaction takes place converting fuel into electrical energy without combustion. Oxygen and hydrogen undergo chemical reactions releasing electricity in water discharging heat. This operating methodology has proved PEMFC to be an efficient and cleanest technology for power generation which is best suited for EV [15].

Table 12.1 summarizes the classification of certain electric vehicles based on their features such as cost, security, complexity, eco-friendliness, fuel efficiency, power system, and applications. Electricity is used for propulsion in all the electric vehicle varieties. However, the electric range and operational phenomenon varies [16]. With further advancements in technology, the carbon footprint of automobiles is reduced significantly and the fuel economy and horsepower are increased. Considering the innovation, regulations, and consumer preferences, more powerful and cleaner vehicles are developed. Further, the Range Extender Hybrid EVs (REHEVs) offer a higher battery range [17].

2 Energy Management and Optimization Strategies

PEMFC and FCHEV are the most suitable type of electric vehicles based on theoretical aspects. However, there are certain challenges in the practical application and commercialization of this type of automobile [18]. The maximum efficiency of PEMFC with the largest energy density is less than 60%, whereas an additional power supply is required under low dynamic response conditions to achieve high dynamic response for meeting the requirements of peak power during acceleration or start of the vehicle. A supercapacitor (SC), Fuel Cell, or battery is used as a power

Table 12.1 Analysis of characteristics of Electric Vehicles

EV Type	Power system	Pros	Cons	Applications
BEV	Li-ion battery	Ecofriendly Quick charging	Low battery life, high cost	Tesla Model 3 BMW i3 Chevy Bolt
HEV	Petroleum and battery	Energy recovery Long battery life	Lack of fuel efficiency, the high initial cost	Toyota Prius Hybrid Honda Civic Hybrid Toyota Camry Hybrid
PHEV	Petroleum and battery	Ecofriendly Quick charging Driving comfort	High cost, complex configuration	Chevy Volt BMW i8 Kia Optima
FCHEV	Fuel cell and battery	Zero-emission Noiseless, smooth Energy efficient	Low security High cost	Toyota FCHV Benz F-Cell

source in hybrid vehicles to overcome the variation in performance of the diverse components. The distribution of power between the battery and fuel cell module imposes several challenges due to their diverse characteristics and nature [19]. The lithium battery offers high output power, long service life, and high energy ratio while the SC offers high efficiency of energy conversion with a large discharge capacity. Within a short time, power is released in response to the relatively low energy density [20].

It is essential to provide a consistent energy management strategy in EVs due to the regenerative braking energy. In a hybrid power systems (HPS), global optimization methods, deterministic rule control, and fuzzy rule control schemes are used for energy management. Real-time optimization issues can be solved strategically using global optimization. Irregular problems can be optimized using this technique in contrast to the classical optimization schemes that solve problems based on rigid rules [21]. The global optimization-based EMS can be solved efficiently using the traditional dynamic programming algorithm. Protection schemes, economic indicators, and system performance are the parameters that affect HPS. The steady-state features and dynamic response are the major characteristics of control performance. Life cycle cost, hydrogen consumption equivalent, and minimization define the economy of the model [22]. Depth of charging and discharging as well as current fluctuation reduction are the features involved in protection [23]. The system energy consumption is only a part of the optimization objective. Control strategy plays a major role in choosing an optimal HPS.

Fuzzy logic uses rule-based techniques for management. The human brain's uncertainties are imitated for reasoning and judgment in fuzzy logic. Based on the system's degree of freedom, the time adjustments of fuzzy logic can be done [24]. The HPS electrical and mechanical design can be performed using fuzzy control. Control experience or expert knowledge are the major limitations of fuzzy rules. Full play may be given to the characteristics of fuzzy logic by combining intelligent schemes with it. The hybrid power transmission systems can be monitored on a real-time basis using deterministic rule-based control strategies. The controller decision-making is entirely dependent on the instantaneous input. Based on the efficiency in meeting the battery and driver requirements using the flow chart and rule table, the performance of the system is calculated. The highest practicability is offered by the deterministic rules based on strategies rather than expert knowledge [25]. Of these systems, we use the global optimization strategy due to its performance and features. The following sections present powertrain system and dynamic programming (DP) schemes for constructing a global optimal EMS that can operate on offline mode in PHEVs using DP.

2.1 *Structuring Your Paper*

In this chapter, we compare both serial as well as parallel powertrain configuration as shown in Fig. 12.2. In the serial configuration, the battery is supplied with electrical energy from Engine Generator Unit (EGU) which is formed by mechanically coupling a diesel engine and an Integrated Starter Generator (ISG). In a parallel configuration, six-speed ratios are available with the Dual-Clutch Transmission (DCT) driven by coaxially connected ISG and engine [26]. Regenerative braking mode, hybrid driving mode, electric driving mode, engine driving mode, and charging mode are the five modes of operation of the PHEV [27]. Table 12.1 provides the comparison of the basic PHEV parameters in serial and parallel configurations. The energy efficiency of the engine generator unit can be estimated using the generator efficiency and engine brake specific fuel consumption (BSFC) (Table 12.2).

The engine model, motor model, battery model, and vehicle model are studied and analyzed for both the configuration. The relationship between the speed, torque, and engine fuel rate is analyzed by building an interpolation model [28]. Calibration is done under diverse speed and torque conditions for obtaining an optimal engine fuel rate. The fuel consumption map of the engine can be obtained by the linear interpolation scheme. The relationship between motor speed, torque, and efficiency can also be described using a spline interpolation scheme [29]. Over 90% efficiency is obtained which is the maximum range for the motor.

The battery performance is characterized using an internal resistance model. Lithium-ion battery is considered for this purpose. Figure 12.3 represents the charging and discharging modes and the open-circuit voltage (OCV) with respect to the internal resistance of the battery. With the variation in the charge state from

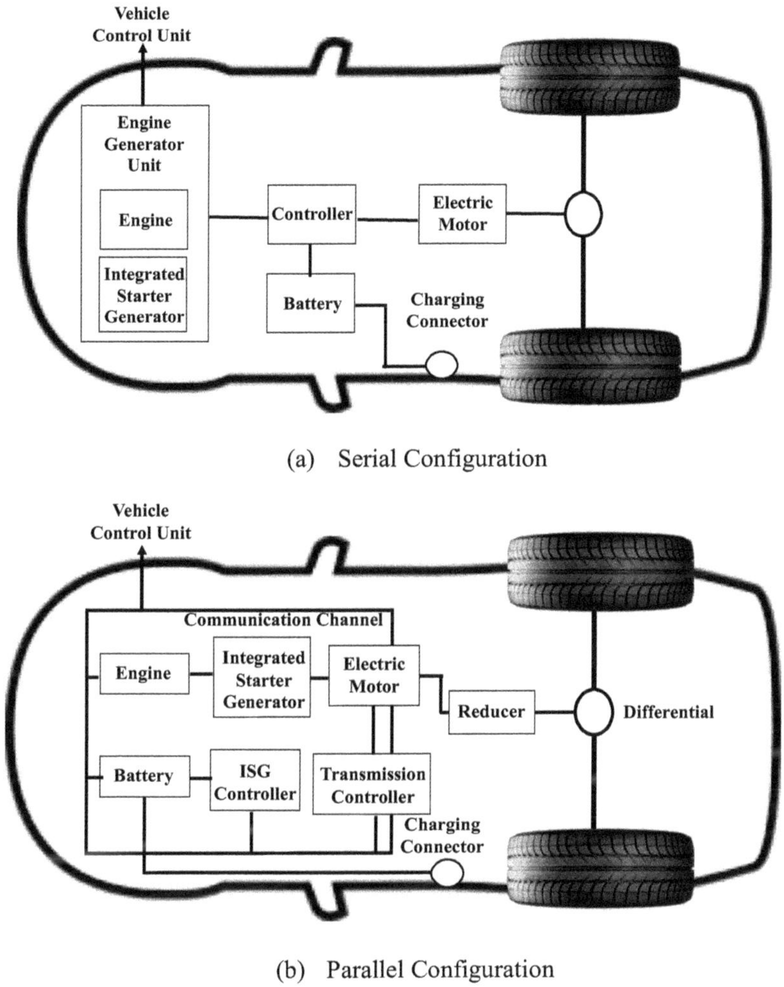

Fig. 12.2 PHEV-based powertrain architecture

Table 12.2 Comparison of PHEV parameters

Item	Parameter	Serial configuration	Parallel configuration
Vehicle	Gear ratio	6.733	6.239
	Mass–full load	1650 kg	1700 kg
Engine	Peak power	110 kW	80 kW
	Peak torque	320 Nm	140 Nm
ISG	Maximum power	90 kW	33 kW
	Maximum torque	330 Nm	140 Nm
Battery	Capacity	180 Ah	38.5 Ah
	Voltage	537.6 V	288 V

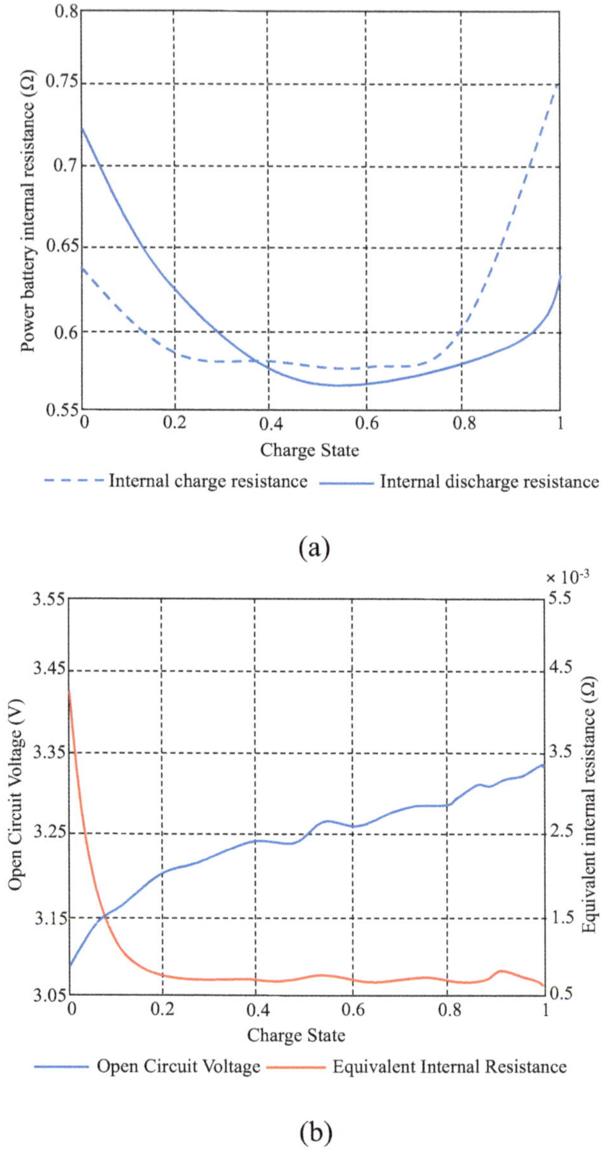

Fig. 12.3 State of charge based (**a**) internal resistance and (**b**) open-circuit voltage of battery cell

0 to 1, the OCV increases gradually. The temperature effects are neglected while estimating the equivalent internal resistance and OCV of the battery [29]. Considering the total battery power to be P_{TOT}, load power consumption as P_C, internal power loss as P_L, electric current I, and equivalent internal resistance R_{EQ}, the battery power balance equation can be calculated using the formula

$$P_{TOT} = P_C + P_L = P_C + I^2 R_{EQ} \tag{12.1}$$

Based on the efficiency of system transmission ηt, rotating mass correction coefficient ε, vehicle speed s, vehicle frontal area A, coefficient of air drag CD, road grade α, coefficient of rolling resistance γ, and mass of the vehicle M, the power required by the vehicle can be estimated using the formula

$$P_v = \frac{1}{3600\eta_t}\left(Mg\gamma\cos(\alpha) + \frac{C_D A s^2}{21.15} + Mg\sin\alpha + \varepsilon M\frac{ds}{dt}\right)s \tag{12.2}$$

2.2 Dynamic Programming

The discretization interval, controlling inputs, and state variables have to be determined for solving the PHEV energy management issue. The PHEV powertrain structure considers the optimal target as cost of fuel consumption and penalty function as frequent transmission shifting. At regular intervals, the driving cycle is split into multiple stages. Prediction of the future intervals is performed at every stage of the diving cycle [30]. The future stage optimal control policy is solved by applying dynamic programming (DP). Based on the DP forward optimization and the inverse solution, the global optimal solution can be obtained by setting the control variables, constraints, and cost function. The local control policy is attained by a multi-time-scale prediction scheme analyzed by the future short-period driving cycle. Optimal control sequence and the cost function are achieved using the reverse solution on implementation of DP algorithm [31]. The positive solution is executed based on the system's initial state. Further, the optimal trajectory is obtained by imposing a state with the optimal control policy

The DP parameter settings are analyzed based on multiple parameters. The cost function of the system is calculated using the reverse solution for the N − 1 stage using the expression

$$O*(x(N-1)) = \min_{u(N-1)}[I(x(N-1), u(N-1))] \tag{12.3}$$

k stage ($0 \le k \le N-2$):

$$\begin{cases} O*(x(k)) = \min_{u(N-1)}[I(x(k), u(k)) + O*(x(N-1))] \\ x(k+1) = f(x(k), u(k)) \end{cases} \tag{12.4}$$

Here, when optimal control is applied to the system, the optimal cost function is given by $O^*(x(k))$ which transfers $x(k)$ to $x(N)$. At the k stage, the instantaneous cost is given by $I(x(k), u(k))$.

3 Optimization and Performance Analysis

Aging of the battery influences the energy management of the EV model. It is essential to consider the aging characteristics while designing the control strategy [32]. Two thresholds namely $\delta 1$ and $\delta 2$ are considered to represent the high and low states of charge, respectively. The engine power PE is denoted by Eq. (12.5) where PDM is the electric motor power demand, P_L and P_H are the engine's high-efficiency range threshold values, P_{DE} is the optimizable engine power, Pmax_batt is the battery pack limitation, η_{APU} is the efficiency of the assistance power unit. The optimal control policy of the battery is obtained under the varying state of health (SOH), z represents the optimization horizon of SOC of the battery pack.

$$P_E = max\left\{\psi_E, \frac{P_{DM} - P_{\max\_batt}(z, SOH)\eta_{batt}}{\eta_{APU}}\right\} \tag{12.5}$$

where

$$\psi_E = \begin{cases} P_{DE}, if \dfrac{P_{DM}}{\eta_{APU}} \in \left[P_H, P_{e\_\max}\right] \\ P_{opt}, if \dfrac{P_{DM}}{\eta_{APU}} \in \left[P_L, P_H\right] \\ 0, if \dfrac{P_{DM}}{\eta_{APU}} \in \left[0, P_L\right] \end{cases} \tag{12.6}$$

Particle swarm optimization (PSO) algorithm is used for implementing the energy management and perform model simulation using inputs from the driving cycles of the power demand. The particle position X of the algorithm is represented by

$$X = P_H P_L P_{DE} \tag{12.7}$$

The optimal control policy u is denoted by

$$u^* = \begin{bmatrix} P_E \\ P_{Batt} \end{bmatrix} = f(X^*(SOH)) \tag{12.8}$$

The PHEV optimal energy management may also be affected by the driving distance as well as the driving condition. Under diverse driving conditions, the target PHEV fuel economy can be improved by incorporating real-time blended EMS with

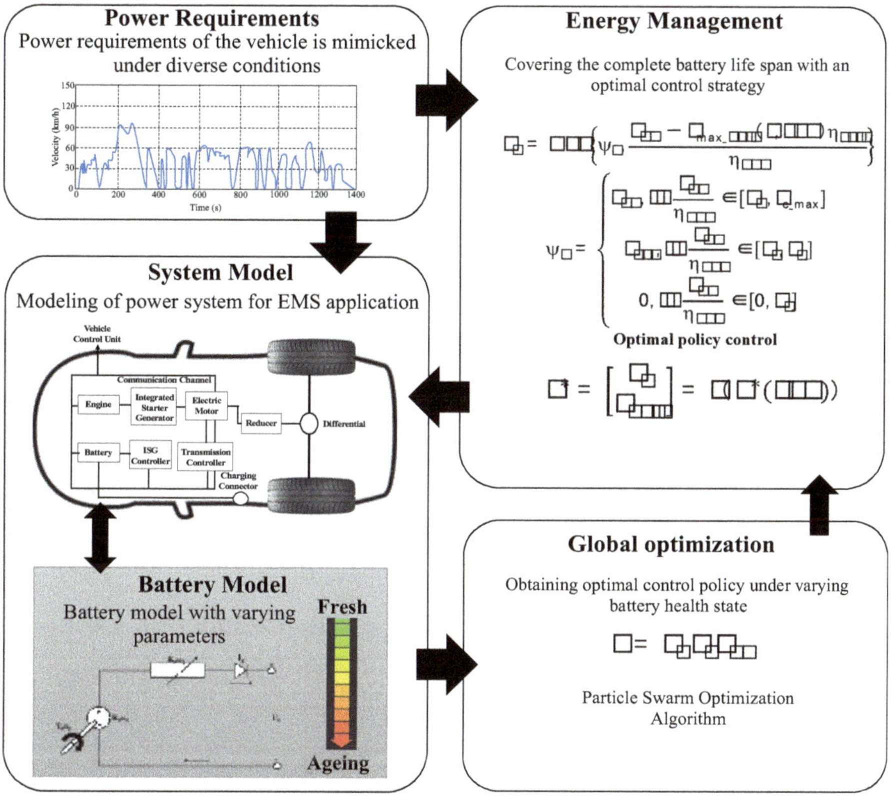

Fig. 12.4 Energy management strategy schematic

GPS and an algorithm for identifying the driving condition. Figure 12.4 provides the complete framework of the EMS which is comprised of four major segments namely power requirements, system model inclusive of the battery model, global optimization, and energy management. This helps in analyzing the driving conditions, battery charging, and discharging status and provides an optimal control policy. The optimal solution can be obtained offline using DP. Aging affects the performance parameters of the battery significantly. The control parameters have to be adjusted dynamically as the battery ages for establishing a globally optimal strategy [33]. Online energy management can be achieved by incorporating real-time driving data along with the control rules.

3.1 Driving Condition

EMS design partially depends on the driving conditions. The driving conditions are categorized based on the congestion status and average speed as represented in Fig. 12.5. With the decrease in congestion, there is a gradual increase in the average

Fig. 12.5 Analysis of driving conditions under (**a**) 75% congestion, (**b**) 60% congestion, (**c**) 50% congestion and (d) 20% congestion

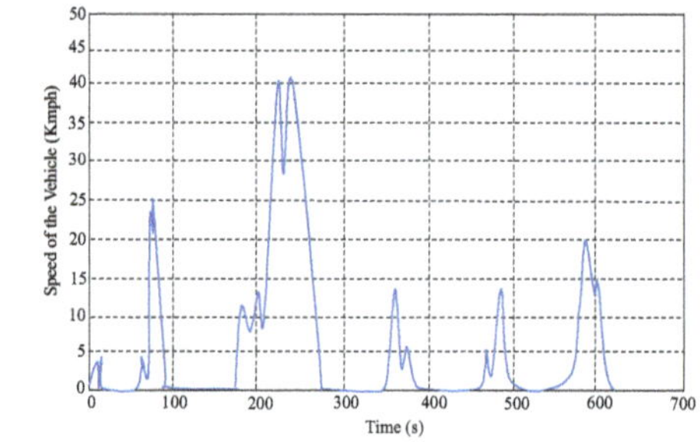

(a)

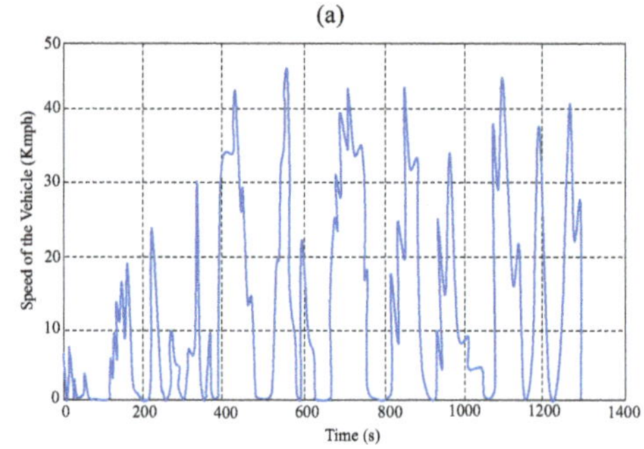

(b)

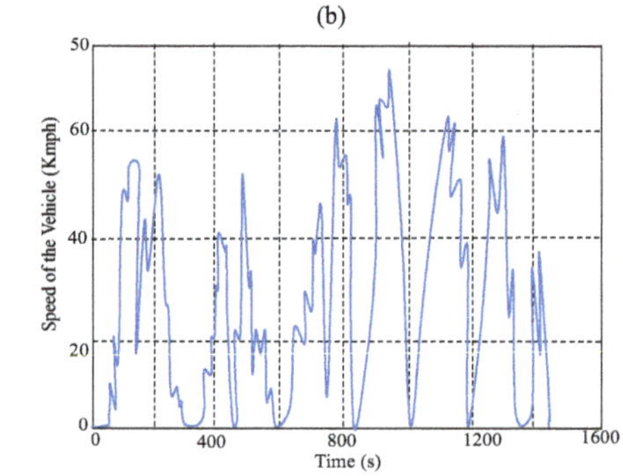

(c)

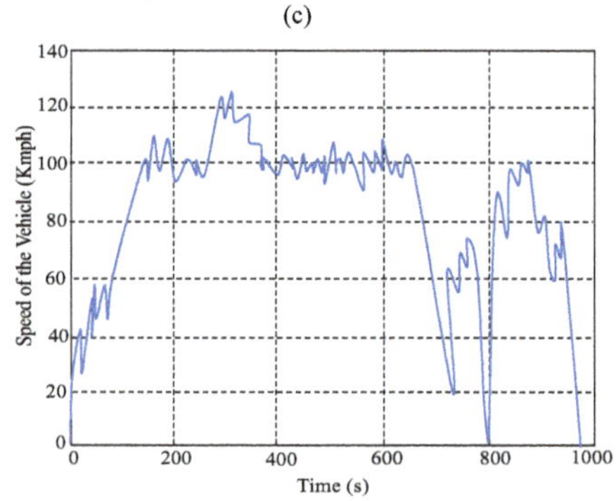

(d)

velocity of the vehicle. The blended EMS design is also influenced by the driving distance [34]. However, the all-electric range of the battery cannot be determined with only the driving distance. The battery's initial SOC is also a crucial factor of analysis. L_{eq}, the equivalent distance factor is considered for analyzing the influence of the distance of driving. This aids in planning the trajectory of SOC decline.

$$\begin{cases} L_{eq} = \dfrac{SOC_{rem}/SOC_{tot}}{L_{rem}/L_{max}} \\ SOC_{rem} = SOC_{all} - SOC_{min} \\ SOC_{tot} = SOC_{max} - SOC_{min} \end{cases} \tag{12.9}$$

here, the remaining distance of driving is represented as L_{rem}, the remaining charge state capacity is given by SOC_{rem}, the maximum and minimum SOC limits are denoted by SOC_{max} and SOC_{min}, the upper and lower limits of mileage data is L_{max} and L_{min}, and the total range of variation is SOCtot. The geographical information system (GIS), global positioning system (GPS), and intelligent transportation system (ITS) module help in obtaining real-time information during driving. Incorporation of this system ensures flexibility and strong adaptation of the model in real-time driving conditions as well as efficient energy management of PHEVs.

3.2 Energy Management

In PHEV as well as traditional HEVs, in order to improve the fuel economy, EMS is implemented. The advantages of the hybrid system can be maximized by focusing on optimization techniques. PSO, GA, DP, and such advanced optimization algorithms are used for achieving optimal control performance. Motor and engine torque distribution, gear shifting rules, engine start power, and such basic rules are defined for efficient energy management of the EV. The high efficiency of the motor can be obtained by appropriate engagement of the engine [35]. The PHEV can be controlled using the engine start and stop schemes. The energy source of the PHEV and its operating state can be adjusted effectively using the engine torque distribution and transmission gearshift ratio schemes. This enables improvement of the operating efficiency of the motor as well as the engine. The equivalent driving distance coefficient, driving pattern, and such driving information are obtained for enabling blended strategy. Figure 12.6 provides a sample driving cycle equivalent driving distance coefficient matrix.

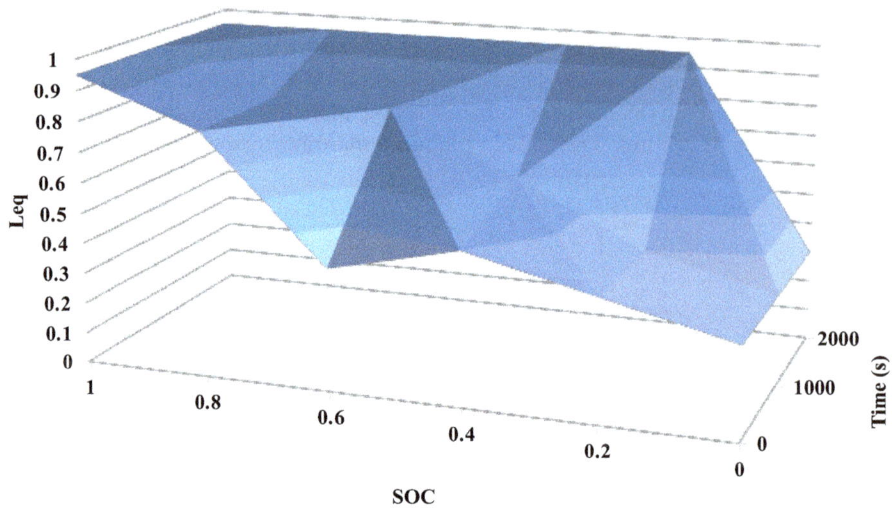

Fig. 12.6 Coefficient matrix of equivalent travel distance

4 Simulation, Validation, and Application

The optimal energy management scheme aids in the investigation of the driving conditions and their influence over the battery, the built strategy effectiveness, and other significant parameters of the vehicle. The PHEV sub-optimal energy management is obtained by applying the proposed energy management scheme under varied driving conditions. An optimal solution is obtained using the dynamic programming technique and the performance of the proposed scheme is evaluated. Different driving distances and conditions are considered for the optimization of PHEV and to perform energy management. 0.9, 0.7, 0.4, and 0.1 are the equivalent travel distance coefficients considered for this analysis. These equivalent factors induce optimal controlling parameters that are compared.

Figure 12.7 provides the scatter diagram of the equivalent driving distances along with the engine operating state for the coefficients ranging between 0 and 1. The start and stop state of the engine is the control decision for the vehicle power demand. The stop state values are marked in red (✦) and the start state values are represented in blue (o). With the decrease in L_{eq}, there is a gradual decrease in the engine starting power. The engine optimal starting decision can be estimated using these values. Based on the largest average power required, the engine starting power varies as represented in Table 12.3.

Based on the optimization results, the driving cycles and their shift schedules are estimated in the electric driving model. The shifting schedule is not affected by the driving conditions despite the wide regulation range of the speed of the motor. The distribution of speed ratio is analyzed and the optimized speed ratio is identified based on the upshift curve of the shift schedule. Except for the motor operating

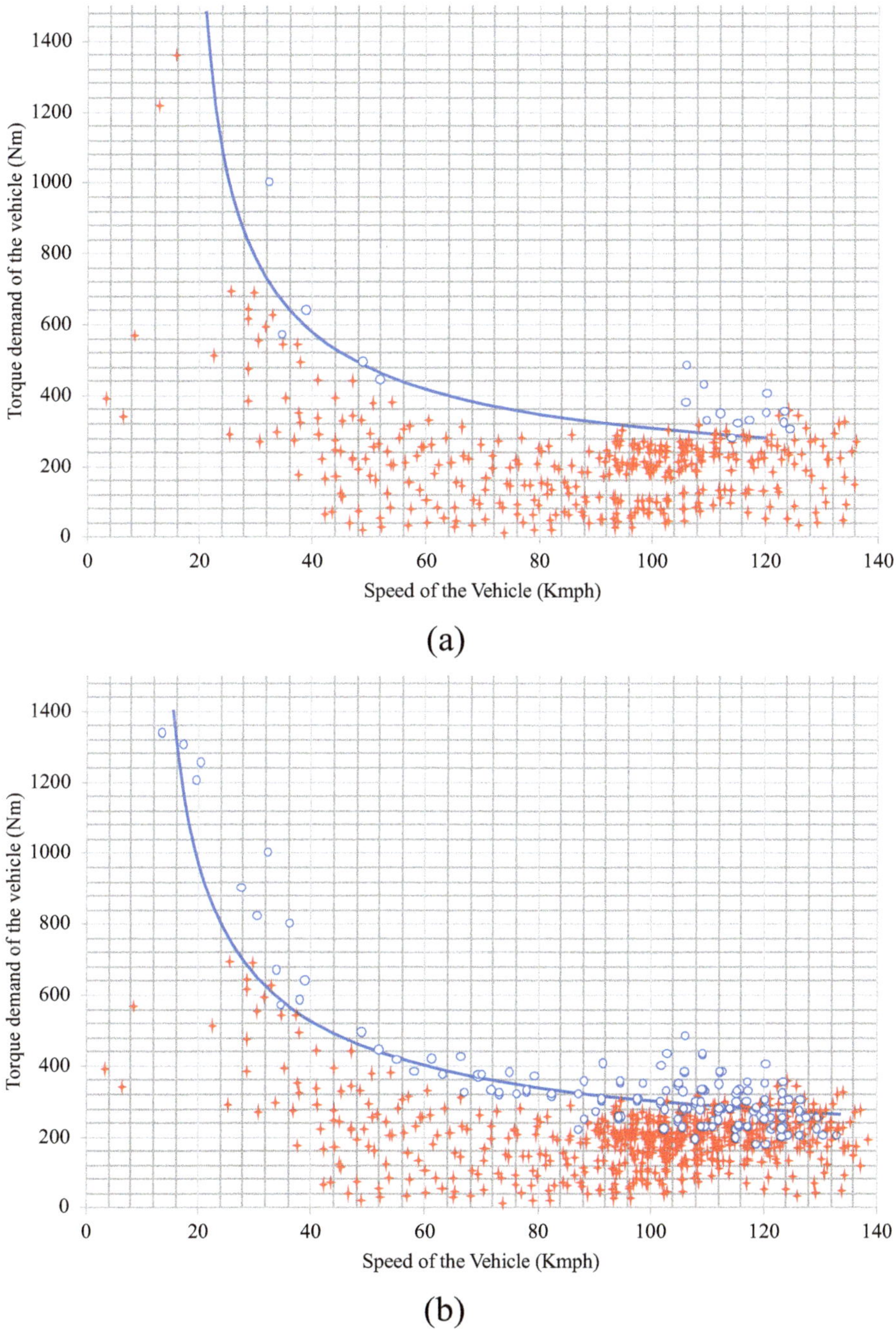

Fig. 12.7 Equivalent driving distances along with engine operating state

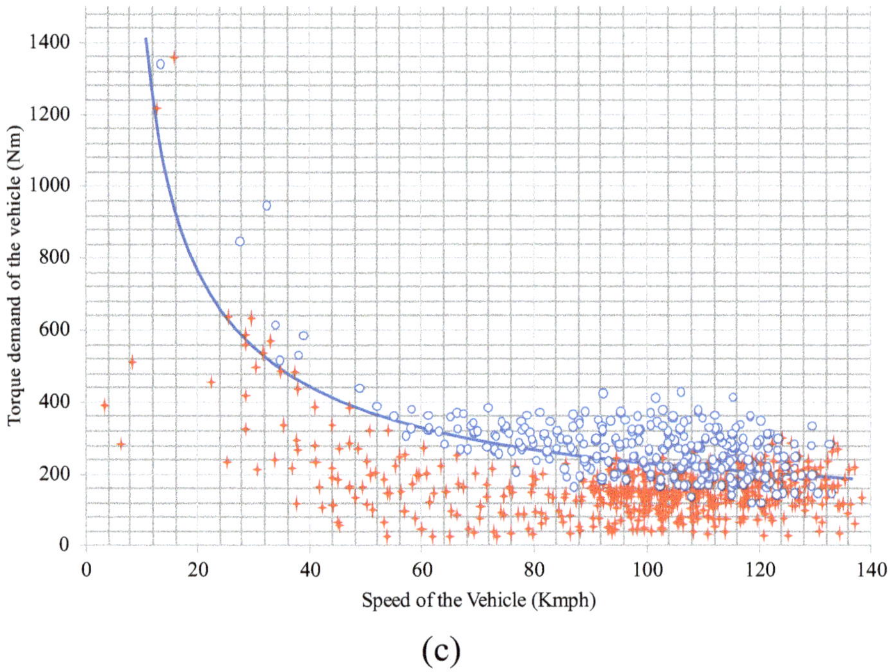

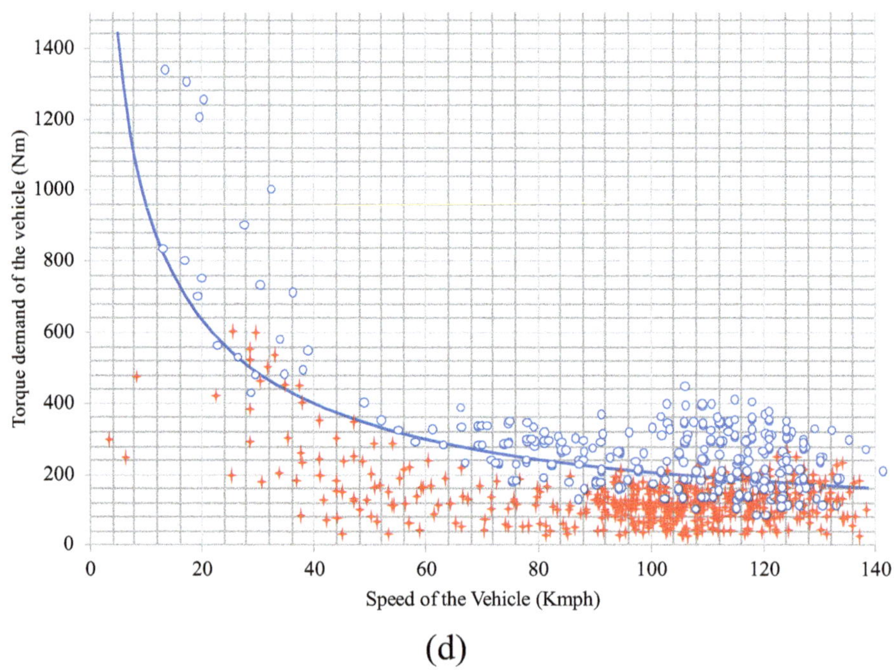

Fig. 12.7 (continued)

Table 12.3 Driving cycle tests under different coefficients and engine starting power

L_{eq}	Test 1	Test 2	Test 3	Test 4
0.9	7	12	13	30
0.7	6	8	9	25
0.4	5	7	6	18
0.1	3	2	2	8

points, there is no significant variation in the shifting schedules. The fitted torque distribution ratio curve can be used as a strategy for the distribution of engine torque as it does not influence Leq. The driving condition affects the optimal torque distribution ratio significantly.

4.1 *Life Cycle Impact*

Life cycle impact analysis is a critical tool for the assessment of the impact of usage of the vehicle. Aging of the battery has to be analyzed accurately to avoid the inefficiency of energy management. It is essential to adjust the parameters of energy management at varying SOH of the battery. When the energy cost of the engine system is large during long trips, the battery aging impact is insignificant. However, in short trips, the factor is critical for analysis. Energy management approach cannot eliminate the aging factor of the battery. The energy costs and their negative impact on PHEV can be reduced partially with efficient EMS. The entire life cycle of the vehicle is analyzed based on its energy consumption, environmental impact, cost, and other information through simulation. This data is compared for observing the energy consumption and life cycle impact for improving the performance of the vehicle further. With respect to emission and economy, optimization of efficiency, investment cost, and driving condition can be analyzed using the cost function equation represented as follows:

$$F_0 = \sum_f C_{fc} + \sum_f C_{em} + \sum_f C_b +$$

where C_{fc}, C_{em}, and C_b represent the cost of the fuel cell, electric motor, and battery, respectively. The impact of vehicle manufacturing, energy production, and vehicle energy utilization is calculated with a view to reduce the emission of greenhouse gases.

4.2 *Comparative Analysis*

In order to verify the performance of the proposed scheme with respect to the existing state-of-the-art models, the DP, Dyna, and genetic algorithms are compared.

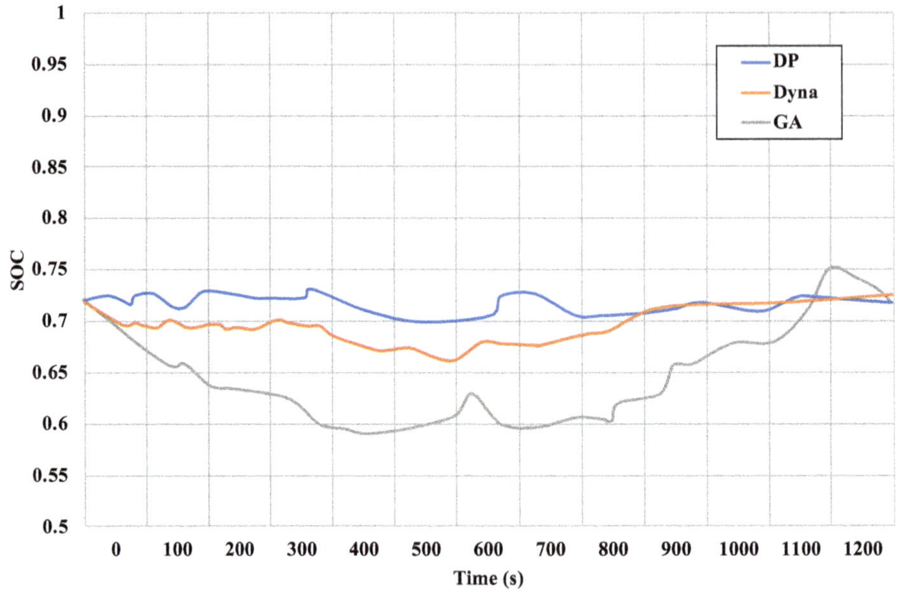

Fig. 12.8 Comparison of control strategies

Table 12.4 Comparison of control strategies

Technique	Initial SOC	Fuel consumption (L)	Electricity consumption (kWh)	Final SOC	Computation time (s)
DP	0.9	4.5325	55.2345	0.2966	125
Dyna		4.0321	52.2343	0.3012	1254
GA		3.9953	51.5664	0.3252	145
DP	0.7	9.2458	34.1245	0.3125	462
Dyna		9.1485	35.3345	0.3632	345
GA		9.0341	36.2543	0.3653	1463
DP	0.4	14.9472	14.3345	0.2857	135
Dyna		14.2745	14.7322	0.2947	256
GA		14.2034	15.2367	0.3014	732
DP	0.1	17.1249	4.6729	0.2946	257
Dyna		16.0465	4.3502	0.3017	422
GA		17.2529	5.2958	0.3590	1465

A driving cycle of 1200 seconds is considered along with a 10 Hz sampling frequency. The differential position function is used along with global navigation satellite system. The discount factor is set at 0.95 and the learning rate at 0.1. The three control strategies and their corresponding SOC trajectories are represented in Fig. 12.8. Table 12.4 provides the comparative analysis of the corresponding values. Based on the initial SOC, the values are categorized.

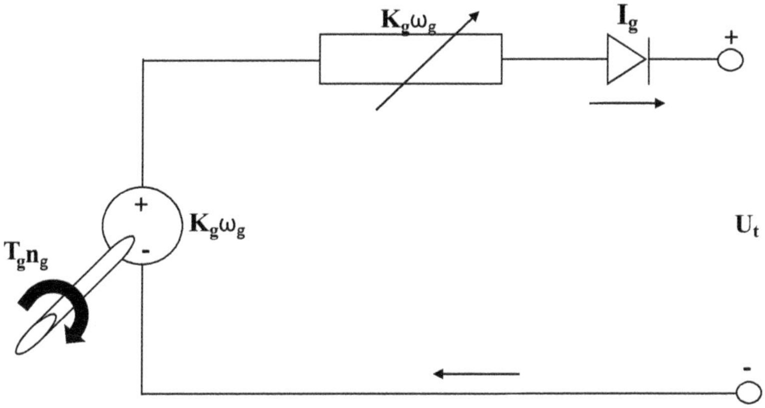

Fig. 12.9 Engine generator set battery model

The deep reinforcement learning algorithms enable the formulation of energy management through heuristic planning [36]. The model is trained with high-precision real-time driving conditions. It enables low consumption of fuel and fast training speed. The model is adaptable to other driving schedules. The Engine Generator Set (EGS) battery model enabling efficient conversion of fossil energy to electrical energy is as represented in Fig. 12.9. The output voltage is given by Ut, kg, and ωg are force, coefficient, and generator speed, respectively. The generator output current is Ig and electromagnetic torque is Tg.

The prediction method is used for analyzing the driving cycle of the battery. Markov chain may be used for modeling the vehicle velocity. The nearest neighborhood and maximum likelihood estimator can be used for calculating the velocity transition probability. The driving cycle is estimated for prediction of the driving cycle using the single-time-scale prediction technique as represented in Fig. 12.10. The prediction horizon can be set at different time limits and further analyzed.

When compared to ICE mode, PHEV sustains for a longer time duration in EM mode. This makes it more suitable on highways as well as the city. The application type, desired performance, weight, and size of the vehicle are some of the preliminary information that has to be understood while selecting a suitable energy management strategy. Driving experience, GPS position, weather forecast, and weather conditions will be considered in high-level applications for power management.

4.3 Output Power Balance

Under certain conditions, an unexpected low utilization rate is obtained due to the overloading of certain components. The simulation results of the power distribution process are represented in Fig. 12.11. Maximum output power of 180 kW is considered wherein during the complete working state, the FC is operational without

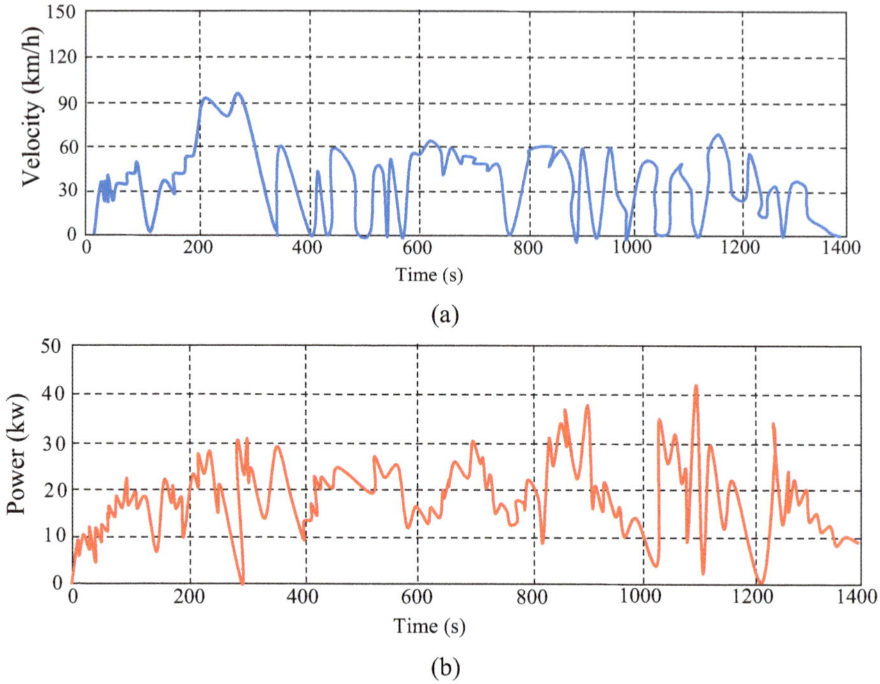

Fig. 12.10 Power demand and driving schedule

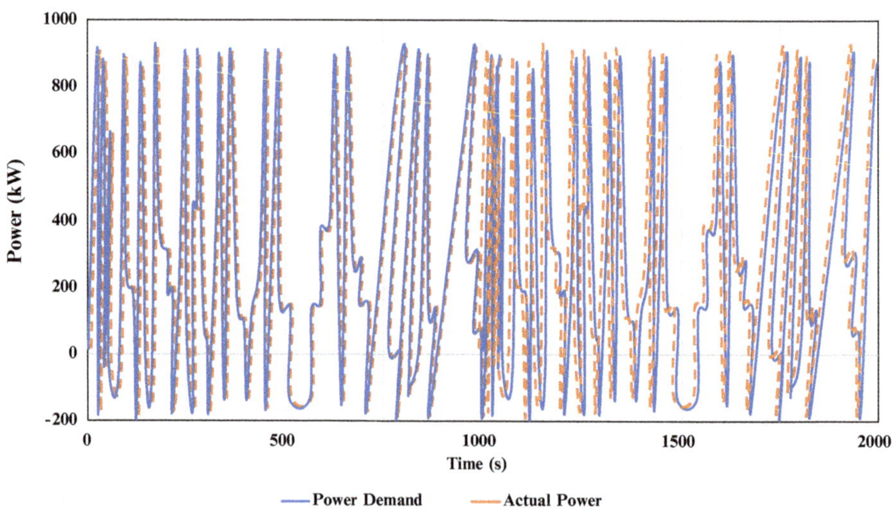

Fig. 12.11 FC output power

being shut down. The train operation energy distribution can be met by the EMS. Inefficient utilization of regenerative braking energy is done due to the small storage battery energy absorption rate and energy output during the traction stage. This results in the loss of renewable energy in large amounts and inefficient recovery of regenerative braking energy in the braking stage. Hence the dynamic system has to be optimized.

Despite serious battery aging, optimal energy efficiency is maintained and a system-level compensation is provided to partially overcome the negative influence of battery aging. The adaptability and effectiveness of the proposed model have been evaluated using MATLAB/Simulink environment. Power distribution simulation and vehicle demand power tracing are performed efficiently using the powertrain model. In the prediction horizon, the power split is optimized using the DP algorithm.

5 Conclusion and Future Scope

5.1 Conclusion

The ICE vehicles are gradually replaced by EVs due to the serious emission issues and depletion of fossil fuel. Hydrogen based FCs act as an ideal source of energy as it provides higher energy on combustion when compared to gasoline while causing zero pollution. Security risks such as explosion, limited hydrogen source, high cost, and such drawbacks exist while developing such FCs. This paper provides a complete analysis of EMS using the global optimization scheme considering the powertrain system, dynamic programming, and deep learning schemes. Simulation is carried out under diverse driving conditions considering the state of charge and the battery aging factors. SC, battery, and FC-based HPS have unique features and advantages that aid in the promotion of HEVs. An efficient EMS is required to determine the battery life, energy consumption, and cost. Various driving cycles are considered for the implementation of EMS with DP to achieve a global optimal solution.

5.2 Challenges of PHEV

Despite the various controversies and challenges in FCHEV commercialization, various researches are carried out to optimize the solution and solve the environmental and energy crisis in the automobile industry. The major criteria in deciding the efficiency of automobiles are their powerful performance. With the increase in features, the FC and battery service life is affected despite taking the power distribution issues into consideration. The distribution scheme, aging of the battery, driving conditions, and several parameters affect the battery performance. The total cost of the vehicle life cycle is affected due to the negligence of these factors. Fuel or hydrogen consumption is considered by certain researchers to provide a solution to the optimization issues. Real-time strategies offer better solutions compared to the random variables used in offline strategies. The proportional–integral–derivative (PID) controller is popularly implemented in control devices.

5.3 Future Direction

Improvement of practicability, verification of experimental platform, solution for oxygen deficiency in system operation, maintenance cost reduction, energy consumption optimization, and FC efficiency improvement are performed. The market of automobile energy has large requirements for clean fuel resources like hydrogen. However, the explosive and flammable nature, as well as high utilization value of hydrogen, causes safety concerns and controversies. Implementing safe operation of hydrogen is essential. The hydrogenation standards, operation management specification, construction, the design formulation, and acceleration of hydrogen facilities are key research areas. Interaction of information and energy, integration of vehicle network technologies, billing equipment, metering, monitoring, network connection, and charging facilities of EVs as well as the integration of these techniques without loss of energy or information is essential

References

1. Mohd, T. A. T., Hassan, M. K., Aris, I., Azura, C. S., & Ibrahim, B. S. K. K. (2017). Application of fuzzy logic in multi-mode driving for a battery electric vehicle energy management. *International Journal on Advanced Science, Engineering and Information Technology, 7*(1), 284–290.
2. Moreno, J., Ortúzar, M. E., & Dixon, J. W. (2006). Energy-management system for a hybrid electric vehicle, using ultracapacitors and neural networks. *IEEE Transactions on Industrial Electronics, 53*(2), 614–623.
3. Alloui, H., Becherif, M., & Marouani, K. (2013, June). Modelling and frequency separation energy management of fuel cell-battery hybrid sources system for hybrid electric vehicle. In *21st Mediterranean Conference on Control and Automation* (pp. 646–651). IEEE.
4. Xiong, R., Chen, H., Wang, C., & Sun, F. (2018). Towards a smarter hybrid energy storage system based on battery and ultracapacitor-A critical review on topology and energy management. *Journal of Cleaner Production, 202*, 1228–1240.
5. Hu, Y., Li, W., Xu, K., Zahid, T., Qin, F., & Li, C. (2018). Energy management strategy for a hybrid electric vehicle based on deep reinforcement learning. *Applied Sciences, 8*(2), 187.
6. Williamson, S. S. (2013). *Energy management strategies for electric and plug-in hybrid electric vehicles*. Springer.
7. https://www.ev-volumes.com/country/total-world-plug-in-vehicle-volumes/
8. Ding, N. (2020). *Energy management system design for plug-in hybrid electric vehicle based on the battery management system applications*. Doctoral dissertation, Auckland University of Technology.

9. Mozhi, S. A., Raja, S. C., Saravanan, M., & Nesamalar, J. J. D. (2020). Energy management of hybrid energy storage system in PHEV with various driving mode. *Artificial Intelligent Techniques for Electric and Hybrid Electric Vehicles, 103*.
10. Zhou, Y., Ravey, A., & Péra, M. C. (2019). A survey on driving prediction techniques for predictive energy management of plug-in hybrid electric vehicles. *Journal of Power Sources, 412*, 480–495.
11. Komarappa, N. B., & LakshmiKantha, B. R. (2018). Power management strategy for hybrid energy storage system in hybrid electric vehicles. *International Journal of Intelligent Engineering and Systems, 12*, 12–21.
12. Huang, Y., Wang, H., Khajepour, A., He, H., & Ji, J. (2017). Model predictive control power management strategies for HEVs: A review. *Journal of Power Sources, 341*, 91–106.
13. Kamal, E., & Adouane, L. (2017, July). Reliable energy management optimization in consideration of battery deterioration for plug-in intelligent hybrid vehicle. In *International conference on informatics in control, automation and robotics* (pp. 150–173). Springer, Cham.
14. Demircalı, A., Sergeant, P., Koroglu, S., Kesler, S., Öztürk, E., & Tumbek, M. (2018). Influence of the temperature on energy management in battery-ultracapacitor electric vehicles. *Journal of Cleaner Production, 176*, 716–725.
15. Nyong-Bassey, B. E., Giaouris, D., Patsios, C., Papadopoulou, S., Papadopoulos, A. I., Walker, S., et al. (2020). Reinforcement learning based adaptive power pinch analysis for energy management of stand-alone hybrid energy storage systems considering uncertainty. *Energy, 193*, 116,622.
16. Martinez, C. M., Hu, X., Cao, D., Velenis, E., Gao, B., & Wellers, M. (2016). Energy management in plug-in hybrid electric vehicles: Recent progress and a connected vehicles perspective. *IEEE Transactions on Vehicular Technology, 66*(6), 4534–4549.
17. Liu, T., Hu, X., Hu, W., & Zou, Y. (2019). A heuristic planning reinforcement learning-based energy management for power-split plug-in hybrid electric vehicles. *IEEE Transactions on Industrial Informatics, 15*(12), 6436–6445.
18. Hussain, S., Ali, M. U., Park, G. S., Nengroo, S. H., Khan, M. A., & Kim, H. J. (2019). A real-time bi-adaptive controller-based energy management system for battery–supercapacitor hybrid electric vehicles. *Energies, 12*(24), 4662.
19. Xu, N., Kong, Y., Chu, L., Ju, H., Yang, Z., Xu, Z., & Xu, Z. (2019). Towards a smarter energy management system for hybrid vehicles: A comprehensive review of control strategies. *Applied Sciences, 9*(10), 2026.
20. Le Rhun, A. (2019). *Stochastic optimal control for the energy management of hybrid electric vehicles under traffic constraints*. Doctoral dissertation, Université Paris-Saclay.
21. Manzali, C., & Bertolini, F. (2020). *Energy management strategy for a high-performance hybrid vehicle*. Doctoral dissertation, Politecnico di Torino.
22. Alam, M. S., & Arefifar, S. A. (2019). Energy Management in power distribution systems: review, classification, limitations and challenges. *IEEE Access, 7*, 92,979–93,001.
23. Xie, S., Hu, X., Qi, S., & Lang, K. (2018). An artificial neural network-enhanced energy management strategy for plug-in hybrid electric vehicles. *Energy, 163*, 837–848.
24. Zhou, D., Al-Durra, A., Gao, F., Ravey, A., Matraji, I., & Simoes, M. G. (2017). Online energy management strategy of fuel cell hybrid electric vehicles based on data fusion approach. *Journal of power sources, 366*, 278–291.
25. Xie, S., Hu, X., Qi, S., Tang, X., Lang, K., Xin, Z., & Brighton, J. (2019). Model predictive energy management for plug-in hybrid electric vehicles considering optimal battery depth of discharge. *Energy, 173*, 667–678.
26. Hu, X., Zou, C., Tang, X., Liu, T., & Hu, L. (2019). Cost-optimal energy management of hybrid electric vehicles using fuel cell/battery health-aware predictive control. *IEEE Transactions on Power Electronics, 35*(1), 382–392.
27. Tie, S. F., & Tan, C. W. (2013). A review of energy sources and energy management system in electric vehicles. *Renewable and sustainable energy reviews, 20*, 82–102.

28. Brahma, A., Guezennec, Y., & Rizzoni, G. (2000, June). Optimal energy management in series hybrid electric vehicles. In *Proceedings of the 2000 American Control Conference*. ACC (IEEE Cat. No. 00CH36334) (Vol. 1, No. 6, pp. 60–64). IEEE.
29. Serrao, L., Onori, S., & Rizzoni, G. (2011). A comparative analysis of energy management strategies for hybrid electric vehicles. *Journal of Dynamic Systems, Measurement, and Control, 133*(3).
30. Kessels, J. T., Koot, M. W., Van Den Bosch, P. P., & Kok, D. B. (2008). Online energy management for hybrid electric vehicles. *IEEE Transactions on Vehicular Technology, 57*(6), 3428–3440.
31. Sun, C., Hu, X., Moura, S. J., & Sun, F. (2014). Velocity predictors for predictive energy management in hybrid electric vehicles. *IEEE Transactions on Control Systems Technology, 23* (3), 1197–1204.
32. He, H., Xiong, R., Guo, H., & Li, S. (2012). Comparison study on the battery models used for the energy management of batteries in electric vehicles. *Energy Conversion and Management, 64*, 113–121.
33. Kennel, F., Görges, D., & Liu, S. (2012). Energy management for smart grids with electric vehicles based on hierarchical MPC. *IEEE Transactions on Industrial Informatics, 9*(3), 1528–1537.
34. Sabri, M. F. M., Danapalasingam, K. A., & Rahmat, M. F. (2016). A review on hybrid electric vehicles architecture and energy management strategies. *Renewable and Sustainable Energy Reviews, 53*, 1433–1442.
35. Gonder, J., & Markel, T. (2007). Energy management strategies for plug-in hybrid electric vehicles (No. 2007-01-0290). *SAE Technical Paper. 10.*
36. Meenakshi Ammal, S., Kathiresh, M., & Neelaveni, R. (2021). Artificial intelligence and sensor technology in the automotive industry: An overview. In M. Kathiresh & R. Neelaveni (Eds.), *Automotive embedded systems. EAI/Springer innovations in communication and computing*. Springer. https://doi.org/10.1007/978-3-030-59897-6_7

Chapter 13
A Novel Sensing Technique for Continuous Monitoring of Volume in an Automobile Fuel Tank

P. Rajalakshmy, P. Subha Hency Jose, K. Rajasekaran, R. Varun, and P. Sweety Jose

1 Introduction

Fuel-level sensors are primarily used to monitor the consumption of fuel by measuring the level of fuel in a vehicle fuel tank or a storage tank. Fuel sensors generally consist of two main components, viz., the primary sensing element and the indicating element. The measuring system comprises a float level sensor, variable potentiometer, and a moving wiper. The float moves in accordance with the fuel level variations in the tank, which in turn causes the wiper to move across a variable resistor (potentiometer) causing a change in voltage. However, the sensing mechanisms may vary based on the techniques used for measurement in order to attain better accuracy and speed of measurement [1]. In general, the automobile tanks are

P. Rajalakshmy
Department of Robotics Engineering, Karunya Institute of Technology and Sciences, Coimbatore, Tamil Nadu, India
e-mail: rajalakshmy@karunya.edu

P. Subha Hency Jose (✉)
Department of Biomedical Engineering, Karunya Institute of Technology and Sciences, Coimbatore, Tamil Nadu, India
e-mail: hency20002000@karunya.edu

K. Rajasekaran
Department of Biomedical Engineering, KPR Institute of Technology, Coimbatore, Tamil Nadu, India

R. Varun
Microchip Technology, Chennai, India

P. Sweety Jose
Department of Electrical & Electronics Engineering, PSG College of Technology, Coimbatore, Tamil Nadu, India
e-mail: psj.eee@psgtech.ac.in

M. Kathiresh et al. (eds.), *E-Mobility*, EAI/Springer Innovations in Communication and Computing, https://doi.org/10.1007/978-3-030-85424-9_13

irregular in shape and it is difficult to measure the volume of the fuel in the tank. In this chapter, a measurement system has been developed using a potentiometric sensor to accurately determine the fluid volume in irregular-shaped automotive fuel tanks. The device for measuring and monitoring the volume of liquid in an irregular-shaped tank from the level measurement sensor output comprises an adapter for mounting the device in an opening at the top of the tank. The adapter supports a potentiometric level sensor going all the way to the bottom of the tank.

2 Existing Technology and Its Review

Capacitive sensors are progressively becoming an alternate for automatic devices in industrial and automotive applications, as they provide the benefits of long-term consistency even in aggressive environments. There are no moving parts in capacitive sensors. The level of any fluid is measured by the changing capacitance value. The capacitance is dependent on the dielectric constant, the area of the conducting plate, and the separation distance of the plates. The capacitance value is proportional to the dielectric constant of the material that separates the two conducting plates Therefore, any change in the fluid level will correspond to a change in the dielectric constant, hence the capacitance value. A fuel level measurement system involving a capacitive sensor is implemented with Neural Network using Matlab software [2]. A capacitive sensor for measuring the liquid level is implemented in inkjet printing technology [3]. The effects of dielectric materials between the capacitance plates are discussed in [4]. The application of fiber optic technology is used in the measurement of volume in a container [5]. The charge transfer technique in the capacitive sensor is used to measure the liquid volume [6]. The capacitive sensor is used for the measurement of pressure, level, and an ultrasonic sensor is employed for the concentration measurement [7]. An ultrasonic sensor is used to measure the liquid volume in automobile tanks [8]. The bending loss phenomenon of fiber optics is employed in volume measurement [9]. Microcontroller-based Ultrasonic sensor is used to measure the water level without measuring it [10]. The measurement of liquid level in four tanked systems using Micro electro mechanical System (MEMS) sensor is discussed [11]. The application of various sensing technologies for the measurement of liquid levels in molten metal is elaborated [12]. The concept of capacitive sensing, its technology, and its applications are well discussed [13].

3 Proposed Methodology

The measurement System is connected to the potentiometric float level sensor to convert the level readings into corresponding voltage levels. A processor is connected to the Voltage Measurement System to take input data to estimate the volume of liquid in the tank. During the calibration operation, a measurement functional relationship is derived to correlate the Voltage Measurement System readings to the volume of liquid in the tank. The calibrated functional relationship is stored in the processor which takes data from the Voltage Measurement System. This device can be used in any irregular-shaped tanks in which the level of volume relationship is nonlinear in nature and also easily retrofitted to any existing tank with level sensors.

The measurement functional relationship is derived using an Adaptive Neuro-Fuzzy Inference System (ANFIS). ANFIS is a machine learning algorithm, which is a synergistic integration of the adaptive nature of ANNs and the reasoning ability of Fuzzy Inference Systems. It is basically a hybrid learning algorithm that adapts its parameters according to the training data. The algorithm uses the gradient descent technique for tuning the nonlinear antecedent parameters and the least square technique to tune the linear consequent parameters of the IF–THEN rule structure -Sugeno type.

4 Description

A device for measuring and monitoring the volume of liquid in an irregular-shaped tank from the level measurement sensor output comprises an adapter for mounting the devices in an opening at the top of the tank. The adapter supports a potentiometric float level sensor for continuous level measurement in the tank. A voltage divider circuit is connected to the potentiometric float level sensor to convert the level data into corresponding voltage levels. A virtual instrument is connected to the voltage divider to take input data to estimate the volume of liquid. While calibrating, a measurement functional relationship is derived to correlate the Voltage to the volume of liquid in the tank. The calibrated functional relationship is stored in the system, comprising voltage measurement system, and calculates the actual volume of liquid in the container and displays the **measured volume**. The device of the present invention can be used in any irregular-shaped tanks in which the level to volume relationship is nonlinear in nature and also easily retrofitted to any existing level sensors. Figure 13.1 represents the block diagram of the proposed volume measurement system.

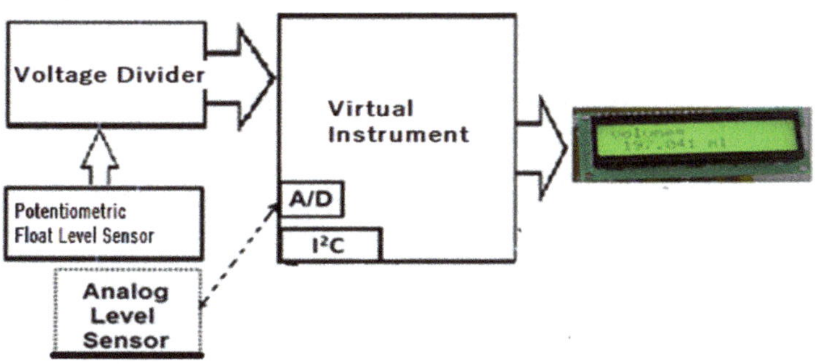

Fig. 13.1 Block diagram of the volume measurement system

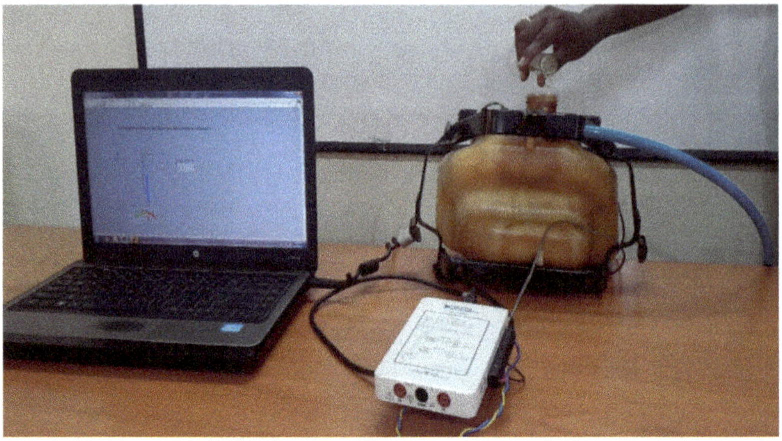

Fig. 13.2 Experimental setup of the volume measurement system

The potentiometric-based volume measuring unit comprises the level sensor immersed in a tank and connected to a display unit. The tank can be any irregular tank such as mounted fuel tanks in automobiles. The level sensor can be mounted near the top of the tank. The special feature of this measurement unit is that it can be applied to any type of level sensor. Calibration of the sensing unit is performed using Artificial Intelligence. Figure 13.2 is the Experimental setup of the volume indicator for an irregular-shaped tank.

The learning model proposed in this system is the Adaptive Neuro-Fuzzy Inference System. ANFIS stands for Adaptive Neuro-Fuzzy Inference System. These are a class of adaptive networks that are functionally equivalent to fuzzy inference systems. The architecture of these networks is referred to as ANFIS or semantically equivalently, adaptive neuro-fuzzy inference system. The objective of an ANFIS is to integrate the best features of Fuzzy Systems and Neural Networks. ANFIS is one of the best trade-offs between neural and fuzzy systems, providing smoothness, due to the Fuzzy Control (FC) interpolation and adaptability due to the Neural Network Backpropagation.

It is one of the data learning methods. The given input is converted to a target output by employing the Fuzzy Inference Model. It belongs to the family of Artificial Neural Network. This is used for the prediction employing functions, operators, and rules. There are two types of Fuzzy Inference Systems—Mamdani and Takagi-Sugeno. It employs the concept of both neural networks and fuzzy logic principles. This can be used to approximate the nonlinear functions. It is more efficient and can be considered as a universal Estimator. This is employed for estimation, modeling, stability analysis, and so on.

ANFIS algorithm estimates the correlation function through its input data collected during the calibration mode. The density of data point distribution required in order to achieve sufficiently high calibration accuracy depends on the extent of irregularity in the tank shape. In other words, the smaller the incremental change in fuel added, the denser the data point distribution that implies finer level sensor calibration. The configured model thus derived can be used for all identical tanks.

The advantages of this method are as follows:

- Improved measurement accuracy.
- Can be operated with any type of level sensor
- The measurement system developed can be further refined to detect abrupt changes in fuel volume that may be caused due to theft and trigger a suitable alarm.

The readings obtained can also be used to indicate of an unusually quick loss of fuel due to a leak

The proposed measurement system comprises of a potentiometric float level sensor, a voltage divider circuit, and a processor with an artificial intelligence algorithm, working in two modes of operation viz.

1. Calibration Mode
2. Measurement Mode

4.1 Calibration Mode

In Calibration Mode, the ANFIS model is trained using known volume readings as the target and the corresponding voltage readings as the input. A fixed volume of the fuel is added at periodic incremental levels to the irregular-shaped tank. The float level sensor in the tank indicates the level of fuel in the tank in terms of voltage measured from the potentiometric voltage divider circuit. These voltage levels are calibrated against the known values of volume. This procedure is repeated for various levels of incremental change and the values are used to train the ANFIS model. The training is done for various trials of incremental volume and the corresponding ANFIS model is saved as the reference model. The reference model is fixed based on the minimum error obtained during training. The algorithm is implemented using LabVIEW software with a Matlab script integrated within the VI. The LabVIEW implementation of Calibration Mode is shown in Fig. 13.3.

The trained model is further validated by presenting the voltages corresponding to a known set of volume readings. The volume readings obtained from the reference model are validated against the expected volume readings and the error is calculated. If the error exceeds the acceptable limits, the training process is repeated until the validation error falls within acceptable levels of tolerance.

In measurement mode, the voltage levels obtained from the potentiometric float level sensor are presented to the trained ANFIS model which in turn returns the corresponding volume based on the reference model.

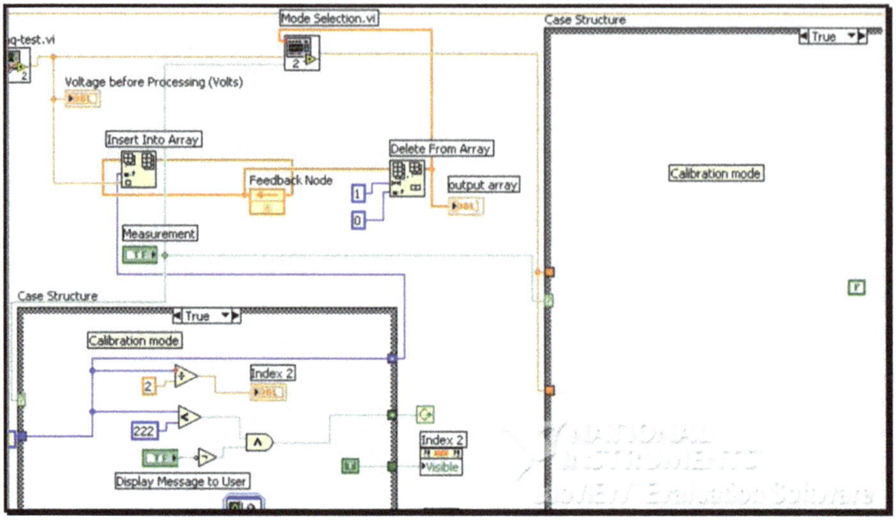

Fig. 13.3 Calibration mode of ANFIS

4.2 Measurement Mode

In measurement mode, the trained and validated ANFIS model is presented with voltage values from the potentiometric float level sensor, and the corresponding volume readings are obtained as output. The LabVIEW implementation of the measurement mode is shown in Fig. 13.4.

The volume output from the ANFIS model is indicated by interfacing a display unit with the VI using a DAQ card. The measurement unit with the display module is shown in Fig. 13.5 whereas the measurement unit along with the display module and the controller is shown in Fig. 13.6.

Finally, the LabVIEW front panel for the fuel indication is shown in Fig. 13.7. The calibration and measurement modes can be selected using the selector switch placed on the front panel. The proposed system can also be used to detect fuel theft from the tank when there is an indication of abnormal variation in the fuel volume which can be indicated to the user.

5 Findings

A nonlinear curve fitting is done to correlate the amount of fuel in the tank and the corresponding reading obtained from the level sensor. The curve fit thus obtained is a mathematical representation of the shape of the tank. The initial calibration is carried out with an empty tank and the sensor in place. A known quantity of fuel is incrementally added into the tank and the corresponding level sensor readings are

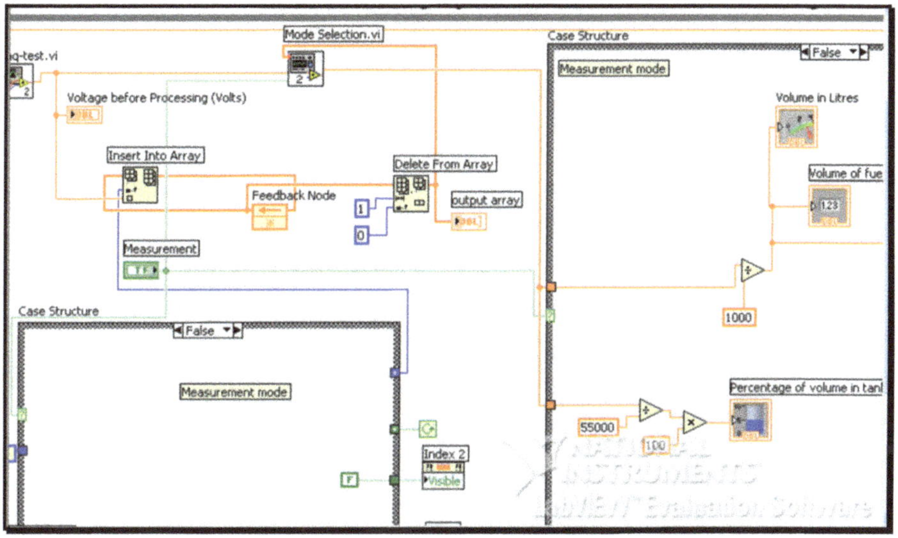

Fig. 13.4 Measurement mode of ANFIS

Fig. 13.5 Display module of the irregular tank

Fig. 13.6 Measurement unit with controller

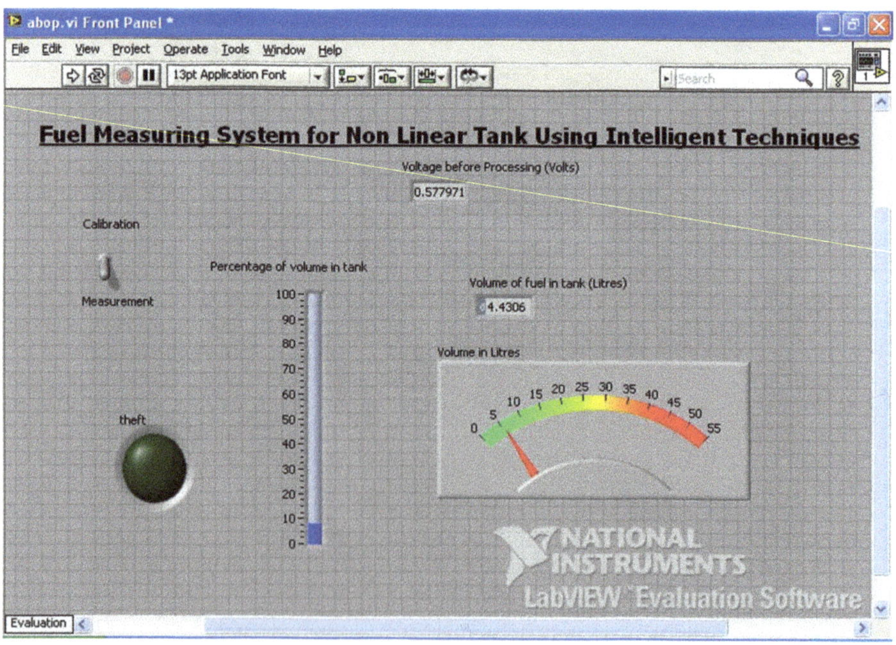

Fig. 13.7 LabVIEW front panel of the measurement module

Table 13.1 Voltage developed for every addition of 100 mL

Voltage	Volume 100 (mL)
0.4	0
0.4	100
0.4	200
0.4	300
0.4	400
0.41	500
0.41	600
0.42	700
0.43	800
0.43	900
0.38	1000
0.37	1100
0.36	1200
0.36	1300
0.36	1400
0.36	1500
0.37	1600
0.28	1700
0.28	1800
0.28	1900
0.28	2000
0.28	2100
0.28	2200
0.28	2300
0.28	2400
0.28	2500
0.19	2600
0.19	2700
0.19	2800
0.19	2900
0.19	3000
0.19	3100

noted for each increment. The inbuilt level sensor is employed and the increment quantity of 100 mL, 200 mL, and 500 mL was added and the voltage developed is portrayed in Tables 13.1, 13.2 and 13.3, and the calibrated response is plotted in Figs 13.6, 13.7, 13.8, respectively. From the graph, it is inferred that there is a discrete change in four levels. The response is not continuous.

From the above graphs and tables, it is observed that the voltage levels are discrete. This is due to the discrete steps in which the potentiometric float level sensor is designed. Since this setup is not suitable to generate sufficient data set to train the ANFIS model, the existing potentiometric float measurement system is replaced with a multi-turn pot. This results in the generation of continuous output

Table 13.2 Voltage developed for every addition of 200 mL

Voltage	Volume 200 (mL)
0.43	0
0.42	200
0.41	400
0.41	600
0.41	800
0.37	1000
0.37	1200
0.28	1400
0.28	1600
0.28	1800
0.28	2000
0.28	2200
0.19	2400
0.19	2600
0.19	2800

Table 13.3 Voltage developed for every addition of 500 mL

Voltage (V)	Volume (mL)
0.41	0
0.41	500
0.37	1000
0.28	1500
0.28	2000
0.19	2500

voltage values that can be used for training the ANFIS model. The voltage values obtained after incorporating the multi-turn pot are shown in Table 13.4 and the corresponding graph is shown in Fig. 13.11.

It is concluded from the above figure that the training error decreases and converges as the number of epochs increases.

From the above figure, it is evident that the training data and the ANFIS Output converge. Table 13.5 shows the ANFIS training and Testing dataset

6 Inferences

A novel method to calibrate a nonlinear fuel tank has been experimentally implemented and validated. The entire process consists of two modes of operation: Calibration Mode and Measurement Mode. The training dataset is generated with step-sizes of 100, 200, and 500 mL in both discrete mode and continuous mode. The nonlinear curve fitting is done using Adaptive Neuro-Fuzzy Inference System. The calibration curves obtained are shown in Figs. 13.8, 13.9, 13.10, 13.11, 13.12, 13.13.

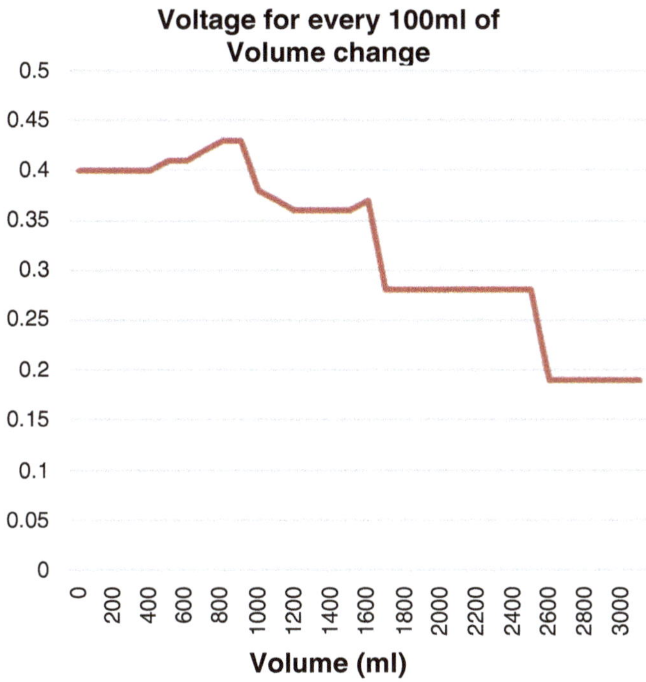

Fig. 13.8 Calibration graph for every addition of 100 ml of fuel

Table 13.4 Continuous voltage level using multi-turn pot

Voltage (V)	Volume (mL)
0.359	0
0.359	200
0.359	400
0.359	600
0.355	800
0.353	1000
0.351	1200
0.348	1400
0.344	1600
0.342	1800

The nonlinear characteristics of the model are well depicted from the graphs. The readings obtained in the measurement mode are depicted in Table 13.5. During this mode, the trained model of the nonlinear system estimates the exact volume of fuel that is poured into the fuel tank. From the observations, it is clear that the estimation error is less than 1% which is well within the acceptable limits.

Table 13.5 ANFIS training and Testing data set

Training		Testing		
Volume (mL)	Voltage (V)	Voltage (V)	Volume (mL)	Error %
100	2.32	2.28	156.3	0.563
200	2.22	2.20	190	−0.05
300	2.16	2.17	278	−0.07
400	2.11	2.12	401	0.0025
500	2.07	2.08	577	0.154
600	2.07	2.08	577	−0.038
700	2	2.01	750	0.071
800	2	2.01	750	−0.0625
900	1.95	1.96	896	−0.004
1000	1.88	1.89	1036	0.036
1100	1.85	1.89	1036	−0.0581
1200	1.83	1.84	1186	−0.0116
1300	1.76	1.76	1345	0.0346
1400	1.76	1.76	1345	−0.039
1500	1.69	1.69	1504	0.0026
1600	1.64	1.64	1663	0.039
1700	1.64	1.64	1663	−0.021
1800	1.56	1.56	1793	−0.003
1900	1.48	1.49	1936	0.0189
2000	1.48	1.49	1936	−0.032
2100	1.4	1.41	2079	−0.01
2200	1.32	1.32	2212	0.005
2300	1.24	1.24	2279	−0.0091

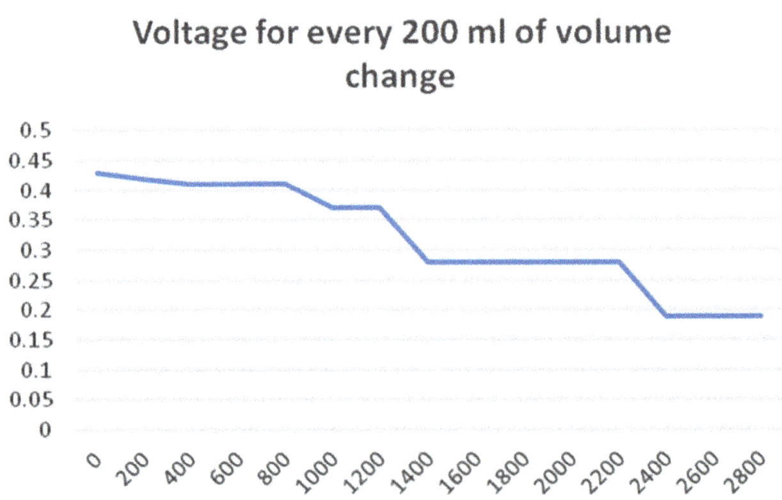

Fig. 13.9 Calibration graph for every addition of 200 mL of fuel

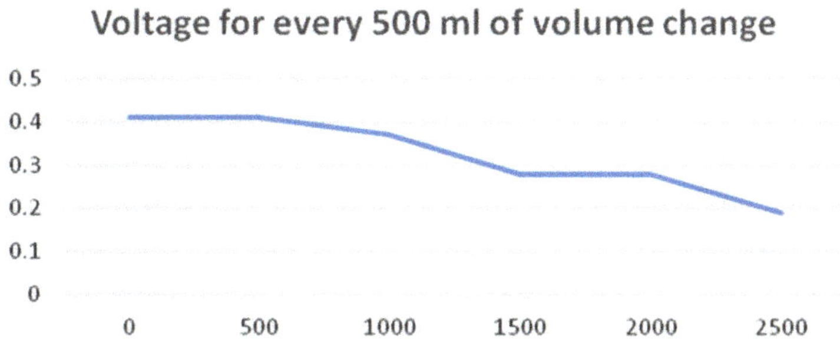

Fig. 13.10 Calibration graph for every addition of 500 mL of fuel

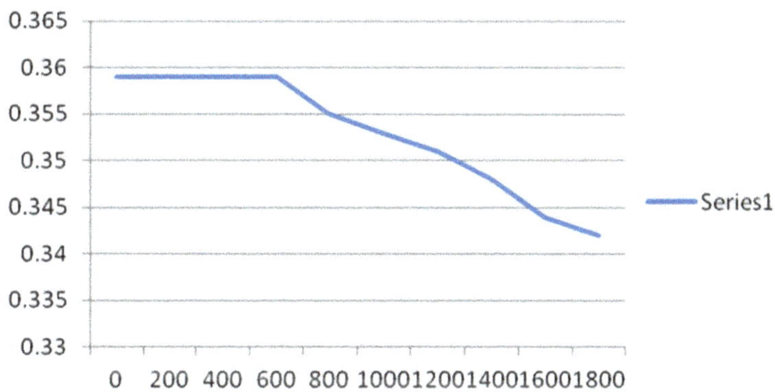

Fig. 13.11 Continuous voltage-level measurement

7 Conclusion

The proposed research works well for all irregular-shaped tanks provided, and it is calibrated for the specific tank. The results obtained are quite promising and well within the tolerable limits of error. Hence, the proposed methodology proves to be an effective method for volume measurement not only in vehicle fuel tanks but also in any nonlinear device. The measurement system developed can be further improved by including a tilt sensor to compensate for the error due to vehicle tilt. The pilot system developed can be made a standalone embedded system by incorporating the algorithm for linearizing in an embedded microcontroller. The authors would like to acknowledge the infrastructural and academic support extended by Karunya Institute of Technology and Sciences towards the successful implementation of the research work presented in this article.

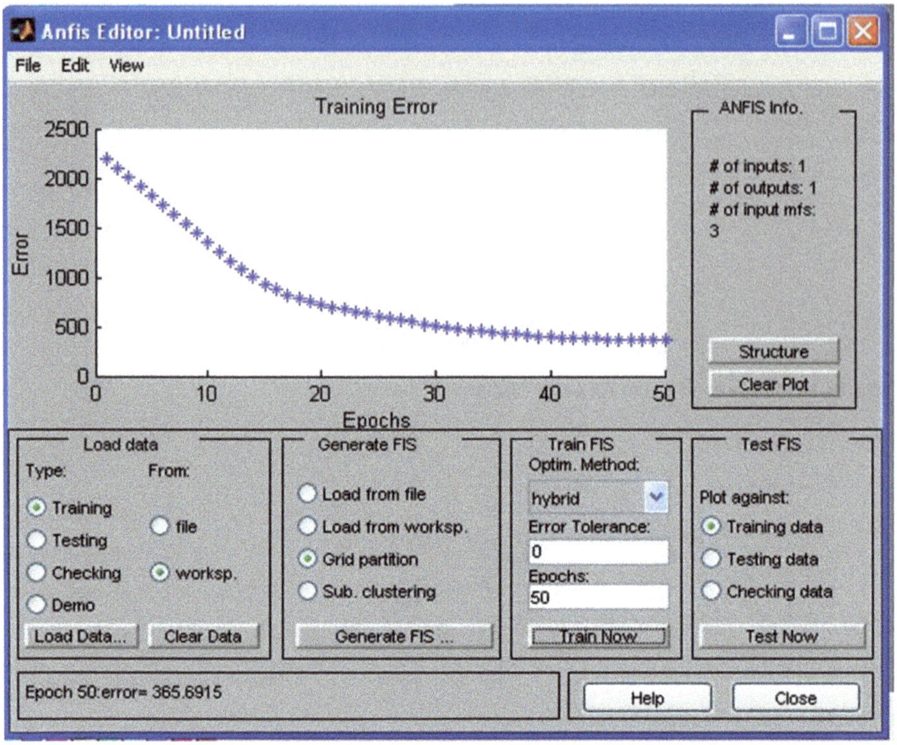

Fig. 13.12 Error graph of ANFIS implementation

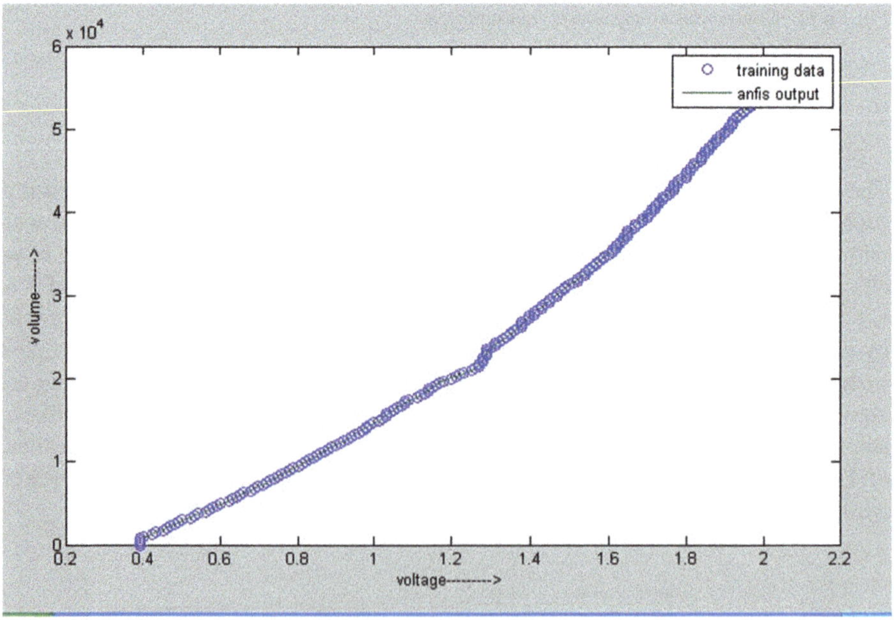

Fig. 13.13 ANFIS output and the training data

References

1. Divakar, V. (2014). Fuel gauge sensing technologies for automotive applications. *International Journal of Advanced Research in Computer Engineering & Technology (IJARCET), 3*(1), 2278–1323.
2. Terzica, E., Nagarajahb, C. R., & Alamgir, M. (2010). Capacitive sensor-based fluid level measurement in a dynamic environment using neural network. *Engineering Applications of Artificial Intelligence, 23*(4), 614–619.
3. Paczesny, D., Tarapata, G., Marzęcki, M., & Jachowicz, R. (2015). The capacitive sensor for liquid level measurement made with ink-jet printing technology. *Procedia Engineering, 120*, 731–735. https://doi.org/10.1016/j.proeng.2015.08.776
4. Bande, V., Pop, S., Pitica, D., & Ioan, C. (2013). Electrical model of a capacitive based level sensor. In *36th International Spring Seminar on Electronics Technology*. doi: https://doi.org/10.1109/ISSE.2013.6648268
5. Sarma, P., Singh, H. K., & Bezboruah, T. (2018). Fiber optic sensor for liquid volume measurement. *IEEE Sensors, 14*(4), 935–936. https://doi.org/10.1109/JSEN/2013/2295433
6. Yoshizaki, R., & Bertemes-Filho, P. (2015). Water volume measuring system using the capacitive sensor technique. *IEEE Latin America Transactions, 13*(12), 3803–3806. https://doi.org/10.1109/TLA.2015.7404911
7. Santhosh, K. V., Joy, B., & Rao, S. (2020). Design of an instrument for liquid level measurement and concentration analysis using multisensor data fusion. *Research Article, 2020*, 4259509. https://doi.org/10.1155/2020/4259509
8. Husni, M., Siahaan, D. O., Ciptaningtyas, H. T., Studiawan, H., & Aliarham, Y. P. (2015).' Liquid volume monitoring based on ultrasonic sensor and Arduino microcontroller, IOP Conference series: Materials Science and Engineering (Vol. 128). *International conference on innovation in engineering and vocational education* 14 November 2015, Bandung, Indonesia.
9. Singh, H. K., Basumatary, T., Chetia, D., & Bezboruah, T. (2014). Fiber optic sensor for liquid volume measurement. *IEEE Sensors Journal, 14*(4).
10. Mohammed, S. L., Al-Naji, A. A., & Farjo, M. (2019). Highly accurate water level measurement system using a microcontroller and an ultrasonic sensor. In *2nd International conference on sustainable engineering techniques (ICSET 2019)* (Vol. 518). doi: https://doi.org/10.1088/1757-899X/518/4/042025
11. Jose, P. S. H., Thomas, J. J., & Rajalakshmy, P. (2019). Design and implementation of stable PID controller for a quadruple tank system. *Indian Journal of Environmental Protection IJEP, 39*(5), 472–478.
12. Rajalakshmy, P., Jayanthi, G., & Jose, P. S. H. (2021). Role of sensor technology in liquid level measurement of molten metal. *Materials Today: Proceedings, 45*, 2524–2527.
13. Murawska, M. (2019), Vision sensor for measuring liquid volume. In *Proceedings of the Conference on Integrated Optics, Sensing Structures and methods*. doi: https://doi.org/10.1117/12.2536534

Chapter 14
Internet of Vehicles

G. Santhakumar and Ruban Whenish

1 Introduction

Traffic management is a well-known global issue due to the increment of vehicles on the road. The movement of vehicles is increasing which led to severe traffic issues in the metro and large cities. Bogota in Columbia topped as the most congested Latin American city with drivers losing 191 h which is nearly 8 days. Indian cities such as Bengaluru, Mumbai, and Delhi are facing the worst traffic congestion in the world with losing 243 h, 209 h, and 190 h, respectively. Drivers in the United States lose 97 h per year due to traffic issues. The lost hours due to traffic issues directly affected fuel wastage, environment, and indirectly affect productivity, economy, and business [1, 2]. Apart from these road traffic accounted for 40% of accidents, 1.3 million casualties, and 20–30 million injuries annually according to the WHO report, 2016. Many countries have developed traffic management systems dealing with signal timing procedures. Timing-based traffic signals controlled by pneumatic actuators are used to control the traffic, yet it could not have adequate control over the heavy traffic scenarios in modern smart cities. It is essential to have better communications and connections within the vehicles as well as surroundings due to their mobility [3].

The enlargement of Information and communication technologies (ICT) is used to extract traffic information and direct the drivers to take traffic less routes. IoT can play a key role by restricting traffic congestion, promoting safety, and coordinating vehicle mobility on the road. IoT has smart sensors, computing devices are interacting together and the ability to receive, process, and transfer data with no human participation [4]. IoT can convert into smart roads by embedding sensory systems and providing information regarding traffic situations. One of the highly

G. Santhakumar
Sri Krishna College of Technology, Coimbatore, Tamil Nadu, India

R. Whenish (✉)
Vellore Institute of Technology, Vellore, Tamil Nadu, India

M. Kathiresh et al. (eds.), *E-Mobility*, EAI/Springer Innovations in Communication and Computing, https://doi.org/10.1007/978-3-030-85424-9_14

attractive applications of IoT in intelligent transportation system (ITS) where traffic management is ensured well by smarter use of transport networks enable safety, route optimization, delay reduction, and pollution control. IoT was introduced in connected vehicles in 1998 applied with radio frequency identification (RFID) sensor tags. Tracking of vehicle mobility, as well as connectivity, becomes more convenient in ITS, after adopting technologies such as global positioning system (GPS), bluetooth, WiFi, Zigbee, and so on [5, 6]. These technologies brought a new environment by communicating between the vehicles as well as controlling vehicle mobility. Vehicular communication systems, such as V2V, V2I, V2P, and vehicle to everything (V2X) allow a vehicle to communicate the circumstances with short-range wireless signals. It can widen the vehicle recognition events aided by smart sensors and detectors to provide necessary guidelines for drivers to take smart and safe decisions. For example, these technologies provide traffic flow parameters, road conditions, short cuts/alternate routes, atmospheric conditions, and so on can be shared by drivers of other connected vehicles developing networks called VANETs [24]. VANET converts all the vehicles as a wireless router or mobile node which takes part in the VANET network, in that way it enables vehicles to be connected inside the network. In case a vehicle moves out of the VANET network, other vehicles can join inside and create a mobile network. VANET is limited with some connected vehicles which cover only a small mobile network. VANET cannot provide sustainable services for broader areas so it is difficult to manage several situations such as heavy traffic jams, larger city areas, driver misbehaviors, complex road networks, and so on. The usage of VANET is local and discrete, temporary and unstable; therefore, usage of VANET is becoming stagnate [7, 8].

Contrary to the VANET network, IoV has brought forth two major technologies such as vehicle networking and vehicle intelligence by integrating communication advancements with connected vehicles. By using techniques such as deep learning, artificial intelligence (AI), cognitive computing, and so on with intelligent integration of humans, vehicles, environment, and things as a whole broader network. IoV is not short and unstable like VANET, but it is an open network system with integrated services that can provide even to a whole country. The benefits of IoV are highly controllable, manageable, operationalized, and credible. IoV network is composed of multiple vehicles, multiple networks, multiple users, and multiple things. IoV can govern and quantify huge complex data with the intention of improving the sustainability of information, communication, and complex network systems [9].

The essential ideology of IoV is to comprehend thoroughly the unification of vehicle–human–environment to nurture efficient transportation, reduce the cost factors, avoid collisions and fatalities, ensure the safety prospects of humans as well as enjoy their rides [10]. VANET is a vehicle interconnection network that can act as a subnetwork of IoV. IoV has another subset that is vehicle Telematics, which can transfer and interchange the electronic data and location-based information between connected vehicles. The information such as remote diagnostics, navigation, entertainment content, and so on comes under Vehicle Telematics, which can align with intelligent transportation systems considered as an application of IoV, for example, traffic guidance system, intelligent vehicle control, safe navigation. VANET, vehicle Telematics, and other connected vehicles have failed to handle global information due to their lack of processing capacity which is limited to short-term applications. The inception of IoT, cloud computing, deep learning, AI, and big data has evolved into IoV, to handle and compute/process global information. IoV has characteristics such as the trajectory of vehicles which is subject to the road distributions of the city, integration of humans and vehicles. Based on the network model, service model and human–vehicle behavior model will be developed. IoV interconnects intelligent systems of vehicles, humans, cyber-physical systems of the surrounding environment and integrates along with sensors, mobile devices into a global network [11–15]. IoV has a combination of inter-vehicular networks, intra-vehicular networks, and vehicular mobile internet. Thus, it is possible to build a global network in IoV, with multilevel collaboration with existing multi-vehicle, multiple users, multiple networks, and multiple things. Sensors are smart devices in IoV, playing a vital role in terms of feeding information and behaving wisely based on the need. IoV has external sensors, internal sensors, and measurable sensors. External sensors primarily providing information on GPS (global positioning system), LIDARs, cameras, and so on. Internal sensors such as automotive sensors (brakes, accelerator, etc.) and cockpit sensors (alertness, the health status of the driver, etc.). Social media, phone texts, tweets will be considered as measurable sensor outputs to understand the state of the driver.

Profitability from connected cars and their drive technology would generate up to 81 million USD income annually. The effective usage of traveling time is the prime goal of IoV. IoV has a potential market opportunity by monetizing the time wasted by traveling in the coming years. Even 5 min saved globally by IoV are expected to generate 25 million Euros per year by 2030. European Union made an initiative to develop next-generation Cooperative Intelligent Transportation Systems (C-ITS). Various reports from the United Kingdom, United States, and Australia suggested the positive impact of connected vehicles. In the United States, security chips were equipped in vehicles to define an identity for every entity and vehicle tracking could be done on the internet. In Delhi, all registered vehicles and metros were installed with GPS and Wireless Fidelity (Wi-Fi). Google is working along with certain automotive industries and IT companies to develop an Android system for connected vehicles; "Carplay" was developed by Apple which enables the driver to access services of iPhone through the display of car with a voice support feature. All such efforts are the roadmap toward the design and development of IoV.

By considering all these facts, the proposed chapter has the following sections. Section 2 provides basic network architecture and its elements of IoV. Section 3 presents four layers (4L), five layers (5L), and seven layers (7L) of IoV architecture and its overview. Various types of communications are also discussed here. Section 4 gives architecture analysis and protocols. Finally, Section 5 brings toward the future direction of IoV and its transformation with conclusions.

2 Network Architecture of IoV

Numerous cycles engaged with network communications in OSI or TCP/IP model like packets directing and bundles conveying. Agreeable C-ITS parts that structure the IoV ecological system depend on numerous components, and they generally include various gadgets or infrastructures. Moreover, being a specific MANETs (Mobile Ad hoc Network) network classification, VANETs, V2I, V2X, and thus IoV sending is unpredictable and needs an unprecedented effort and thought because about their characteristics, for instance, a huge degree of versatility and dynamic change in the geography, which produces dispersed organizations. For example, ETSI in Europe design is portrayed by Fig. 14.1.

The principal stands out from the conventional TCP/IP or OSI model is:

- The occupancy of a facilities layer which is liable for VANETs affiliated functions.
- The blend of the network and transport layer falls under a single layer.
- Incorporation of two uncommon layers: Management and Security.
- The occupancy of ITS committed stack which coordinates the geo networking addressing.

Notwithstanding, such designs are just identified with the inward structure and capacities of one gadget engaged with the interchanges without explaining the

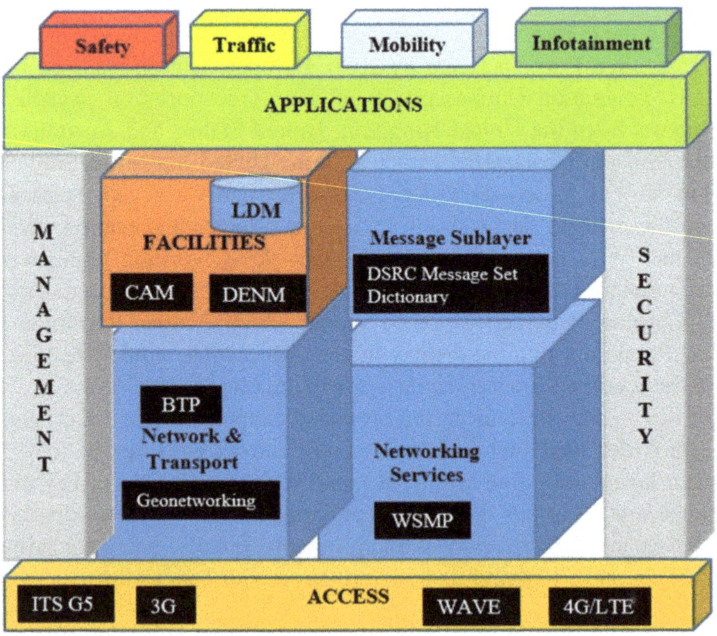

Fig. 14.1 ETSI architecture

communication cycle if there should arise an occurrence of handover or different systems that include outside gadgets cooperation. The game plan of IoT in the VANETs setting engaged promising courses of action and applications (for instance, persistent applications for autonomous driving, road traffic the board applications and comfort applications), which were delivered purported as IoV. IoV biological system is essentially shaped by four sections which are:

End centers: vehicles, PDAs, sensors, and related contraptions
Establishment: Roadside units (RSUs), Wi-Fi hotspots, cell associations (3G/LTE) base stations
Activities, for example, methodology prerequisite, stream-based organization, security and assessing and
Administrations, for instance, public cloud for enrollment-based organizations, private cloud, attempt cloud for huge business data, voice or video, and so on.

These days, few suitable association structures between these IoV segments are suggested. This offer would like to IoV arrangement probability despite the RSU sending delay since its costly execution, which is assessed to €660 Million from 2020 to 2026. Multiple specifications and normalization exercises in IoV space are still in development, a survey on network design in IoV found that may be of extraordinary significance and will assist the scientists to know and refreshed to what in particular is as of now completed in IoV. This persuades us to commit to the IoV research network.

3 IoV Architectures Overview

As of now, various endeavors and the academic world examiners are giving in-solid thought to novel association plans that could capably permit the IoV sending and associated business market models. In [16], Bonomi from Cisco has depicted 4L-based plans as showed up in Fig. 14.2. The suggested model has four phases that each IoV correspondence reliably incorporates with Embedded systems and sensors, Multiservice Edge, Core, Datacenter, and Cloud, as seen in Fig. 14.2.

In [17], authors likewise presented a five-layered engineering, outlined in Fig. 14.3, which is made out of the accompanying layers:

- Perception: The connection between the vehicle and its present condition is depicted in this layer. Devices kept inside the vehicle, for instance, sensors, actuators, singular devices, and those presented over the road, RSU to collect appropriate information to be used in vehicle's comments.
- Coordination: this layer is primarily trustworthy on interoperability, controlling, and report transportation security.
- AI: This is the center layer where choices part undertakings must be executed. This layer predominantly centers around large information examination, information mining, distributed computing, and master frameworks-based choice.

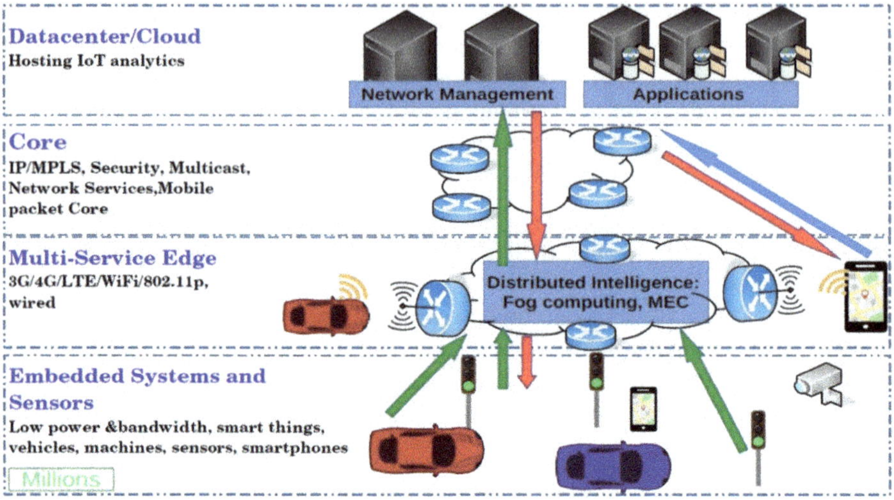

Fig. 14.2 Four layers architecture of IoV

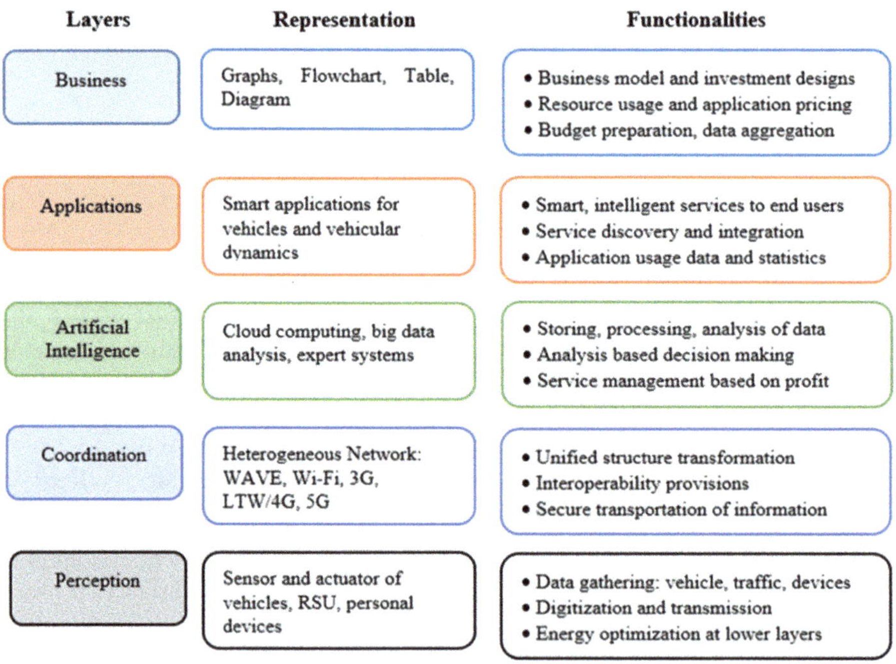

Fig. 14.3 Five layers architecture of IoV

- Application: This layer concerns such an organization and pre-necessities present in the structure.
- Business: the part depicts which kind of associations the IoV deface kit will offer to customers.

To propose a solid controlling show for IoV climate [18] designing given in past works, by planning the software-defined networks (SDN) perspective which involves secluding the association traffic light plane and the data transfer plane. Accordingly, the presented architecture with six layers (6L) which are perception layer, correspondence layer, application layer, cost layer, security layer, and a layer for law, ethic, private life, and legal use. The SDN perspective is applied in the correspondence layer in which they decide an SDN steering convention (Control plane + Data plane) sub-layer and radio access technologies (RAT) types (homogenous or heterogeneous) sub-layer [19].

In [20], the authors proposed a 7L-based design, which showed up in Fig. 14.4. They arranged this 7L plan by decreasing the layers functionalities' multifaceted

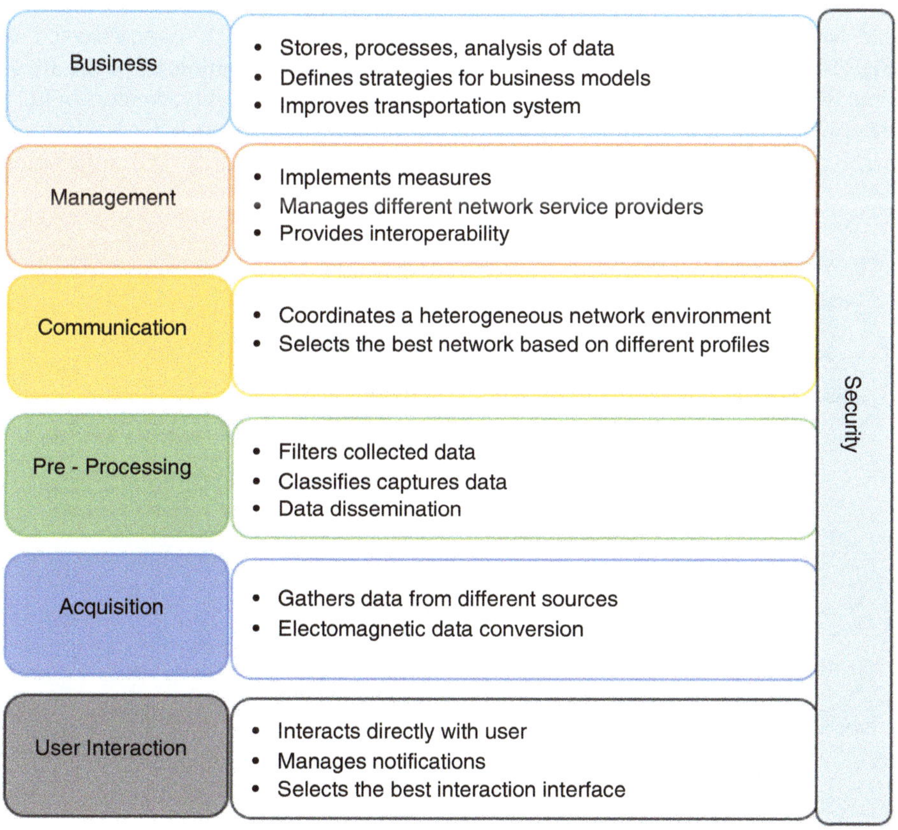

Fig. 14.4 Seven layers architecture of IoV

nature and by social event the essentially equivalent to limits in same and fitting layers, as needs are, making straightforward its utilization.

The principle goal of this design is the enhancement of the number of layers by upgrading the differentiability among layers. This enhancement must be additionally sent as much productively as conceivable to accomplish the organization qualities and necessities which are mostly: interoperability, dependability, versatility, particularity, straightforwardness, and combination adaptability with the internet. As it is mentioned in Fig. 14.4, a layer for customer collaboration which exchanges with the UI, a layer for data making sure about, a pre-taking care of layer in which assembled data must be pre-arranged before employed in the correspondence layer, which encourages the heterogeneous organization surroundings. From that point forward, they incorporate a layer for interoperability and organization specialist providers which are called management. At long last, they recommended a business layer and a security-related layer.

They additionally acquaint a gadget with gadget (D2D) correspondence approach which may be an encouraging and presumably utilized arrangement in the following years in machine to machine (M2M) interchanges setting.

An altogether and comprehensive audit on the gadget to gadget correspondences can be discovered in [21]. The D2D design approach in IoV is represented in Fig. 14.5. Considering the moving issue toward asset distribution to assure Real-time (RT) traffic in IoV and to update the asset utilization capability, creators of [22], determined an IOV design, regardless, they moreover suggested a model for asset allotment and improvement by following the effortlessly and solicitation

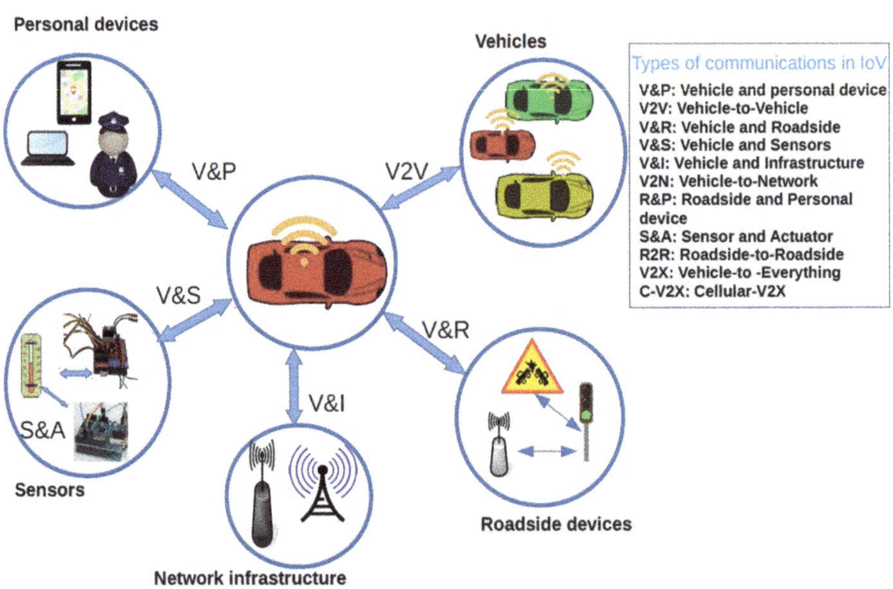

Fig. 14.5 Different types of D2D communications

approach and utility limit. The recommended plan is a different leveled IoV designing that involves three layers which are: a data gathering cloud, a web-access cloud, application cloud. In their architecture, they additionally thought to be four organizations which are: The on-vehicle Sensor (OVS) organization, V2Vorganization, close to street V2I organization, and a V2P organization.

Authors of [23] presented a fog enlisting RT-Based ITS Big Data Analytics (RITS-BDA) designing in IoV conditions, which is made out of a three-estimation structure configuration including the components of IoV, smart figuring, and consistent huge data assessment. RITS-BDA is then multi-dimensional layered designing which is made of the going with layers: 4L in the astute enrolling estimation (three various leveled Fog figuring layers, circulated processing layer), three layers (3L) for the progressing enormous data assessment estimation (serving layer, group layer, speed layer), and 6L for IoV estimation. Their designing way to provide the real execution of persistent ITS colossal data applications and is loosened up from a nonexclusive consistent enormous data getting ready to plan considered lambda designing that was introduced in [24].

4 Protocols Stack and Architecture Analysis of IoV

A conventional stack is given for every design which contains particular of the utilitarian demands of each plan layer by figuring out the proper existing shows. For example, VANETs standards, 3GPP rules, and so on. For the 5L design [17], a convention stack (shown in Fig. 14.6) is formed by four planes which are:

The board plane, activity plane, security plane, and layer plane. In any case, [21] proposed a show store of two planes: an operational plane and a security plane appeared in Fig. 14.7.

CALM-SL = CALM Service Layer
OMA-DM = Open Mobile Alliance Device Management
6LoWPAN = IPv6 over Low-Power Wireless Personal Area Networks
RPL = Routing Protocol for Low-Power and Lossy Networks
IP = Internet Protocol
ROLL = Routing Over Low Power and Lossy Networks
XMPP = Extensible Messaging and Presence Protocol
CoAP = Constrained Application Protocol
HTTP REST = Hypertext Transfer Protocol Representational State Transfer
MQTT = Message Queuing Telemetry Transport
LLAP = Lightweight Logical Automation Protocol
LoRaWAN = Low-Power Wide Area Network
OTrP = Open Trust Protocol
S-MIB = Security Management Information Base
HSM = Hardware Security Module
S-IC = Security Information Connector

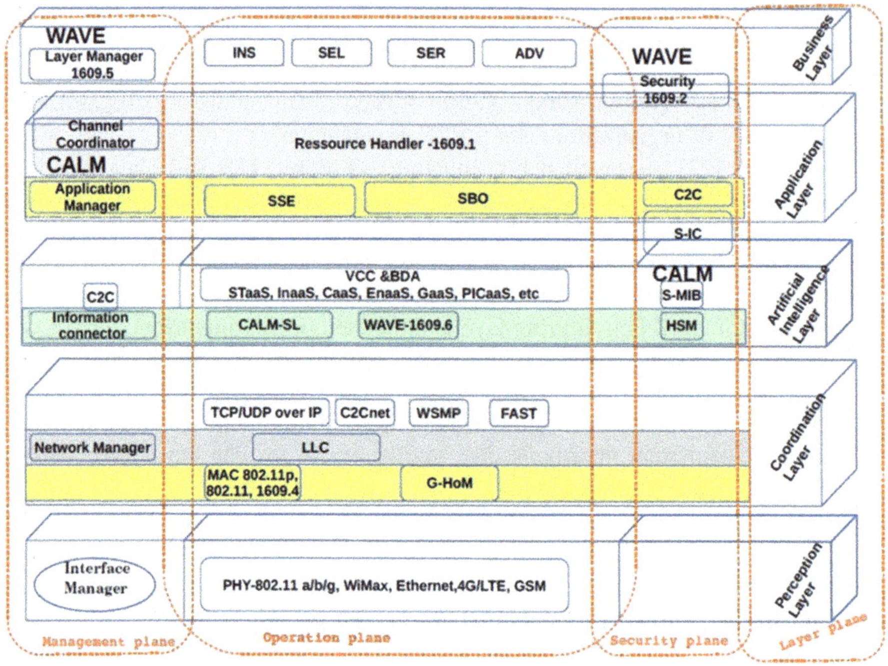

Fig. 14.6 Protocol stack of the IoV architecture with five layers

The coordination calculation control layer is utilized for acknowledging processing and control on human, vehicle, climate equivalently. It can control single and multitude through registering singular data coordinately. Besides, it can finish some deal with task IoV. This layer can accomplish the supportability of administration necessities employing processing and control on-request assets by planning coordinately. As per various requests from IoV, shall separate the coordination model into two classes that are individual coordination model and multitude coordination. The individual model is utilized for tackling the harmonization issues among humans and vehicles and individual article and multitude objects. This model comprises the human (driver) and the driving vehicle. The multitude object comprises all objects of IoV aside from the individual item. Human and vehicle communication understands the tight coupling and finishes the immediate connection through an in-vehicle organization. To settle the bottleneck of correspondence, a clever picture is needed for going about as a specialist for the knowledge of driver and vehicle. The specialist can finish the harmonization of the items through an examination of the individual conduct. The multitude model is utilized for tackling the harmonization issue among humans, vehicles, and climate in IoV from the point of view of collaboration administrations. In this model, humans, vehicles, and climate arrange with one another. The human remembers driver, traveler, and human in climate (rider, person on foot). The vehicle incorporates driving vehicles and leaving vehicles. Similarly, the climate incorporates charging heaps, ecological

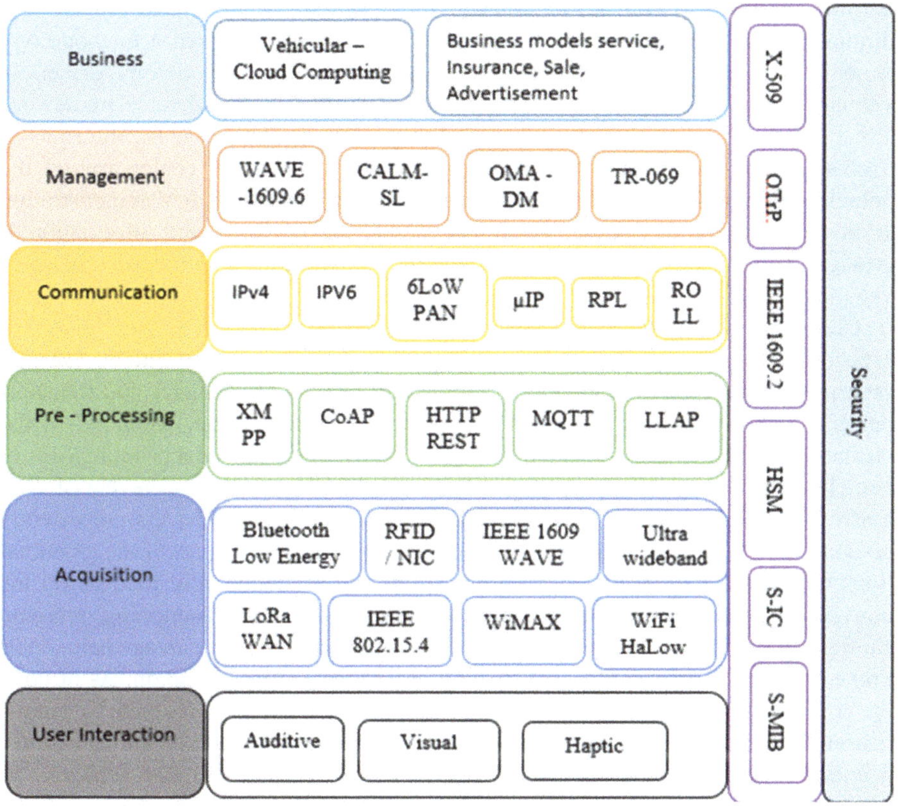

Fig. 14.7 IoV Protocol stack with seven layers

checking, and data channel administration passageways. Attributable to the nonsymmetric of figuring capacities among human, vehicle, and climate, the all-encompassing detecting of IoV depends on the keen vehicles, yet additionally on the multitude coordination detecting. Humans and vehicles play out a functioning part in the coordination registering through cell phones and vehicles. Notwithstanding, they partake in detachment through wise electronic stations, which is conveyed by ITS. Along these lines, the multitude co-appointment model of IoV comprises humans, vehicles, climate, insightful pictures, and administrations. To adapt to the difficulties of IoV, the knowledge of humans and vehicles should be displayed on the internet. The vehicle and street to fairly see the city traffic information and registered the traffic to acquire the geography of city traffic organization and its attributes, whereupon better IoV administrations can be given. The vehicle cooperation issue to the vehicle speed harmonization issue by figuring if the vehicle direction converges. Vehicles can organize at the crossing point by taking care of the harmonization issue. Ghaffarian et al. changed the traffic issue into a whole number programming issue by examining the two-way single path between segment structures. They tackled the harmonization issue with no significant traffic signals between area climate and diminished the normal deferral adequately. Milan et al. proposed a methodology dependent on V2V and V2I to facilitate processing. In this methodology, driverless vehicles with a few distinctive correspondence guidelines and diverse framework structures can speak with one another, and they settle on choices as per information exchange. For the current exploration, they overwhelmingly center around the vehicle's harmonization, especially in VANET. Be that as it may, IoV administration additionally requires correspondence assets, calculation assets, and information to arrange in IoV. Besides, there are just a couple of studies that emphasize human–vehicle–climate harmonization.

Closed services are administrations that focus on explicit ventures or stages, especially benefits profoundly associated with the transportation of vehicles. Driving security is one of the overwhelming application records for vehicles and transportation, for instance, dynamic street wellbeing can diminish the likelihood of car crashes and advance transportation wellbeing. In light of shared data about places of vehicles and convergences, speed, and driving vehicle separation, the event of a car crash might be anticipated by V2V and V2I correspondence. Drivers consistently respond rapidly to stay away from auto collisions. The greater part of the examinations on dynamic street wellbeing applications pre predominantly focused on the convergence strife admonishing, surpassing cautioning, crash cautioning, rear-end cautioning, chain impact admonishing, crisis vehicle cautioning, salvage help, crisis brake, petty criminal offense admonishing, traffic state sees, and so on. As of now, the crash admonishing procedures are autonomous program bundles gave by unique gear makers as the premise of frameworks. They give notification of auto collisions, a notice of street conditions (e.g., dangerous asphalt), and rear-end vehicles. To evade a mixture impact, Colombo et al. have shown a technique, which uses the vehicle's dynamic model to take care of the vehicle booking issues. Most of the functions in ITS are closed service as it is a complex circulated framework joined with cutting-edge innovations in zones of correspondence, detecting, versatile situating, information bases, smart data handling, and programmed control. ITS dominatingly contains six essential subsystems. They are progressed city the board framework, progressed route framework, progressed vehicle control framework, business vehicle the executives, progressed public transportation framework, and progressed metropolitan transportation framework. The insightful traffic the board likewise incorporates ETC, which guarantees vehicles to cross street and extension cost stations at ordinary speed for decreasing the likelihood of blockage. Considering processing assets and capacity assets brought about by huge scope versatile clients, novel metropolitan traffic the executives' framework is proposed for taking care of the issue of deficient registering and capacity assets dependent on keen transportation cloud [25].

Open administrations in IoV are mostly given by the outsider to clients, which are grouped into on the web and disconnected stream media and human–machine intelligent administrations, including video gathering, climate data, information transmission, web administrations, music downloads, game intelligence, and side of the road administrations. Future portable internet providers are reached out to vehicular administrations, which offer types of assistance for vehicles, for example, Apples "CarPlay." For the most part, the administrations accommodated clients incorporate two perspectives, specifically customized amusement administrations and transportation administrations. Clients' amusement benefits dominatingly center around those that can be acquired from the network or different vehicles. For instance, a notice of focal points, nearby online business, and media downloads. In any case, the focal point of customized transportation administrations is prevalently centered around transportation information that the clients ought to recover from networks, for example, way route and HD (high definition) maps for mechanized driving. Telematics additionally includes some open administrations like telematics in taxicabs, which may gather direction information, break down traffic state, and offer opening types of assistance. Telematics is a combination of media transmission and informatics. In addition, it is a helpful framework, which gives data across internet innovation, to be specific vehicular PC frameworks, remote correspondence innovation, satellite route gadgets, trade messages, and voices. Telematics gives capacities about security applications, crisis salvage, guard against burglary, and distant analysis. Telematics can contact with administration focus through remote correspondence to find the flaw precisely, give the shortcoming causes and analysis to look after staff, and guarantee the vehicles travel all the more securely. Using cloud administrations, telematics synchronizes information with other electronic gadgets, gives continuous street condition data, and chooses the best course. Telematics can refresh the most recent guide data to keep up the guide information precise and state-of-the-art. At that point, it can question the data about encompassing offices, stopping plots, shops, and administrations. It can likewise give the elements of phone administrators and one-contact calling focus to diminish activity and encourage clients. All the previously referred administrations center around upgrading the driving well-being and some normal applications. Not many of them join worldwide traffic data and the driver inclination to offer customized assistance, which could turn out to be increasingly more significant in the future. Subsequently, we ought to investigate novel IoV administration.

In IoV, vehicles and foundations access networks utilizing an assortment of remote access advancements. Nonetheless, there exist huge contrasts between various advancements. Subsequently, a transmission control network is needed for protecting these distinctions, which infer that the heterogeneous organizations' mix is inescapable with the improvement of IoV. To understand the transmission control organization, the heterogeneous organizations must be incorporated with serious level, which will draw in numerous difficulties. Consequently, the mix of heterogeneous organizations has become a hot examination field. SDN can control network traffic deftly through isolating the organization gadget control and information. In SDN, like a pipeline, the network turns out to be wiser, and it can understand the organization's transmission and control. As indicated by the distinctions created in preparing the vehicular information utilizing diverse correspondence advances (e.g., cell organization and DSRC), a way to deal with shield these distinctions. A novel vehicle correspondence design dependent on SDN. In this engineering, the distinctions of various heterogeneous access advancements could be protected through the SDN trade interface. To dispose of the distinctions of cell organization and broadcast organization, a multi-radio organization joining approach dependent on substance appropriation organization. This combination organization can fulfill ser-indecencies of sound and video. Distributed computing has the upsides of incredible figuring, dynamic booking of the asset, giving on-request benefits, handling huge data productively, and incorporating the executive's instruments. These favorable circumstances can be utilized for taking care of the issues of data sharing and transmission delay in IoV. Henceforth, joining the cloud and vehicle is an altogether significant improvement of IoV. Vehicloud, which is an engineering dependent on distributed computing, and tackles vehicle correspondence precariousness issues through moving the conventional vehicular organization to support-based design. A novel engineering consolidated vehicle with distributed computing, named V-Cloud, for tackling the correspondence inadequacies issue of V2V and V2I for current 3G/4G. MEC (mobile edge computing) coordinates the Internet and remote organization viably, and it builds the elements of figuring, stockpiling, and information preparing in the remote organization. Also, it assembles an open stage for embedded functions and opens the data communication between remote organizations and administration workers through a remote application interface. MEC coordinates the remote organizations and administrations, and it overhauls the conventional base station to a shrewd base station. For future organization transmission and control, MEC will likewise assume a significant job. Likewise, with the approach of the 5G time, the correspondence postpone will be extraordinarily decreased, and the street data will be instantly sent to the information stage. Thus, the plat structure can control the traffic all the more precisely and actualize V2X correspondence applications [26]. All the previously mentioned advancements don't consider the vehicle highlight (e.g., vehicle speed and data transmission) for vehicle access and transport, which are needed in IoV and significant for vehicles to associate organizations.

As of now, the organization access innovations of IoV shall be characterized into between vehicle network access advancements and versatile Internet access advances. The entrance advances of bury vehicle networks incorporate DSRC (Dedicated Short Range Communications) and WAVE (Wireless Access in the Vehicular Environment). Also, the portable Internet access innovations incorporate LTE (Long Term Evolution) and WiMAX-WLAN. Vehicles depend on these remote correspondence advancements to get to networks, which can understand the correspondence among vehicles and organizations. DSRC is a kind of effective remote correspondence innovation, which bolsters the moving objective acknowledgment and two-path correspondence with rapid movement in a particular region (generally many meters). DSRC embraces the correspondence standard IEEE 802.11a. DSRC has two working modes: one is to set up the association among vehicles, which is utilized for upgrading the traffic security through being careful separation and cautioning auto collisions. The other is to build up the association among vehicles and streets, which is utilized for facilitating traffic pressure through the ideal course. At present, WAVE innovation has gotten one of the principal access advances to associate with the organization. The WAVE is utilized for taking care of the direct obstructing issue in the actual layer when vehicles access the organization. A multi-need conveyed channel blockage control approach dependent on IEEE 802.11p. The object of this methodology is to guarantee the low crash rate and most extreme transmission likelihood of the high need data [27]. In the investigation of LTE, a methodology of vehicle access network dependent on 4G and LTE-A (LTE–Advanced) [28]. The test outcomes uncover that the methodology can be worked in the vehicle with a speed of 140 km/h. The creators initially summed up the connected examination on LTE research organizations and industry. At that point, they examined the difficulties in the current examination about this issue and anticipated the advancement heading of LTE in IoV [29]. The focus has been given to the foundation of correspondence engineering and model in the LTE-A framework. Additionally, the presentation distinction is somewhere in the range of 2D and 3D channels. Additionally, HUAWEI dispatched LTE-V (LTE–Vehicle) for notice and controlling vehicle impact [30]. In light of the LTE and WLAN, portable Internet can give correspondence among vehicle and vehicle and vehicle and organization. LTE is the third versatile age correspondence standard created by the association venture association. The engineering of LTE is more straightforward, and it can diminish network hubs and complex framework degrees, which lessens the framework delay. Additionally, it decreases the expense of organization sending and support. WiMAX (Worldwide interoperability for Microwave Access) and WLAN (Wireless Local Area Network) are two remote correspondence innovations dependent on IEEE 802.11. Since these kinds of remote access advancements are integral, significant examinations consolidate the two innovations to enable the vehicle to interface with the organization. The chance of joining WiMAX and LTE-A, and analyzed the throughput and deferral of V2I utilizing two innovations. The performance indicated that the innovation can upgrade the correspondence proficiency of V2I utilizing the blend of WiMAX and WLAN. A half-breed of WiMAX and WLAN, named Carlink, can give the principles of vehicle correspondence and security of route frameworks [31].

It is to be seen that each plane collaborates with all the layers in its engineering. For additional insights regarding convention stack functionalities and portrayal, perusers are urged to allude to the comparing articles in [32]. By breaking down these previously mentioned proposed structures in the IoV area, we discovered numerous perspectives that demonstrated that IoV is still in its beginning phase of normalization and shows numerous chances and difficulties for both scholarly world and enterprises specialists, IT engineers, internet suppliers, and so on. This is wonderful particularly while considering the IoV observation from various examinations, regardless of whether it is from modern or scholastic analysts. The thought about perspective used to propose and plan these structures are extraordinary and at times layers are compatible.

- In Fig. 14.3, the discernment layer functionalities relate to the functionalities introduced in the inserted frameworks and sensors layer in Fig. 14.2. A similar layer is a bit of two layers (2L) (e.g., client cooperation and obtaining) in Fig. 14.4.

- The coordination layer in Fig. 14.3 is known as a multi-organization edge in Fig. 14.2, while it is known as a correspondence layer in Fig. 14.4.
- Datacenter/cloud layer in consideration, 2 is parceled along with 3L (artificial understanding layer, application layer, business layer) in Fig. 14.3, while it is disconnected into 2L (Management layer and Business layer) in Fig. 14.4.

Likewise, remark on an issue in the layer's structure between Figs. 14.3 and 14.4. In Fig. 14.4, there is a preplanning layer, which looks at counterfeit information in Fig. 14.3, going before the correspondence layer. Nevertheless, in Fig. 14.3, the arrangement which occurs in the fake information layer comes after the coordination layer. Another perspective to be examined is the presence of the security submitted layer in Fig. 14.2 which was not present in the 5L-based plans in Fig. 14.3. The correlation may be long while differentiating these models independently, from 3L-based plan to a 13L-based plan.

5 IoV Applications and Resource Management System

Connected vehicles in the system can share a variety of available information to make wise decisions. For example, a cloud-based VANET architecture facilitates the identification of accessible resources in real-time. It can be applicable for cloud-based IoV applications like real-time video sharing, complex computation, dynamic bandwidth sharing, enhanced resource management, and so on. IoV is an emerging technology that integrates multiple sensors, which are placed on roads, vehicles, and devices worn by pedestrians to ensure safe driving and secure vehicle-to-vehicle communication (V2V). Sensors in vehicles collect information such as GPS location, vehicle health conditions, surrounding environment, conditions of the road, and upload the collected data in the cloud. The data will be processed and provide optimized outcomes to the user with the intention of vehicle performance enhancement and safety of drivers, vehicles, and pedestrians [33]. Cloud-based VANET architecture consists of a local VANET cloud network (LVCN), wide VANET cloud network (WVCN), and central VANET cloud network (CVCN). LVCN shares its connected vehicle resources, computational, storage within a mile's range. Vehicles across LCVN will lose their connectivity from the cloud network. Vehicles inside LCVN can obtain the available mined information such as traffic congestion, traveling time, short routes based on safe route optimization. Each LCVN has dedicated servers attached with other cloud networks in the VANET environment. WVCN has more resources in the cloud network which offers vehicles to be connected and communicated within the WVCN cloud site. WVCN establishes interconnection with sets of LCVN via Wi-Fi, internet, and DSRC. CVCN is like a cellular communication that has a group of WVCN and its servers are connected through the internet. CVCN has a resource-rich cloud network for superfast computations, information transfers which facilitates vehicles to make decisions. Vehicles in CVCN have guided well to face complex issue [34]. Cloud services are offered

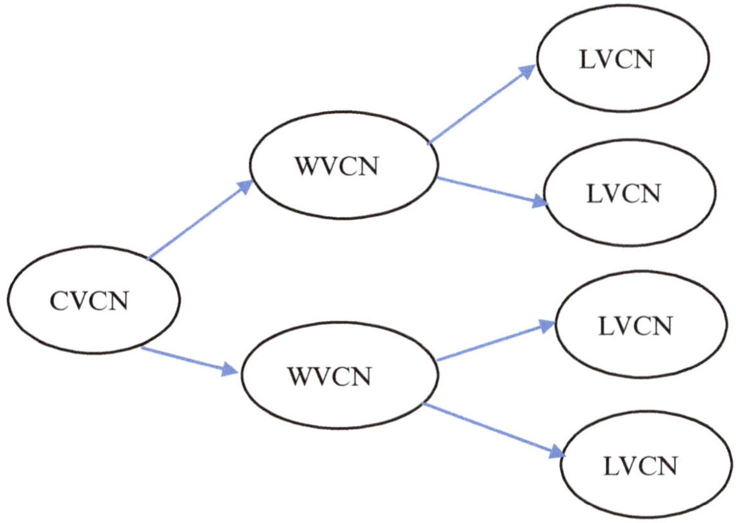

Fig. 14.8 VANET architecture with cloud based

commercially by several software platforms such as IBM, Google, Oracle, SAP, and so on. The process consists of RT data gathering, transmission, and analysis of data [35]. Cloud-based VANET architecture is described in Fig. 14.8.

Cloud-based VANET architecture facilitates fast, reliable, and efficient V2V, V2I, and V2X communication with routing strategy and optimal resource allocation. Thus, it equips rapid implementation of smart cities and capable to manage under cloud-based VANET architecture. The advantages of using a cloud server are capable of fast computing, processing, and analyses in real time. Based on this strategy, some ideal systems were developed under this cloud-based network integrated with IoV, to handle global vehicle management as follows,

- Smart traffic management
- Smart accident warning system
- Privacy and security
- Smart city design
- Cost-effective web services

Considering one of the IoV applications, designing a smart city requires certain key technologies which are mentioned in Fig. 14.9. These technologies effectively monitor and control vehicle mobility and improvise the safety and time savings.

The cloud-based architecture is required to manage all the resources to avoid data collision and provides global communication. It has three stages which are mentioned in Fig. 14.10.

The design of cloud-based architecture needs to be directed potentially with the following functional parameters which are mentioned in Table 14.1.

Fig. 14.9 Key technologies for smart city design

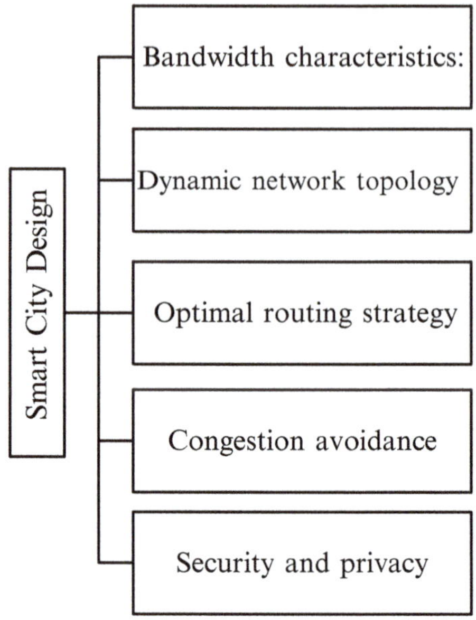

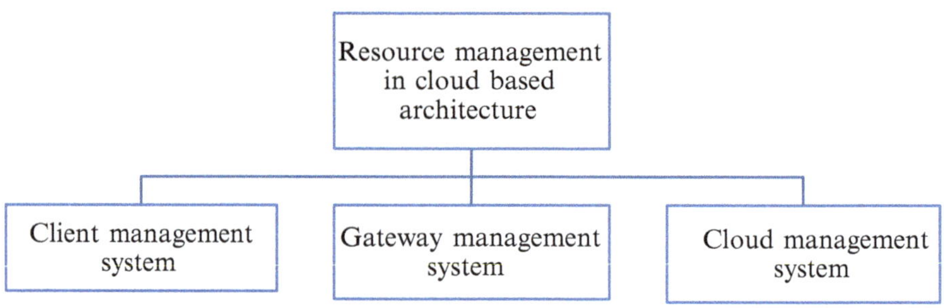

Fig. 14.10 Resource management in cloud-based architecture

RT three-dimensional vehicle tracking: In a cloud-based VANET network, IoV could be applied effectively for traffic management analysis through RT 3Dvehicle tracking. IoV application can provide the 3D map, street, and intersection views with optimal routing strategy to avert traffic obstruction. A vehicle in the network selects the route, will be updated to the driver with RT traffic conditions on the road for the whole trip. This IoV function permits every vehicle in the network, to stock the required route map and can stake their trips with family, friends, and social networks.

IoV has other applications like bandwidth resource sharing (vehicles in the cloud network can stake their bandwidth resources as per their requirements), Enhanced resource management (Intelligent Disaster Management Reinforcement System such as Emergency Medical Services for people affected by the earthquake, flood can be

Table 14.1 Key design characteristics of resource management system

Key design characteristics	Purpose
Resource monitoring	It is the key important characteristic in an exceptionally complex network. The combination of service-oriented architecture and task-driven infrastructure is a promising approach due to the close relationship between monitoring of resources, processing, and control. Resource monitoring includes other external resources such as material, energy, and personality.
Inter compatibility	The system must avoid disruption, but compatible with existing functionalities and interfaces.
RT integration and operation	To identify tasks, information flows, and operational needs with quality outputs. To find out optimal system performance with novel technologies that can satisfy performance requirements.
Infrastructure development management	Design of infrastructure and its reliability, long run, adding recent technologies and provides advanced functionalities. It requires to meet the performance indices.
System management	Millions of devices using in autonomous vehicle environments have different hardware and software configurations. It is mandated to maintain, upgrade, and reconfigure system functions with flexibility.
Adaptability support	Adaptability support is considered for static and mobile infrastructure in autonomous vehicle environments during the interaction between the mobility of devices, users, and the environment.
RT information processing	It is a challenging characteristic of resource management, to collect and process the real-time information from sources like sensors and transfers the outcome to the end user. RT information requires efficient algorithms and effective collaborative approaches.
Scalability	To add scale-up (vertical scalability) and scale-out (horizontal scalability) features for complex systems such as autonomous vehicles at all levels to maintain the bandwidth, processing power, and energy availability.
System simulation feasibility	To pick out the possible conflicts and side effects before implementing the complex system. Assist the system design, development, and operating of future infrastructure.
Unique resource identification	A standard universal identification with a unique mechanism capable to identify and address the resources even for a complex system.

rescued), monitoring, storing, and sharing of trip videos (cloud-based network can store, monitor, and retrieve HD videos in the cloud instead of storing in a large volume of a hard disk). Unique Vehicle Identity (UVI) harmonizes vehicle smart information such as auto insurance, rescue operations, and vehicle remote inspections.

6 Future Directions and Conclusions

With rapid development in internet and communication technologies, traditional vehicles are becoming smart vehicles. IoV has emerged as a global player which interconnects people, vehicles, and things inside the cloud network. The advancements that occurred in areas such as IoT, VANET, and software tools evolved together and bring forth new technology called "IoV." IoV is a multiplex combined network model which mainly focused on vehicular computing especially cloud computing, intelligent geographic information system, ITS. IoV has major benefits in particular safe driving, traffic administration, active information service, and minimize traffic obstruction. The smart integration of people, vehicles, and things under IoV in addition to focuses safety prospects and also concentrates advanced transport-related services such as autonomous driving and green driving. Communications made in IoV are implemented by Standard development organizations (SDO) like IEEE, 3GPP, and ETSI. This chapter investigates the need for IoV, network paradigms and architectures, resource management, and novel applications. The future direction and prospects of IoV play a huge role shortly by adopting advancements in computing, softwarization technologies, sensoric advancements [36, 37]. Few future IoV implications listed for readers knowledge is as follows:

- In the future, IoV shall add significant features and adopt advanced technologies, for example, parking slot availability data analytics. SMARTPARK—an algorithm is used for the selection of parking slots, which minimizes the traveling time and maximizes the possibilities to secure a parking spot [38].
- Image and command hybrid model for autonomous vehicles to detect obstacles using speech recognition or video cameras mounted on the car [39].
- Data mining technology for driving behavior and safety-relevant applications [40, 41].
- In smart cities, vehicle tracking and traffic control can be controlled by video camera sensor networks [42].
- Prediction of vehicle trajectories and future locations can be identified by algorithm-based models [43].
- Quality of service (QoS) is another future concern which depends on performance metric such as low delay, reliability, centralized control server adjusts routing decisions [44].
- During the absence of infrastructure support, data exchange, self-configure, and other services can be functionalized by using Fog computing, data distribution of IoV, 5G network functions (expecting 6G to arrive soon), which are the prospects of future IoV [45, 46].
- Efficient hierarchical clustering protocol (EHCP) for multihop communication in VANETs that offers efficient handling of resources and maintains the multi-local networks in the vehicle movement environment [47–50].

References

1. Joy, J., Rabsatt, V., & Gerla, M. (2018). Internet of vehicles: Enabling safe, secure, and private vehicular crowdsourcing. Internet Technology Letters, 1(1), e16. doi: 10.1002/itl2.16.
2. Vijayaraghavan, V., & Leevinson, J. R. (2020). Intelligent traffic management systems for next generation IoV in smart city scenario. In *Connected vehicles in the internet of things* (pp. 123–141). Springer. https://doi.org/10.1007/978-3-030-36167-9_6
3. da Silva Barbosa, F. E., de Mendonça Júnior, F. F., & Dias, K. L. (2020). A platform for cloudification of network and applications in the internet of vehicles. *Transactions on Emerging Telecommunications Technologies, 31*(5), e3961. https://doi.org/10.1002/ett.3961
4. Ghanshala, K. K., Sharma, S., Mohan, S., & Joshi, R. C. (2018, November). Cloud-based cognitive radio adhoc vehicular network architecture: A next-generation smart city. In *2018 IEEE 9th annual information technology, electronics and Mobile communication conference (IEMCON)* (pp. 145–150). IEEE. doi: https://doi.org/10.1109/IEMCON.2018.8614995.
5. Sharma, S., Muhammad, A. W. A. N., & Mohan, S. (2018). U.S. patent application no. 15/705,542.
6. Guerrero-Ibáñez, J. A., Flores-Cortés, C., & Zeadally, S. (2013). Vehicular ad-hoc networks (vanets): Architecture, protocols and applications. In *Next-generation wireless technologies* (pp. 49–70). Springer. https://doi.org/10.1007/978-1-4471-5164-7_5.
7. Sharma, S., & Mohan, S. (2016, November). Cognitive radio adhoc vehicular network (CRAVENET): Architecture, applications, security requirements and challenges. In *2016 IEEE international conference on advanced networks and telecommunications systems (ANTS)* (pp. 1–6). IEEE. doi: https://doi.org/10.1109/ANTS.2016.7947842.
8. Sharma, S., & Mohan, S. (2020). Cloud-based secured VANET with advanced resource management and IoV applications. In *Connected vehicles in the internet of things* (pp. 309–325). Springer, Cham. . https://doi.org/10.1007/978-3-030-36167-9_11.
9. Yang, F., Wang, S., Li, J., Liu, Z., & Sun, Q. (2014). An overview of internet of vehicles. *China Communications, 11*(10), 1–15. https://doi.org/10.1109/CC.2014.6969789
10. Mahmood, Z. (2020). Connected vehicles in the IoV: Concepts, technologies and architectures. In *Connected vehicles in the internet of things* (pp. 3–18). Springer. https://doi.org/10.1007/978-3-030-36167-9_1
11. Mayilsamy, K., Ramachandran, N., & Raj, V. S. (2018). An integrated approach for data security in vehicle diagnostics over internet protocol and software update over the air. *Computers & Electrical Engineering, 71*, Elsevier, Print ISSN: 0045-7906 E-ISSN:1879-0755.
12. Kathiresh, M., & Varun, C. (2014). Automotive ethernet in on-board diagnosis (Over IP) and in-vehicle Networking, *International Conference on Embedded Systems*, Amrita Vishwa Vidyapeetham, Coimbatore and published in IEEE Explorer. ISBN 978-1-4799-5026-9.
13. Kathiresh, M., Neelaveni, R., Benny, M. A., & Moses, B. J. S. (2021). Vehicle diagnostics over internet protocol and over-the-air updates. In M. Kathiresh & R. Neelaveni (Eds.), *Automotive embedded systems. EAI/Springer innovations in communication and computing*. Springer. https://doi.org/10.1007/978-3-030-59897-6_5.
14. Jadhav, A. (2021). Automotive cybersecurity. In M. Kathiresh & R. Neelaveni (Eds.), *Automotive embedded systems. EAI/Springer innovations in communication and computing*. Springer. https://doi.org/10.1007/978-3-030-59897-6_6.
15. Janeera, D. A., Gnanamalar, S. S. R., Ramya, K. C., & Kumar, A. G. A. (2021). Internet of things and artificial intelligence-enabled secure autonomous vehicles for smart cities. In M. Kathiresh & R. Neelaveni (Eds.), *Automotive embedded systems. EAI/Springer innovations in communication and computing*. Springer. https://doi.org/10.1007/978-3-030-59897-6_11.
16. Flavio, B. (2013). *The smart and connected vehicle and the internet of things*. WSTS.
17. Kaiwartya, O., Abdullah, A. H., Cao, Y., Altameem, A., Prasad, M., Lin, C. T., & Liu, X. (2016). Internet of vehicles: Motivation, layered architecture, network model, challenges, and future aspects. *IEEE Access, 4*, 5356–5373. https://doi.org/10.1109/ACCESS.2016.2603219

18. Alouache, L., Nguyen, N., Aliouat, M., & Chelouah, R. (2018, April). Toward a hybrid SDN architecture for V2V communication in IoV environment. In *2018 fifth international conference on software defined systems (SDS)* (pp. 93–99). IEEE. https://doi.org/10.1109/SDS.2018.8370428
19. Vinoth, K. K. (2021). Model-based automotive software development. In M. Kathiresh & R. Neelaveni (Eds.), *Automotive embedded systems. EAI/Springer innovations in communication and computing*. Springer. https://doi.org/10.1007/978-3-030-59897-6_4.
20. Contreras-Castillo, J., Zeadally, S., & Guerrero Ibáñez, J. A. (2017). A seven-layered model architecture for internet of vehicles. *Journal of Information and Telecommunication, 1*(1), 4–22. https://doi.org/10.1080/24751839.2017.1295601
21. Gandotra, P., Jha, R. K., & Jain, S. (2017). A survey on device-to-device (D2D) communication: Architecture and security issues. *Journal of Network and Computer Applications, 78*, 9–29. https://doi.org/10.1016/j.jnca.2016.11.002
22. Kang, L., Li, J., Li, X., Wang, W., Zhang, Y., & Poslad, S. (2015). A traffic-aware resource allocation scheme for internet of vehicles: A supply and demand function approach. *J. Computational Information Systems, 11*(3), 987–994. https://doi.org/10.12733/jcis13115
23. Darwish, T. S., & Bakar, K. A. (2018). Fog based intelligent transportation big data analytics in the internet of vehicles environment: Motivations, architecture, challenges, and critical issues. *IEEE Access, 6*, 15679–15701. https://doi.org/10.1109/ACCESS.2018.2815989
24. Marz, N., & Warren, J. (2015). *Big data: Principles and best practices of scalable real-time data systems*. Manning Publications Co.. Print ISBN: 9781617290343.
25. Kathiresh, M., Neelaveni, R., & Abinesh, S. 2014. Analysis of multi-core architecture for automotive applications. In *International conference on embedded systems*, Amrita Vishwa Vidyapeetham, Coimbatore and published in IEEE Explorer. ISBN 978-1-4799-5026-3.
26. Zhou, H., Xu, W., Chen, J., & Wang, W. (2020). Evolutionary V2X technologies toward the internet of vehicles: Challenges and opportunities. *Proceedings of the IEEE, 108*(2), 308–323.
27. Shen, X., Cheng, X., Zhang, R., Jiao, B., & Yang, Y. (2013). Distributed congestion control approaches for the IEEE 802.11 p vehicular networks. *IEEE Intelligent Transportation Systems Magazine, 5*(4), 50–61. https://doi.org/10.1109/MITS.2013.2279176
28. Mosyagin, J. (2010, June). Using 4G wireless technology in the car. In *2010 12th international conference on transparent optical networks* (pp. 1–4). IEEE. doi: https://doi.org/10.1109/ICTON.2010.5549094.
29. Araniti, G., Campolo, C., Condoluci, M., Iera, A., & Molinaro, A. (2013). LTE for vehicular networking: A survey. *IEEE Communications Magazine, 51*(5), 148–157. https://doi.org/10.1109/MCOM.2013.6515060
30. Thota, J., Almesaeed, R., Doufexi, A., Armour, S., & Nix, A. R. (2015, May). Infrastructure to vehicle throughput performance in LTE-A using 2D and 3D 3GPP/ITU Channel models. In *2015 IEEE 81st vehicular technology conference (VTC spring)* (pp. 1–5). IEEE. doi: https://doi.org/10.1109/VTCSpring.2015.7145880.
31. Yang, Y., Hu, H., Xu, J., & Mao, G. (2009). Relay technologies for WiMAX and LTE-advanced mobile systems. *IEEE Communications Magazine, 47*(10), 100–105. https://doi.org/10.1109/MCOM.2009.5273815
32. Sukuvaara, T., & Pomalaza-Ráez, C. (2009). Vehicular networking pilot system for vehicle-to-infrastructure and vehicle-to-vehicle communications. *International Journal of Communication Networks and Information Security, 1*(3), 1.
33. Naboulsi, D., & Fiore, M. (2013, July). On the instantaneous topology of a large-scale urban vehicular network: The cologne case. In *Proceedings of the fourteenth ACM international symposium on mobile ad hoc networking and computing* (pp. 167–176). doi: https://doi.org/10.1145/2491288.2491312.
34. Ghaffarian, H., Fathy, M., & Soryani, M. (2012). Vehicular ad hoc networks enabled traffic controller for removing traffic lights in isolated intersections based on integer linear programming. *IET Intelligent Transport Systems, 6*(2), 115–123. https://doi.org/10.1049/iet-its.2010.0207

35. Abbas, M. T., Jibran, M. A., Afaq, M., & Song, W. C. (2020). An adaptive approach to vehicle trajectory prediction using multimodel Kalman filter. *Transactions on Emerging Telecommunications Technologies, 31*(5), e3734. https://doi.org/10.1002/ett.3734
36. Antony, M. M., & Whenish, R. (2021). Advanced driver assistance systems (ADAS). In M. Kathiresh & R. Neelaveni (Eds.), *Automotive embedded systems. EAI/Springer innovations in communication and computing*. Springer. https://doi.org/10.1007/978-3-030-59897-6_9.
37. Liu, K., Xu, X., Chen, M., Liu, B., Wu, L., & Lee, V. C. (2019). A hierarchical architecture for the future internet of vehicles. *IEEE Communications Magazine, 57*(7), 41–47.
38. Atif, Y., Kharrazi, S., Jianguo, D., & Andler, S. F. (2020). Internet of things data analytics for parking availability prediction and guidance. *Transactions on Emerging Telecommunications Technologies, 31*(5), e3862. https://doi.org/10.1002/ett.3862
39. Malik, K. R., Ahmad, M., Khalid, S., Ahmad, H., Al-Turjman, F., & Jabbar, S. (2020). Image and command hybrid model for vehicle control using internet of vehicles. *Transactions on Emerging Telecommunications Technologies, 31*(5), e3774. https://doi.org/10.1002/ett.3774
40. Sun, Y., Bi, Y., Han, Y., Xie, D., & Li, R. (2020). Research on safe driving behavior of transportation vehicles based on vehicle network data mining. *Transactions on Emerging Telecommunications Technologies, 31*(5), e3772. https://doi.org/10.1002/ett.3772
41. Hamsini, S., & Kathiresh, M. (2021). Automotive safety systems. In M. Kathiresh & R. Neelaveni (Eds.), *Automotive embedded systems. EAI/Springer innovations in communication and computing*. Springer. https://doi.org/10.1007/978-3-030-59897-6_1.
42. Naeem, M., Ejaz, W., Iqbal, M., Iqbal, F., Anpalagan, A., & Rodrigues, J. J. (2020). Efficient scheduling of video camera sensor networks for IoT systems in smart cities. *Transactions on Emerging Telecommunications Technologies, 31*(5), e3798. https://doi.org/10.1002/ett.3798
43. Chen, C. M., Xiang, B., Liu, Y., & Wang, K. H. (2019). A secure authentication protocol for internet of vehicles. *IEEE Access, 7*, 12,047–12,057.
44. Shirley, D. R. A., Sundari, V. K., Sheeba, T. B., & Rani, S. S. (2021). Analysis of IoT-enabled intelligent detection and prevention system for drunken and juvenile drive classification. In M. Kathiresh & R. Neelaveni (Eds.), *Automotive embedded systems. EAI/Springer innovations in communication and computing*. Springer. https://doi.org/10.1007/978-3-030-59897-6_10.
45. Ansari, S., Ahmad, J., Aziz Shah, S., Kashif Bashir, A., Boutaleb, T., & Sinanovic, S. (2020). Chaos-based privacy preserving vehicle safety protocol for 5G connected autonomous vehicle networks. *Transactions on Emerging Telecommunications Technologies, 31*(5), e3966. https://doi.org/10.1002/ett.3966
46. Shen, X., Fantacci, R., & Chen, S. (2020). Internet of vehicles. *Proceedings of the IEEE, 108* (2), 242–245.
47. Dutta, A. K., Elhoseny, M., Dahiya, V., & Shankar, K. (2020). An efficient hierarchical clustering protocol for multihop internet of vehicles communication. *Transactions on Emerging Telecommunications Technologies, 31*(5), e3690. https://doi.org/10.1002/ett.3690
48. Tigadi, A. S., Changappa, N., Singhal, S., & Kulkarni, S. (2021). Autonomous vehicles: Present technological traits and scope for future innovation. In M. Kathiresh & R. Neelaveni (Eds.), *Automotive embedded systems. EAI/Springer innovations in communication and computing*. Springer. https://doi.org/10.1007/978-3-030-59897-6_7.
49. Priyan, M. K., & Devi, G. U. (2019). A survey on internet of vehicles: Applications, technologies, challenges and opportunities. *International Journal of Advanced Intelligence Paradigms, 12*(1–2), 98–119. https://doi.org/10.1504/IJAIP.2019.096957
50. Sharma, S., & Kaushik, B. (2019). A survey on internet of vehicles: Applications, security issues & solutions. *Vehicular Communications, 20*, 100182.

Index

M. Kathiresh et al. (eds.), *E-Mobility*, EAI/Springer Innovations in Communication and Computing, https://doi.org/10.1007/978-3-030-85424-9

D

L

Milton Keynes UK
Ingram Content Group UK Ltd.
UKHW020203211024
449810UK00001B/13